Genomics and Precision Healthcare

Scrivener Publishing
100 Cummings Center, Suite 541J
Beverly, MA 01915-6106

Publishers at Scrivener
Martin Scrivener (martin@scrivenerpublishing.com)
Phillip Carmical (pcarmical@scrivenerpublishing.com)

Genomics and Precision Healthcare

Innovations, Applications and Challenges

Edited by

Swetha Kumari Koduru

Sarad Pawar Naik Bukke

Hussaini Bello

Muhamad Mustafa Mundu

and

Ayush Madan

Scrivener Publishing

WILEY

This edition first published 2026 by John Wiley & Sons, Inc., 111 River Street, Hoboken, NJ 07030, USA and Scrivener Publishing LLC, 100 Cummings Center, Suite 541J, Beverly, MA 01915, USA

For more information about Scrivener publications please visit www.scrivenerpublishing.com.

Wiley Global Headquarters
111 River Street, Hoboken, NJ 07030, USA

For details of our global editorial offices, customer services, and more information about Wiley products visit us at www.wiley.com.

The manufacturer's authorized representative according to the EU General Product Safety Regulation is Wiley-VCH GmbH, Boschstr. 12, 69469 Weinheim, Germany, e-mail: Product_Safety@wiley.com.

Library of Congress Cataloging-in-Publication Data

ISBN 9781394367528

Cover image: Generated with AI using Adobe Firefly
Cover design by Russell Richardson

Set in size of 11pt and Minion Pro by Manila Typesetting Company, Makati, Philippines

Printed in the USA

10 9 8 7 6 5 4 3 2 1

Contents

Preface

Two decades ago, genomics was largely the domain of research laboratories and visionary scientists. Today, it has moved decisively into the clinic, the pharmacy, the newborn nursery, and even the consumer marketplace. A patient with metastatic cancer now routinely receives tumour sequencing to guide therapy; a child with an undiagnosed developmental disorder can obtain answers through rapid whole-genome analysis; an individual considering a new medication may have their genome screened to avoid life-threatening adverse reactions. What was once aspirational is now part of everyday healthcare decision-making.

This book, *Genomics and Precision Healthcare*, is designed to provide a comprehensive yet accessible guide to this rapidly maturing field. It is written for clinicians, genetic counsellors, researchers, students, policymakers, and health-system leaders who need a clear, current, and practical understanding of how genomics is reshaping medicine across the continuum of care.

The journey begins with foundational principles (Chapter 1) and traces the path from gene discovery to bedside impact (Chapter 2). We then explore the core pillars of precision healthcare: tailoring treatments to individual genetic profiles (Chapter 3), revolutionary diagnostic technologies (Chapter 4), and pharmacogenomics (Chapter 5). No contemporary discussion would be complete without addressing the ethical, legal, and social implications that accompany the power of genomic information now confers (Chapter 6).

Subsequent chapters examine genomics in key clinical domains: rare and common non-communicable diseases (Chapter 7), cancer (Chapter 8), infectious diseases (Chapter 9), and prenatal and paediatric conditions (Chapter 10). We address essential infrastructure challenges—integrating genomic data into electronic health records (Chapter 11) and the human elements of genetic counselling (Chapter 12) and patient experience (Chapter 14). Finally, we look to the broader ecosystem: health policy and implementation science (Chapter 13) and the profound global disparities and opportunities that define genomic medicine today (Chapter 15).

Our aim is not only to summarise scientific and technological advances but to equip readers with the knowledge required to navigate real-world complexities: interpreting variants of uncertain significance, counselling families, designing equitable health policies, and ensuring that genomic advances reach all populations, not just the privileged few.

As we stand in late 2025, genomic medicine is no longer emerging, it is here. The question is no longer whether genomics will transform healthcare, but how responsibly, equitably, and effectively we will integrate it into practice worldwide. We hope this book serves as both a roadmap for that integration and an invitation to join the ongoing effort to realise the full promise of precision healthcare for every patient, everywhere.

December 2025

1

The Fundamentals of Genetics and Genomics: A Primer for Healthcare Professionals

K. Haripriya[1]*, Swetha Kumari Koduru[1] and K. Nandini[2]

[1]*Dept. of Biosciences and Sericulture, Sri Padmavati Mahila Visvavidyalayam, Tirupati, India*
[2]*School of Nursing, Sri Padmavati Mahila Visvavidyalayam, Tirupati, India*

Abstract
Genetics and genomics are two related areas which are highly important in terms of health and disease. Genetics refers to the study of individual genes and how certain types of traits or conditions occur as one generation factors into the proceeding generation. In contrast, genomics refers to the overall study of all genes of such an individual, which includes all the genes of the same. The combination of genetics and genomics is the unit of heredity including all chromosomes and genes that instruct cells of our body how genes function and work together with the environment. Many kinds of modifications in chromosome, DNA, genes, proteins, and vitamins lead to diversity and disease and can be inherited in an autosomal, sex-linked, Mitochondrial, Mutations, Genetic variation mechanisms. The genetics and genomics have numerous applications and technologies in healthcare and this is also extensively covered, including DNA sequencing technologies, genetics testing and screening, genomics editing and gene therapy, and bioinformatics and computational genomics. Genomics will help to drive research innovations and development, genetic disease diagnosis, treatment of various chronic diseases and identify the genetic disorders in newborns, enabling early interventions. Primarily, it will be useful in predicting individual response to some drugs to minimize the occurrence of side effects, and enhance positive treatment response. The genomics and genetics dictate the likelihood of the risk that the disease can be inherited by the future generation. This chapter shall establish the major tenets of

**Corresponding author*: drjharipriya7@gmail.com

Swetha Kumari Koduru, Sarad Pawar Naik Bukke, Hussaini Bello, Muhamad Mustafa Mundu and Ayush Madan (eds.) Genomics and Precision Healthcare: Innovations, Applications and Challenges, (1–38) © 2026 Scrivener Publishing LLC

genetic and genomics as a base to comprehending that the variations of our DNA and its transmission can bring illness. This comprehensive primer provides healthcare professionals with a foundational understanding of genetics and genomics.

***Keywords*:** Genetics, genomics, precision healthcare, genetic counselling, genomic data management

1.1 Introduction

1.1.1 Definition of Genetics and Genomics

Gregor Mendel is largely credited with laying the groundwork for genetics. The fundamentals of heredity were established by his work with pea plants in the mid-19th century. He is frequently called the "father of genetics" [1]. It is also important to remember that William Bateson coined the term "genetics" in 1905 [2]. The study of genes, heredity, and genetic variation in organisms is known as genetics. Individual genes and their transmission from one generation to the next are its main emphasis. In essence, genetics examines how individual genes determine inheritable traits as well as their makeup and function. For instance, genetics may investigate how a certain disease is brought on by a particular gene mutation [3].

The term "genomics" is more recent. It was coined in 1986 by Thomas Roderick, a scientist at the Jackson Laboratory [4]. Genomics as a field was enabled by the technological advancements of the late 20th and early 21st century, primarily the ability to sequence DNA. Therefore, it is not tied to one single person, but to the collective efforts of many scientists and engineers. Genomics is a branch of biology that is interdisciplinary and looks at the structure, mechanism, evolution, and mapping of genomes as well as editing genomes [5]. All the genes of an organism, including its DNA, make its genome. Genomics is a broader discipline that investigates the entire genome and the interactions of genes with one another and with the environment. It aims to understand the complex interplay of all genes in an organism. For example, genomics might study how multiple genes interact to increase a person's risk of heart disease [6].

1.1.2 Importance in Modern Healthcare

Genetics and genomics play a crucial role in modern healthcare by transforming the way diseases are diagnosed, treated, and prevented. These fields enable personalized medicine, improve disease risk assessment, and enhance treatment strategies based on an individual's genetic makeup [7].

1.1.2.1 Disease Diagnosis and Early Detection

Genetic testing helps in the early detection of inherited diseases, allowing for timely interventions. For example, in cancer screening, identifying mutations in genes like BRCA1 and BRCA2 can assess the risk of breast and ovarian cancer [8]. In the case of genetic disorders, testing for conditions such as cystic fibrosis, sickle cell anaemia, and Down syndrome enables early medical planning [9].

1.1.2.2 Personalized and Precision Medicine

Genomic data enable medical practitioners to give medications depending on the genetic profile of the patients, improving efficacy, and reducing side effects. For example, pharmacogenomics determines how individuals metabolize drugs, optimizing medication dosages for better effectiveness (e.g., warfarin, antidepressants) [10]. Targeted therapies for certain cancers, such as lung cancer, use precision drugs to target specific genetic mutations (e.g., EGFR inhibitors) [11].

1.1.2.3 Risk Assessment and Prevention

By analysing genetic markers, individuals can be informed about their susceptibility to diseases and take preventive actions. In cardiovascular risk assessment, identifying genetic predispositions to heart disease allows for lifestyle changes and early intervention [12]. Genetic tests can also indicate the likelihood of developing diseases like Alzheimer's, enabling preventive strategies [13].

1.1.2.4 Reproductive and Prenatal Health

Genetic counselling and prenatal testing help identify potential risks for inherited disorders before birth. Carrier screening enables prospective parents to assess the risk of passing on genetic conditions such as Tay-Sachs or sickle cell disease [14]. In *in-vitro* fertilization (IVF), genetic testing can ensure the selection of embryos without genetic disorders, improving the chance of a healthy pregnancy [15].

1.1.2.5 Gene Therapy and Advanced Treatments

Genetic advancements are leading to groundbreaking treatments. Technologies such as CRISPR and gene editing are used to correct faulty genes responsible for inherited diseases like sickle cell anaemia [16]. Stem cell therapy is also being used to treat genetic conditions such as blood disorders and immune deficiencies [17].

1.1.2.6 Infectious Disease Research and Vaccine Development

Genomics plays a key role in understanding viral mutations and developing vaccines. COVID-19 genomic sequencing helped scientists track virus mutations and design effective vaccines rapidly [18]. HIV and malaria research has increasingly relied on genetic research to come up with more effective preventive measures and treatment options [19].

1.1.3 Historical Perspective

The study of genetics has evolved over centuries, beginning with early observations of inheritance and culminating in modern breakthroughs in DNA sequencing and gene editing. Below is a historical timeline highlighting key milestones in genetics.

1.1.3.1 Early Observations and Theories in Genetics

Early human beings observed that the features could be transferred to the offspring, especially in the process of taming of plants and animals [20]. Early speculative theories of heredity were proposed by ancient Greek philosophers Hippocrates and Aristotle and many others, but they did not have any experimental support [21]. Modern genetics origins are traced to the 19th century when Gregor Mendel, through careful experimentation with the pea plant, set the groundwork. He proposed the ideas of the dominant and recessive traits as well as the laws of inheritance [1]. In 1859, Charles Darwin presented his theory of natural selection, wherein he stressed the relevance of the inheritability of the population variation [22].

1.1.3.2 The Rise of Classical Genetics

This resurgence in genetic research happened in the early 20th century when the work of Mendel was rediscovered. Mendel and his principles were reintroduced with the aid of scientists Hugo de Vries, Carl Correns, and Erich von Tschermak [23]. The origins of its theory were developed when the chromosome theory of inheritance was produced and genes were linked with chromosomes [24]. In 1953, a shortcut was made by James Watson and Francis Crick against the X-ray diffraction pictures of the team led by Rosalind Franklin of a DNA; then everything changed in the perception of the genetic material due to a double-helix structure [25]. The core of the molecular biology, the one-way flow of information, DNA, to RNA, to protein, soon became established [26]. Another milestone in the history of DNA technology was the role played by the development of recombinant DNA technology in the 1970s [27].

1.1.3.3 The Genomic Era and Modern Genetics

The 21st century was the era of rapid innovation in genome. It was in 1983 that scientists were able to identify the gene involved in Huntington's disease; this was the discovery involving a single disease in humans linked to a particular gene [28]. Between 1990 and 2003, the whole human genome (~3 billion base pairs) was fully sequenced in the Human Genome Project (HGP), which developed a point of departure of genomic medicine [58]. In 2012, Jennifer Doudna and Emmanuelle Charpentier developed a gene-editing tool called CRISPR-Cas9 that allows performing highly specific DNA alterations [29].

Gene therapy became one potential solution to genetic diseases like sickle cell anaemia and cystic fibrosis [30]. With the emergence of personalized medicine, it has become possible to customize the treatment, according to the genetic make-up of a person [31]. More recently, the genomics field has been subjected to the use of Artificial Intelligence (AI), which has improved the rate and accuracy of genetic analysis and disease prediction [32].

1.1.4 Precision Healthcare

Precision healthcare, also known as personalized medicine, is a medical approach that adapts prevention, diagnosis, and treatment strategies as per a patient's genetic profile, habits, and environment. It moves away from a one-size-fits-all model and focuses on customized healthcare interventions based on a person's unique biological profile. Precision healthcare heavily relies on genetics and genomics, using DNA sequencing, biomarkers, and data analytics to improve disease prediction, prevention, and treatment [33].

1.1.4.1 Benefits of Precision Healthcare in Genetics and Genomics

Precision healthcare offers a highly individualized approach to managing health. One major advantage is early detection and risk assessment

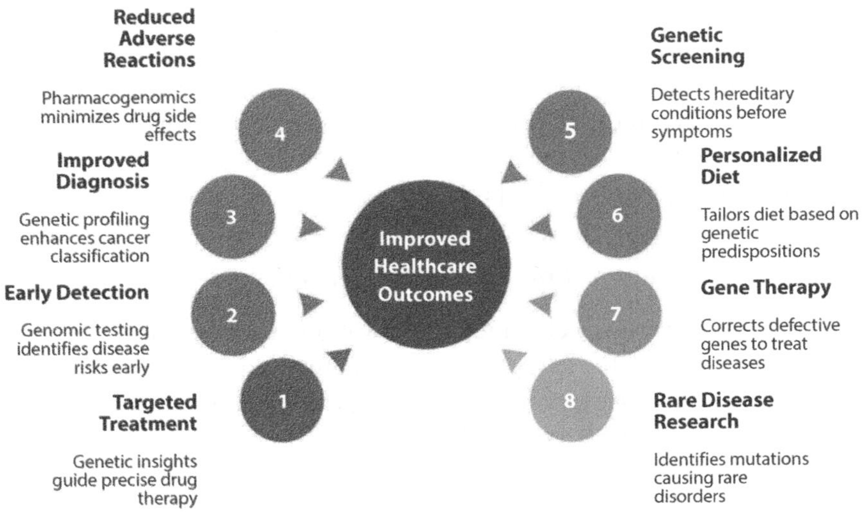

Figure 1.1 Benefits of precision healthcare in genetics and genomics.

for inherited conditions, enabling preventive care and lifestyle modifications [34]. It enables targeted therapies, particularly in oncology, where treatments are matched to specific genetic mutations in tumors, leading to improved patient outcomes and fewer side effects [35].

Another key area is pharmacogenomics, where understanding a patient's genetic profile helps determine the most effective and safe drug and dosage—thereby reducing adverse drug reactions and enhancing therapeutic efficacy [10, 36].

In cases of rare diseases, whole-genome sequencing or exome sequencing can lead to faster and more accurate diagnoses [37]. Precision healthcare also facilitates reproductive planning through carrier screening and prenatal genetic testing, helping families make informed decisions [14].

Overall, precision healthcare promotes more effective, efficient, and personalized medical care, contributing to better patient outcomes, cost-effectiveness, and long-term healthcare sustainability [38] (shown in Figure 1.1).

1.2 Principles of Genetics

Deoxyribonucleic Acid (DNA) carries the genetic code of all living organisms. Its primary role is to preserve and transfer genetic information between generations [39]. DNA serves as a blueprint for cellular growth, function, and reproduction [40]. Structurally, DNA resembles a twisted ladder, known as the double helix, a discovery credited to James Watson and Francis Crick in 1953, with critical contributions from Rosalind Franklin and Maurice Wilkins [25, 41]. The double helix as shown in Figure 1.2 consists of two complementary strands wound around each other [42]. DNA is composed of nucleotides, each containing three components, namely Sugar (Deoxyribose) – a five-carbon sugar forming part of the DNA backbone; Phosphate Group – bonds with sugar molecules to create the backbone of the DNA strand; and Nitrogenous Bases – Adenine (A), Thymine (T), Cytosine (C), and Guanine (G) [43].

DNA follows the principle of complementary base pairing: A pairs with T *via* two hydrogen bonds, and C pairs with G *via* three hydrogen bonds, ensuring replication accuracy [44]. The DNA strands are anti-parallel—one strand runs 5′ to 3′, the other 3′ to 5′. Each turn of the helix contains 10 base pairs, with an inter-turn distance of 3.4 nm, and 0.34 nm per base pair. The helix's diameter is approximately 2 nm. The major groove is wider and more accessible than the minor groove [45].

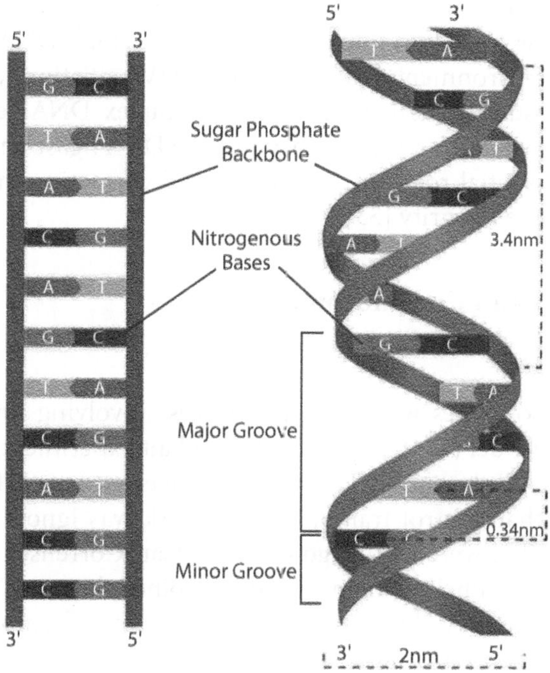

Figure 1.2 Watson and Crick, structure of DNA.

DNA encodes instructions for protein synthesis, which regulates all cell functions. Genes are distinct DNA sequences that encode specific proteins [46]. Before cell division, DNA replication ensures each daughter cell receives an exact copy of the genetic material. This replication is semi-conservative, meaning each new DNA molecule retains one original strand and one new strand, a process facilitated by the enzyme DNA polymerase [47].

DNA directs the synthesis of proteins through two key processes: transcription and translation [39]. During transcription, DNA is transcribed into messenger RNA (mRNA) within the nucleus. The enzyme RNA polymerase binds to the DNA and synthesizes a complementary strand of mRNA based on the DNA template [48]. The mRNA then exits the nucleus and carries the genetic code to the ribosome, where translation occurs. At the ribosome, transfer RNA (tRNA) molecules transport amino acids that correspond to the codons (three-base sequences) on the mRNA strand [49]. Ribosomes facilitate the assembly of amino acids into a polypeptide chain, forming a protein [45].

Mutations—changes in the DNA sequence—introduce variation in genetic information. Some mutations are harmless, while others may result in genetic disorders such as sickle cell anemia or cystic fibrosis [50, 51]. Additionally, certain mutations serve as a source of evolutionary change, introducing new traits into populations over time [52]. DNA damage can be caused by environmental factors such as UV radiation, chemicals, or radiation exposure. Cells have developed complex DNA repair mechanisms to counteract these effects. Enzymes like DNA ligase and DNA polymerase play essential roles in recognizing and correcting these errors to maintain genome integrity [53].

1.2.1 Genes and Chromosomes

1.2.1.1 Genes

The discovery of genes was a gradual process involving multiple scientists. Gregor Mendel (1866) conducted pea plant experiments and formulated the Laws of Inheritance. He proposed the existence of "factors" (now called genes) that control traits [54]. His work was ignored until 1900, when it was rediscovered by Hugo de Vries, Carl Correns, and Erich von Tschermak [55]. Wilhelm Johannsen (1909) coined the term "gene", replacing Mendel's "factors". He introduced the concepts of genotype (genetic makeup) and phenotype (observable traits) [56].

The unit of heredity having instructions in constructing and controlling proteins in living things are genes. These are parts of a DNA (Deoxyribonucleic Acid) that decide characters like eye color, height, and being prone to illnesses. Every gene is a particular arrangement of nucleotides (Adenine, Thymine, Cytosine, and Guanine), and the information contained is translated to either proteins or working RNA molecules [57]. The life blueprint is genes, which control each characteristic, the natural life mechanisms, and the susceptibility to diseases. There is a revolution in genetics brought about by the understanding of the genes resulting in gene therapy, personalized medicine, and biotechnology [58].

Genes are organized into chromosomes, which are found inside the nucleus of cells. The key structural components of a gene include:

i. Promoter Region

- Located at the beginning of a gene.
- Acts as a control switch, determining when and where a gene is turned on or off.
- Contains binding sites for transcription factors that regulate gene expression [59].

ii. Coding Region (Exons and Introns)

- Exons: The sequences that code for proteins.
- Introns: Non-coding sequences that are removed during RNA processing.
- The final messenger RNA (mRNA) formed after processing contains only exons, which are translated into proteins [60].

iii. Terminator Region

- Found at the end of a gene.
- Signals the stop of transcription, ensuring the correct length of RNA is produced [59].

Different functions are exhibited by genes. The simpler examples of how proteins are made are through the two fundamental processes by genes, namely Transcription – the copying of the DNA into RNA (messenger RNA) in the nucleus; and also Translation – the translation of the RNA (messenger RNA) into a protein in the ribosome by the assistance of the transfer RNA (tRNA) and ribosomal RNA (rRNA) [61]. Physical and biological traits are determined by genes that parents pass on to their offspring. All these variations in qualities were due to the different forms of a gene (alleles), which is what led to the variation. Genes oversee the vital processes of cell growth, metabolism, and immune responses. Other genes are switches, and, based on the environmental clues, may switch other genes on or off. Genetic diseases may be caused by a change in the gene sequence (mutations): sickle cell anemia, cystic fibrosis, and Huntington are only a few examples. Certain genes such as oncogenes when mutated can contribute to cancer [40, 62].

1.2.1.1.1 Types of Genes

Genes can be classified based on their function and mode of inheritance:

A. Structural Genes

- Encode proteins that form cell structures and enzymes.
- Example: Hemoglobin gene (HBB), which produces hemoglobin in red blood cells [63].

B. Regulatory Genes

- Control the expression of other genes, determining when and where they are activated.
- Example: Homeobox (Hox) genes, which control body development [64].

C. Housekeeping Genes

- Essential for basic cellular functions and are expressed in all cells.
- Example: GAPDH gene, which is involved in energy metabolism [65].

D. Oncogenes and Tumor Suppressor Genes

- Oncogenes: Promote cell growth; mutations can lead to cancer (e.g., RAS gene).
- Tumor Suppressor Genes: Prevent uncontrolled growth; mutations can lead to cancer (e.g., TP53 gene) [62].

E. Mitochondrial Genes

- Found in mitochondria, inherited only from the mother.
- Example: MT-ND1, involved in energy production [66].

F. Pseudogenes

- Non-functional genes that resemble normal genes but are inactive due to mutations [67].

1.2.1.2 Chromosomes

Chromosomes are like a thread made of DNA and proteins which contain genetic information. They are present in the cell nuclei and correspond to genes, which shape traits and manage the functions of cells. Each chromosome consists of a single, tightly coiled DNA molecule wrapped around proteins called histones, which help in compacting the DNA and regulating gene expression [68, 69].

DNA in chromosomes is wrapped around histone proteins, forming a structure called chromatin. This packaging helps fit DNA into the nucleus and plays a role in gene regulation. Each chromosome has two identical chromatids (sister chromatids) when preparing for cell division. The chromatids are joined at a region called the centromere, which plays a role in chromosome movement during cell division. The ends of chromosomes are called telomeres, which protect DNA from damage and degradation. Telomeres shorten with each cell division, contributing to aging [70, 71].

1.2.1.2.1 Types of Chromosomes in Humans

The human somatic (body) cells contain 46 chromosomes (23 pairs). This is divided into: Autosomes (22 pairs), which govern most of the inherited characteristics including eye colour, height, and metabolism, and occur in both males and females. Sex chromosomes (1 pair) such as XX in females and XY in males define biological sex and affect secondary sexual characteristics [68].

The chromosomes contain genes that store the instructions for making proteins and controlling cellular processes. Chromosomes duplicate during mitosis (for growth and repair) and meiosis (for sexual reproduction). Proper chromosome division ensures the correct number of chromosomes in new cells. Chromosomes help regulate which genes are turned on or off in different cells. This differentiation allows specialized functions (e.g., muscle cells vs. nerve cells). Chromosomes guarantee genetic inheritance from parents to offspring. Genetic diversity, important to evolution, is brought about by recombination that happens during meiosis [72]. Chromosomal disorders are of two types, namely Numerical Chromosomal Disorder and Structural Chromosomal anomalies.

Numerical Chromosomal Disorders occur when there is a missing or extra chromosome (aneuploidy). Normal human cells have 46 chromosomes (23 pairs), but numerical disorders result in 45 or 47+ chromosomes. It includes autosomal Chromosomal Disorders (affect non-sex chromosomes) and Sex Chromosome Disorders (affect X or Y chromosomes) [73].

Autosomal Chromosomal Disorders examples include: Down Syndrome (Trisomy 21) – caused by the extra copy of chromosome 21 (47, XX+21 or 47, XY+21). Symptoms are intellectual disability, flattened facial features, heart defects, short stature. Edwards Syndrome (Trisomy 18) is caused by an extra chromosome 18 (47, XX+18 or 47, XY+18). Symptoms are severe intellectual disability, heart and kidney defects, low birth weight. Life expectancy is very low and most infants do not survive beyond infancy. Patau Syndrome (Trisomy 13) is caused by an extra chromosome 13 (47, XX+13 or 47, XY+13). Symptoms are severe brain and heart defects, cleft lip, extra fingers, or toes. Life expectancy is very low and most infants do not survive beyond a few months [74].

Examples of Sex Chromosome Disorders are as follows. Turner Syndrome (Monosomy X): this disorder is due to the female having only one X chromosome (45, X0 instead of 46, XX). Symptomatic manifestations are short stature, infertility, webbed neck, and heart abnormalities. Klinefelter Syndrome (XXY Syndrome) is a disorder that is caused by the presence of an additional X chromosome (47, XXY versus 46, XY) in men. The signs include tall size, diminished muscle mass, infertile, difficulty in learning. XYY Syndrome (Jacobs Syndrome) comes about because of an additional Y chromosome (47, XYY) in the male person. The symptoms include increased height, learning retardation, normal fertility. In Triple X Syndrome (XXX Syndrome), the females have an additional X chromosome (47, XXX). They include taller height, problems with learning, and normal fertility [75].

The Structural Chromosomal Disorders are caused by alternations in chromosome structure (deletions, duplications, translocations, or inversions). Possible examples are: Cri-du-Chat Syndrome (Cat Cry Syndrome) a deletion of a segment of chromosome 5 (5p deletion); Wolf-Hirschhorn Syndrome, a deletion of section of chromosome 4 (4p deletion); Fragile X Syndrome, repeating mutation CGG repeats in FMR1 gene on X chromosome; and Chronic Myeloid Leukemia (CML), a Philadelphia Chromosome (a translocation between chromosomes 9 and 22 (t9;22)) [76, 77].

1.2.2 Mendelian Inheritance Patterns

The Mendelian inheritance is a system of passing parental genes and traits to their children that are based on what Gregor Mendel found out in the 1860s. Three fundamental laws of inheritance were generated out of his experiments on pea plants. These patterns of inheritance are classified to traits that have dominant and recessive alleles and that are controlled by a single gene (monogenic inheritance) [59].

1.2.2.1 Mendel's Three Laws of Inheritance

A. ***Law of Segregation (First Law)*** - All of us possess two copies (alleles) of a gene, one that is produced by each parent. When we form gametes these two alleles are separated such that each individual gamete (sperm or egg) only has one allele (during gamete formation). Children receive one copy of each of the two parents to restore that couple [60].
B. ***Law of Independent Assortment (Second Law)*** - Genes belonging to different traits are passed on independently provided they are found on different chromosomes. The inheritance of a trait has no effect on the inheritance of another trait [61].
C. ***Law of Dominance (Third Law)*** - There are dominant alleles; so there may be a dominating force that covers up the influence of recessive alleles. A superior gene occurs in a heterozygote (Aa) and a homozygote (AA). Recessive allele is present only in a recessive person (aa) [59, 62].

1.2.2.2 Common Mendelian Inheritance Patterns

The Mendelian inheritance patterns spell out the transfer of genetic characteristics between the parents and the offspring. The main types include autosomal dominant, where one copy of a dominant allele causes the trait (e.g., Huntington's disease), and autosomal recessive, where two copies of a recessive allele are needed (e.g., cystic fibrosis). X-linked dominant and X-linked recessive patterns involve genes on the X chromosome, affecting males and females differently (e.g., hemophilia for X-linked recessive). Y-linked traits affect only males. These different patterns as indicated in

Characteristic	Autosomal Dominant	Autosomal Recessive	X-linked Recessive	X-linked Dominant
Allele Copies Needed	One copy	Two copies	N/A	N/A
Parental Status	At least one affected parent	Unaffected parents (carriers)	N/A	N/A
Chromosome Location	N/A	N/A	X chromosome	X chromosome
Male vs Female	N/A	N/A	Males more affected	N/A
Carrier Status	N/A	Heterozygous individuals	Females can be carriers	N/A
Transmission	N/A	N/A	N/A	Affected males pass to daughters
Examples	Huntington's, achondroplasia	Cystic fibrosis, sickle cell anaemia	Haemophilia, colour blindness	Hypophosphatemic rickets

Figure 1.3 Types of mendelian inheritance patterns.

the Figure 1.3 follow Mendel's laws of segregation and independent assortment, which explain the predictable inheritance of genetic traits across generations [40, 63].

1.2.3 Non-Mendelian Inheritance (Epigenetics, Mitochondrial Inheritance)

Non-Mendelian inheritance refers to patterns of genetic inheritance that do not follow Mendel's laws of dominance, segregation, and independent assortment. Two major types of non-Mendelian inheritance are epigenetics and mitochondrial inheritance [64].

1.2.3.1 Epigenetics: Changes in Gene Expression without DNA Alteration

Epigenetics describes the study of changes in gene expression density that occur without changes in the DNA code. Rather, it influences the process of epigenetic modifications, i.e., turning of genes on or off. Some of the epigenetic regulation mechanisms are the DNA methylation, histone modification, and regulation of non-coding RNA (ncRNA) [65]. DNA methylation involves making the DNA undergo a reaction involving accretion of a methyl (-CH3) unit to the DNA (generally to the base cytosine). In highly methylated cells, the process of silencing of genes takes place and in the cells with little methylation the genes open [66].

DNA methylation can be illustrated by X-chromosome inactivation (inactivation of the X chromosome in a female) [67]. Under histone modification, the DNA molecule is coiled around histone proteins and the activity of the gene is altered through modifications such as acetylation and methylation. Acetylation causes DNA to relax, therefore becoming more active, which in turn activates genes, whereas methylation can suppress gene activity [68]. The ncRNAs (e.g., microRNAs) aid in controlling gene expression by inhibiting the translation of specific messenger RNA (mRNA); for example, miRNA controls the growth of cancer cells [69].

1.2.3.2 Mitochondrial Inheritance: Maternal Inheritance

Mitochondrial inheritance (also called maternal inheritance) refers to the transmission of genes found in mitochondria, which are only inherited from the mother. Unlike nuclear DNA, which follows Mendelian inheritance, mitochondrial DNA (mtDNA) is passed exclusively from mother to offspring [70]. Sperm mitochondria are usually destroyed after fertilization, leaving only the mother's mitochondrial DNA in the embryo [71]. Since mitochondria play a key role in energy production, mutations in mtDNA can lead to diseases affecting high-energy-demand organs like the brain, muscles, and heart [72]. Examples of mitochondrial disorders include Leber's Hereditary Optic Neuropathy (LHON), which causes sudden vision loss due to optic nerve damage; Mitochondrial Myopathy, leading to muscle weakness, fatigue, and exercise intolerance; and Mitochondrial Encephalopathy, Lactic Acidosis, and Stroke-like episodes (MELAS), which affect the brain, muscles, and metabolism causing strokes and seizures [73].

1.3 Genomic Technologies and Applications

1.3.1 Genomic Sequencing and Interpretation

Genomic sequencing refers to the process of ascertaining the entire DNA code of an organism's genome. This entails determining the exact sequence of nucleotide molecules (adenine, thymine, cytosine, and guanine) that form DNA [74]. Next-generation sequencing (NGS) technology has transformed genome sequencing into a faster and more cost-effective process, offering the ability to read large quantities of DNA efficiently [75]. NGS allows scientists to study whole genomes or specific regions of interest. Whole-genome sequencing (WGS) involves sequencing the entire genome of an organism, while whole-exome sequencing focuses on sequencing only the protein-coding regions (exons) [76]. The general workflow involves DNA extraction, DNA fragmentation, library preparation, sequencing, and data analysis [77].

Genomic interpretation involves analyzing the sequenced DNA data to identify and understand the biological significance of genetic variations. This includes determining how these variations may contribute to health, disease, or other traits [78].

1.3.1.1 Key Aspects

Variant Identification: Detecting differences in the DNA sequence compared to a reference genome [79].
Variant Classification: Predicting possible effects of variants, such as pathogenic, likely pathogenic, variants of uncertain significance, benign, or likely benign [80].
Clinical Relevance: Assessing the implications of genetic variations for diagnosis, prognosis, and treatment of diseases [81].
Data Analysis: Using bioinformatics tools to analyze and interpret genomic data [82].

1.3.1.2 Applications

Personalized Medicine: Tailoring medical treatment based on an individual's genetic makeup [83].
Disease Diagnosis: Identifying genetic causes of disease conditions [84].
Cancer Genomics: Understanding the genetic basis of cancer to develop specific treatments [85].
Infectious Disease Surveillance: Monitoring pathogen transmission and evolution in infected patients [86].

1.3.2 Genetic Testing (Types and Purposes)

Genetic testing is a diagnostic tool used to identify changes in genes, chromosomes, and proteins. It helps clinicians confirm or exclude suspected genetic disorders and assess the risk of an individual being affected or being a carrier of inherited health issues. The types of genetic tests are illustrated in Figure 1.4 [87].

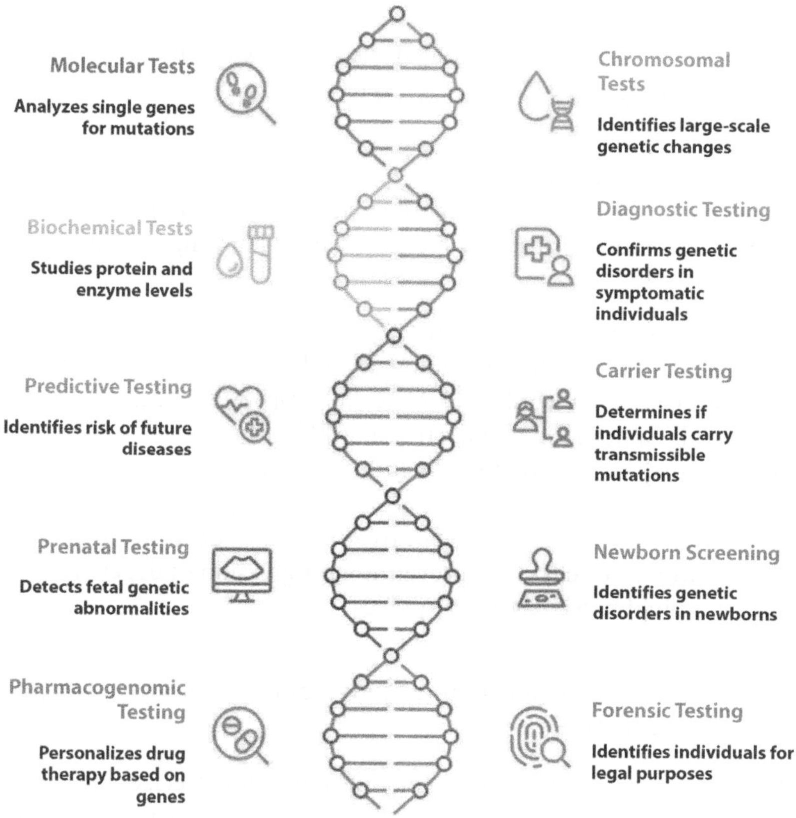

Figure 1.4 Genetic testing framework.

1.3.3 Gene Therapy and CRISPR Technology

In gene therapy, manipulation or introduction of genetic material into a somatic cell is done to achieve treatment or prevention of illness. It aims to correct defective genes responsible for disease development [29].

1.3.3.1 Types of Gene Therapy

1.3.3.1.1 Gene Addition (Replacement Therapy)

A normal gene is added to replace a missing or faulty gene. An example includes adding a functional copy of the *RPE65* gene to treat Leber congenital amaurosis, a rare eye disease [88].

1.3.3.1.2 Gene Editing

Faulty genes are directly corrected within the DNA, often using CRISPR or other gene-editing tools [89].

1.3.3.1.3 Gene Silencing

This approach reduces or turns off the activity of a faulty gene. For example, RNA interference is used to silence genes in Huntington's disease [90].

1.3.3.1.4 Cell-Based Gene Therapy

Patients' cells are modified outside the body and then reintroduced. A well-known example includes CAR-T cell therapy for cancer, where T cells are genetically modified to attack tumors [91].

The purposes of gene therapy include treating inherited disorders (e.g., severe combined immunodeficiency (SCID), hemophilia); treating cancers (e.g., leukemia *via* CAR-T therapy); treating viral infections (e.g., HIV by modifying immune cells); and preventing progression of degenerative diseases (e.g., spinal muscular atrophy) [92].

1.3.3.2 Clustered Regularly Interspaced Short Palindromic Repeats (CRISPR)

CRISPR is a gene-editing technology designed for precise alterations in DNA sequences. The CRISPR-Cas9 tool acts as molecular scissors to cut and paste DNA at desired sites. It has revolutionized gene therapy by offering a more precise and efficient way to edit genes. CRISPR can be used to correct disease-causing mutations directly within the genome, enhance the effectiveness of cell therapies such as those used in cancer treatment, and improve the safety and accuracy of gene delivery [93].

CRISPR is becoming a key tool in advancing gene therapy. It offers the potential to overcome limitations of traditional gene therapy approaches. Researchers are developing CRISPR-based therapies for genetic diseases such as sickle cell disease, cystic fibrosis, and certain cancers [94].

1.3.3.2.1 Key Points

- CRISPR offers a level of precision in gene editing that was previously unattainable [95].
- The combination of gene therapy and CRISPR is an effective approach for treating and potentially curing genetic diseases [96].

- Ethical considerations are crucial in the use of CRISPR technology, especially concerning germline editing [97].
- CRISPR is a powerful tool greatly enhancing the potential of gene therapy [98].

1.3.4 Pharmacogenomics

The study of **pharmacogenomics** investigates how a person's genes affect their response to medications. By tailoring drug therapy to a patient's genetic makeup, pharmacogenomics is central to **personalized medicine** [99]. Genetic variations can influence how drugs are absorbed, metabolized, distributed, and excreted—impacting efficacy and safety. Identifying these genetic variants helps predict drug response and avoid adverse effects [100]. The Table 1.1 lists out the applications areas of pharmacogenomics and its significance.

Genetic profiling allows healthcare providers to forecast individual drug responses, select the most appropriate medication, determine optimal dosages, and reduce the risk of harmful reactions [101]. Many drug-metabolizing enzymes are encoded by genes, and variations in these genes can lead to **slow metabolism** (increased risk of toxicity) or **rapid metabolism** (reduced effectiveness) [102]. Additionally, gene variants in drug targets can alter drug efficacy by affecting how medications interact with their targets as mentioned in the Table 1.2 saying genes involved and their relevance [103].

Table 1.1 Applications of pharmacogenomics.

Application area	Description
Optimizing Drug Therapy	Pharmacogenomics helps identify individuals who are likely to benefit from specific medications and those who are at risk of adverse reactions [104].
Reducing Adverse Drug Reactions	Adverse drug reactions are a major cause of hospitalization and death. Pharmacogenomic testing can identify individuals at increased risk and prevent harm [105].
Improving Drug Development	Genetic markers identified through pharmacogenomics can predict drug response, improving the efficiency and success of drug development [100].

Table 1.2 Pharmacogenomics – examples.

Example	Gene(s) involved	Clinical relevance
Warfarin	*CYP2C9, VKORC1*	Variations affect warfarin metabolism and sensitivity; testing helps determine safe and effective dosage [4, 105, 106].
Cancer Treatment	Tumor-specific mutations	Identifying mutations enables use of targeted therapies, improving treatment outcomes in various cancers [85].
Antidepressants	*CYP2D6, CYP2C19*	Variations impact drug metabolism; testing guides selection and dosing of antidepressants for better results [107].

1.4 Genetic Disorders and their Clinical Relevance

Genetic disorders, which can range from chromosomal abnormalities to single-gene mutations, are illnesses brought on by anomalies in a person's DNA. From minor to fatal illnesses, these abnormalities can result in a broad range of clinical symptoms. For example, Huntington's disease causes neurodegeneration because of a faulty HTT gene, while cystic fibrosis, which is brought on by mutations in the CFTR gene, damages the respiratory and digestive systems. Early diagnosis through genetic testing can guide treatment and reproductive decisions, enhancing patient outcomes. Advances in genomics have enabled personalized medicine approaches for managing genetic disorders [108].

1.4.1 Single Gene Disorders

Mutations in a single gene are the cause of single-gene disorders, also referred to as monogenic disorders. These disorders often follow predictable inheritance patterns. Two well-known examples are cystic fibrosis and sickle cell disease [109].

1.4.2 Chromosomal Abnormalities

Errors in the number or structure of chromosomes, which carry our genetic information, are known as chromosomal abnormalities. Numerous health and developmental issues may result from these anomalies [110].

During the development of egg or sperm cells, chromosomal abnormalities frequently arise. As mothers age, they are more susceptible to certain genetic abnormalities, such as Down syndrome [111]. Karyotyping is one type of genetic testing that can detect chromosomal abnormalities. Depending on the specific chromosomal anomaly, a variety of developmental and health problems may result. Additional instances of chromosomal abnormalities include Edwards Syndrome (Trisomy 18), Patau Syndrome (Trisomy 13), Klinefelter Syndrome (XXY), and Turner Syndrome (Monosomy X) [112].

1.4.3 Multifactorial and Complex Disorders

Multifactorial and complex disorders are caused by a combination of multiple genes and environmental factors. Unlike single-gene or chromosomal disorders, as stated in Figure 1.5, these conditions do not follow simple

Characteristic	Cystic Fibrosis (CF)	Sickle Cell Disease (SCD)
Cause	CFTR gene mutation	HBB gene mutation
Effects	Thick mucus, lung/digestive issues	Blocked blood flow, pain, anaemia
Inheritance	Autosomal recessive	Autosomal recessive
Diagnosis	Newborn screening, sweat/genetic tests	Hemoglobin electrophoresis, genetic testing
Treatment	Manage symptoms, thin mucus	Manage symptoms, blood transfusions

Figure 1.5 Single gene disorders – characteristics of CF and SCD.

Mendelian inheritance patterns and are influenced by lifestyle, environment, and genetic predispositions [113].

1.4.3.1 Key Characteristics

1.4.3.1.1 Multiple Factors

These disorders result from the interaction of several genes and environmental influences, making them complex and challenging to predict [113].

1.4.3.1.2 Genetic Predisposition

Individuals may have a genetic susceptibility to a particular disorder. However, the disorder may not develop unless triggered by environmental factors [114].

1.4.3.1.3 Environmental Influences

Factors such as diet, lifestyle, exposure to toxins, and infections play significant roles. These environmental elements interact with genetic predispositions to influence disease development [115].

1.4.3.1.4 Complex Inheritance

These disorders do not follow simple Mendelian inheritance. Though family history can influence risk, inheritance patterns are often unclear [116].

1.5 Genomics in Healthcare

Genomics in healthcare refers to using a person's genetic information to guide decisions related to prevention, diagnosis, treatment, and management of diseases. It is a cornerstone of personalized (or precision) medicine, tailoring healthcare to the individual.

1.5.1 Applications of Genomics in Healthcare

Genomics has revolutionized healthcare by enabling more precise, predictive, and personalized medical practices. One of its primary applications is in diagnostic genomics, where genome sequencing helps identify genetic mutations responsible for inherited diseases, aiding early diagnosis and treatment planning. In oncology, genomics plays a pivotal role in detecting tumor-specific mutations, allowing the development of targeted therapies and improving patient outcomes. Pharmacogenomics, a branch of genomics, helps tailor drug prescriptions based on a person's genetic makeup, minimizing adverse drug reactions and enhancing drug efficacy (as listed out in Table 1.3). Additionally, genomic screening in newborns can detect conditions that may not be apparent at birth but can benefit from early intervention. Reproductive genomics also supports informed

Table 1.3 Summary table: Examples of multifactorial and complex disorders.

Disorder	Cause	Genetic factors	Environmental triggers	Management
Diabetes Mellitus (Type 2)	Interaction of multiple genes affecting insulin production and resistance; influenced by lifestyle factors.	*TCF7L2, PPARG, FTO, KCNJ11*	High sugar diet, lack of exercise, obesity	Lifestyle modification, medication, insulin therapy (if needed) [117]
Cancer	Mutations in genes related to cell growth and DNA repair; often accelerated by external factors.	*BRCA1, BRCA2, TP53, APC*	Smoking, radiation, chemicals, infections (e.g., HPV)	Depends on type: surgery, chemotherapy, immunotherapy, targeted therapy [118]
Heart Disease	Result of genetic predisposition and environmental risk factors leading to atherosclerosis and cardiovascular issues.	*LDLR, APOB, PCSK9, LPA*	High-fat diet, smoking, hypertension, lack of exercise	Diet and lifestyle changes, statins, antihypertensives, surgical interventions [119]
Asthma	Chronic inflammation of airways, often triggered by allergens and pollutants.	*ORMDL3, IL4, IL13, ADAM33*	Pollution, allergens, respiratory infections, smoking	Inhalers (bronchodilators, corticosteroids), avoiding triggers, immunotherapy [120]
Alzheimer's Disease	Neurodegenerative disorder with progressive cognitive decline.	*APOE ε4, APP, PSEN1, PSEN2*	Aging, head trauma, poor cardiovascular health	No cure; managed with medications (e.g., donepezil), cognitive therapy, support [121]

Table 1.4 Application areas of genomics in healthcare.

Application area	Description	Examples
Disease Diagnosis and Risk Prediction	Enables early and accurate diagnosis of genetic diseases and prediction of inherited conditions.	Cystic fibrosis, Marfan syndrome, BRCA1/ BRCA2 testing, Huntington's predictive testing
Personalized Treatment (Precision Medicine)	Tailors medical treatment based on an individual's genetic profile.	EGFR inhibitors for lung cancer, HER2-targeted therapy in breast cancer
Pharmacogenomics	Uses genetic data to predict individual responses to drugs, improving safety and effectiveness.	CYP2C19 testing for clopidogrel, CYP2D6 for antidepressants
Cancer Genomics	Identifies somatic mutations in tumors for therapy selection and monitors tumor progression or resistance.	Liquid biopsy, KRAS, BRAF, and EGFR mutation analysis
Reproductive Health	Assists in prenatal, preimplantation, and newborn screening to identify or avoid genetic disorders.	PGD during IVF, prenatal testing for Down syndrome, newborn screening for metabolic disorders
Rare and Undiagnosed Disease Diagnosis	Whole genome or exome sequencing helps diagnose unexplained rare diseases, especially in infants.	Pediatric rare disease diagnosis in NICUs using WGS or WES
Infectious Disease Tracking and Control	Monitors mutations in infectious agents to guide public health responses and vaccine design.	SARS-CoV-2 variant tracking, genome sequencing of influenza viruses
Population Health & Preventive Genomics	Screens large groups for genetic risk to enable early interventions and lifestyle changes.	Screening for familial hypercholesterolemia, Lynch syndrome in asymptomatic individuals
Gene Therapy	Repairs or replaces faulty genes using gene-editing technologies.	Zolgensma for SMA, CRISPR-based therapies for sickle cell disease

decision-making through carrier screening and preimplantation genetic testing. Furthermore, population-scale genomic initiatives are enabling predictive risk assessments, identifying individuals at higher risk for common conditions like diabetes or heart disease, which supports preventive healthcare strategies. The integration of genomics with electronic health records is improving clinical decision-making and care management across systems. Overall, genomics is shifting healthcare from a reactive model to a proactive and personalized one, promising better patient outcomes and healthcare efficiency [122]. Figure 1.6 point towards the different applications and impact of genomics in healthcare.

The benefits include more accurate diagnoses, tailored treatments with fewer side effects, early detection and prevention of diseases, better understanding of disease mechanisms and empowered patients with personalized health plans. The Table 1.4 enumerated the application areas of genomics in healthcare.

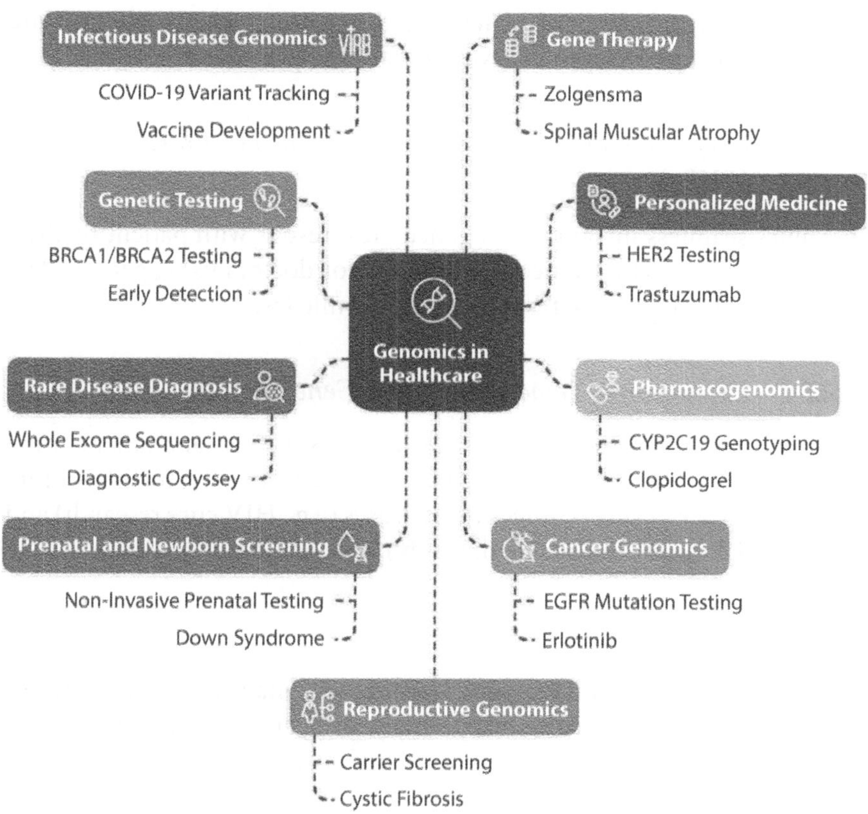

Figure 1.6 Application and impact of genomics in healthcare.

1.5.2 Future Directions for Genomics in Healthcare

Genomics is rapidly advancing, and its integration into mainstream healthcare is transforming how we prevent, diagnose, treat, and manage diseases. The key future directions and trends shaping the future of genomics in medicine include the following:

1.5.2.1 Precision and Personalized Medicine for All

From one-size-fits-all to individualized care based on each person's genetic profile. Widespread use of genomic data to guide medication choices, treatment plans, and disease risk assessments. Example: Tailored cancer therapies based on tumor mutations.

1.5.2.2 Widespread Genomic Screening in Healthy Individuals

Use of whole genome sequencing (WGS) as a routine part of health check-ups. Early identification of genetic predispositions, carrier status, and actionable mutations (even in asymptomatic people).

1.5.2.3 Integration of Genomics into Electronic Health Records (EHRs)

It enables seamless integration of genetic test results with patient medical records, real-time clinical decision support for doctors using genetic data and ensures lifelong value from a single genomic test.

1.5.2.4 Expansion of Gene Editing and Gene Therapy

It can expand to the broader application of CRISPR-Cas9 and other editing technologies. Development of safe, ethical, and effective treatments for rare genetic disorders, cancers, infectious diseases (e.g., HIV cure research) and has potential for *in vivo* (inside the body) gene editing.

1.5.2.5 AI and Big Data in Genomics

Artificial Intelligence and machine learning will analyse massive genomic datasets, predict disease risks, discover new drug targets and enables better interpretation of variants of unknown significance (VUS).

1.5.2.6 Global Genomic Initiatives and Equity

Programs like the UK's Genomics England and All of Us (USA) aim to sequence millions of genomes. It ensure genomic medicine benefits all populations, including underrepresented groups and addresses the ethnic diversity gap in genetic databases.

1.5.2.7 Polygenic Risk Scores (PRS)

Use of combined effect of many genes to estimate risk for complex diseases like Type 2 diabetes, coronary artery disease, and Alzheimer's and enables preventive strategies tailored to genetic risk.

1.5.2.8 Advances in Reproductive Genomics

Non-invasive prenatal testing (NIPT) with full fetal genome sequencing will be improved. It enables better embryo selection using PGT (Preimplantation Genetic Testing) in IVF and new ethical frameworks around embryo gene editing.

1.5.2.9 Ethics, Regulation, and Data Privacy

As genomics becomes more widespread, stronger data protection laws (e.g., GDPR, HIPAA), ethical policies for germline editing and incidental findings and focus on informed consent and genetic literacy would be accomplished.

1.5.2.10 Preventive and Predictive Genomics in Everyday Healthcare

Routine use of genomics to predict disease before it appears could be done. Custom health recommendations like diet & nutrition (nutrigenomics), Exercise plans and screening schedules can be sorted.

1.6 Challenges and Limitations of Genomics and Precision Healthcare

1.6.1 Data Management and Analysis

Genomic sequencing generates vast datasets—whole-genome sequencing yields hundreds of gigabytes per individual—that require robust storage, transmission, and bioinformatics pipelines [123]. Hospitals must adopt scalable cloud infrastructure and secure data warehouses. Analyses often involve complex variant calling, annotation, and filtration, necessitating high-performance computing and reproducible workflows [124]. Machine learning offers promise but introduces issues of interpretability and bias [125].

1.6.2 Interpretation and Clinical Utility of Genomic Data

Interpreting variants remains a bottleneck. Many detected variants are classified as variants of uncertain significance (VUS), complicating clinical decisions [37]. Even pathogenic variants may have variable penetrance: for instance, BRCA1 mutations convey distinct risks based on age and family history [126]. Clinical utility depends on actionable findings that guide prevention or treatment—yet only a minority of genomic findings meet this threshold [127].

1.6.3 Ethical, Legal, and Social Implications (ELSI)

Genomic data raises privacy concerns: risk of re-identification even from de-identified data exists, and unauthorized sharing can lead to discrimination [128]. Ethical challenges include deciding whether and how to return incidental findings unrelated to the primary clinical question [129]. Legally, laws like GINA in the U.S. protect against genetic discrimination—gaps remain around life insurance, disability, and global contexts [130]. Socially, concerns about stigmatization and psychological impact of "genomic knowledge" loom large [131].

1.6.4 Access and Equity in Genomics and Precision Healthcare

Precision healthcare uptake is uneven. High-income and urban areas benefit, but underserved communities, especially in low- and middle-income countries, lag due to infrastructure limitations, cost barriers, and underrepresentation in genomic studies [132]. For example, GWAS datasets are ~80% European in ancestry, limiting risk predictions for other groups [133]. Efforts like NIH's All of Us and H3Africa aim to broaden inclusion [134].

1.6.5 Regulatory Frameworks and Policy Challenges

Regulatory oversight of genomic tests has evolved. In the U.S., the FDA regulates certain high-risk tests, while many labs operate under CLIA and

CMS guidance—leading to inconsistencies in oversight and test quality [135]. Globally, harmonizing standards is a challenge: the EU IVDR has introduced stricter approval pathways but varies in implementation [136]. Policies must balance innovation and patient safety while adapting to new technologies like gene editing.

1.7 Role of Healthcare Professionals in Genetics and Genomics

1.7.1 Genetic Counseling and Patient Communication

Genetic counselors are essential in translating complex genomic information into patient-friendly terms. They assess family and medical history, communicate genetic risk, obtain informed consent, and discuss test limitations and psychosocial impact [137]. Training in communication skills and cultural competency is critical as patients may have strong emotional and ethical concerns [138].

1.7.2 Integrating Genomics into Clinical Practice

Primary clinicians increasingly order genetic tests, necessitating genomic literacy. Clinical decision support (CDS) tools, integrated into electronic health records, help providers interpret results and recommend management [139]. Multidisciplinary tumor boards, pharmacogenomics advisory groups, and joint clinics (e.g., cardio-genetics) foster collaboration [140]. Continuous education and point-of-care resources are vital for safe implementation.

1.7.3 Future Trends in Personalized Medicine

Advances in polygenic risk scores (PRS) are enabling risk stratification for common diseases, though their predictive ability varies by ancestry [141]. Direct-to-consumer (DTC) testing is democratizing access but raises concerns about result validity and follow-through with healthcare providers [142]. The rise of multi-omic profiling (genome, epigenome, transcriptome, microbiome) promises nuanced patient stratification and identification of novel therapeutic targets [143].

1.8 Conclusion

1.8.1 Case Studies and Examples

Pharmacogenomics: CYP2C19 genotyping before prescribing clopidogrel reduces adverse cardiovascular events in high-risk individuals [144].
Oncology: The National Lung Matrix trial matches patients to targeted therapies based on tumor genomics with improved outcomes [145].
Rare disease diagnostics: WES in pediatric cohorts achieves diagnostic yield up to 40%, ending diagnostic odysseys and enabling tailored management [146].

1.8.2 Summary of Key Takeaways

1. **Data infrastructure** is critical to manage the scale and complexity of genomic information.
2. **Interpreting variants** remains challenging; most results lack clinical actionability.
3. **ELSI concerns**—privacy, consent, equity—must be addressed proactively.
4. **Healthcare professionals** need genomic training and decision-support tools to guide testing and interpretation.
5. **Policy and regulation** must evolve with technology to ensure test quality and patient safety.

1.8.3 Future Directions in Genetics and Genomics

Precision healthcare is moving toward integrating real-time multi-omic data, wearable biosensors, and AI-driven analyses to personalize prevention and therapy. Improving global genomic representation will enhance equity and accuracy. Enhanced AI transparency and shared decision-making models are key for patient trust and engagement. Finally, learning healthcare systems that iteratively refine practice based on genomic and clinical outcomes will close the loop between discovery and patient benefit.

AI disclosure statement: The author acknowledges the use of AI-based tools for minor language editing or grammatical correction. However, all intellectual contributions, analyses, and conclusions presented in this chapter are entirely the author's own.

References

1. Mendel, G., *Experiments in Plant Hybridization*, 1865, [English translation by W. Bateson, 1901].
2. Bateson, W., *Mendel's Principles of Heredity: A Defence*, Cambridge University Press, Cambridge, UK, 1909.
3. Alberts, B., *et al.*, *Molecular Biology of the Cell*, 6th ed., Garland Science, New York, NY, 2015.
4. Pennisi, E., A brief history of genomics. *Science*, 291, 5507, 1195–1200, Feb. 2001.
5. National Human Genome Research Institute, Genomics vs. Genetics, [Online]. Available: https://www.genome.gov/genetics-glossary/Genomics. [Accessed: Jul. 24, 2025].
6. Collins, F.S., Green, E.D., Guttmacher, A.E., Guyer, M.S., A vision for the future of genomics research. *Nature*, 422, 6934, 835–847, Apr. 2003.
7. Green, E.D. and Guyer, M.S., Charting a course for genomic medicine from base pairs to bedside. *Nature*, 470, 204–213, Feb. 2011.
8. Domchek, M.M., *et al.*, BRCA1 and BRCA2: Risks and response to therapy. *Annu. Rev. Med.*, 64, 131–143, 2013.
9. Gregory, J.W., *et al.*, The role of genetics in disease susceptibility and clinical decision making. *Clin. Med.*, 16, 5, 441–447, 2016.
10. Roden, D.J., *et al.*, Pharmacogenomics: Challenges and opportunities. *Ann. Intern. Med.*, 145, 10, 749–757, 2006.
11. Sequist, L.V., *et al.*, Molecular predictors of response to epidermal growth factor receptor antagonists in non-small-cell lung cancer. *J. Clin. Oncol.*, 25, 5, 587–595, 2007.
12. Knowles, J.A. and Ashley, D.E., Cardiovascular disease: The rise of the genetic risk score. *PLoS Med.*, 15, 3, e1002546, 2018.
13. Green, R.C., *et al.*, Disclosure of APOE genotype for risk of Alzheimer's disease. *N. Engl. J. Med.*, 361, 3, 245–254, 2009.
14. Ioannidis, A.E., *et al.*, Carrier screening for recessive disorders in clinical practice. *Am. J. Med. Genet. Part A*, 173, 10, 2710–2721, 2017.
15. Wells, S.K., *et al.*, Clinical utility of preimplantation genetic testing. *Nat. Rev. Genet.*, 22, 9, 587–602, 2021.
16. Neguembor, V.A., *et al.*, CRISPR-based technologies in gene therapy and biomedical research. *Nat. Rev. Mol. Cell Biol.*, 24, 1, 21–40, 2023.
17. Trounson, A. and McDonald, C., Stem cell therapies in clinical trials: Progress and challenges. *Cell Stem Cell*, 17, 1, 11–22, 2015.
18. Oude Munnink, R.D., *et al.*, The next phase of SARS-CoV-2 surveillance: Real-time genomic monitoring. *Lancet Infect. Dis.*, 21, 5, e76–e84, 2021.
19. Mbogo, J.C. and Mburu, K.M., Genomic insights into HIV and malaria: Implications for vaccine development. *Infect. Genet. Evol.*, 91, 104821, 2021.
20. Ruse, M., *Monad to Man: The Concept of Progress in Evolutionary Biology*, Harvard University Press, 1996.

21. Shipley, A.E., The early theories of heredity. *J. Hist. Biol.*, 4, 1, 1–22, 1971.
22. Darwin, C., *On the Origin of Species by Means of Natural Selection*, John Murray, London, 1859.
23. Bowler, P.J., *The Mendelian Revolution: The Emergence of Hereditarian Concepts in Modern Science and Society*, Johns Hopkins University Press, 1989.
24. Morgan, T.H., The mechanism of Mendelian heredity. *Science*, 37, 959, 871–873, 1913.
25. Watson, J.D. and Crick, F.H.C., Molecular structure of nucleic acids: A structure for deoxyribose nucleic acid. *Nature*, 171, 737–738, 1953.
26. Crick, F.H., Central dogma of molecular biology. *Nature*, 227, 5258, 561–563, 1970.
27. Berg, P., *et al.*, Potential biohazards of recombinant DNA molecules. *Science*, 185, 4148, 303–305, 1974.
28. Bird, A.P., DNA methylation and the frequency of CpG in animal DNA. *Nucleic Acids Res.*, 9, 4, 989–997, 1981.
29. Doudna, J.A. and Charpentier, E., The new frontier of genome engineering with CRISPR-Cas9. *Science*, 346, 6213, 1258096-1–1258096-9, 2014.
30. Cavazzana-Calvo, M., *et al.*, Gene therapy of human severe combined immunodeficiency (SCID)-X1 disease. *Science*, 288, 5466, 669–672, 2000.
31. Topol, E., *The Patient Will See You Now: The Future of Medicine is in Your Hands*, Basic Books, 2015.
32. Zou, J., *et al.*, A primer on deep learning in genomics. *Nat. Genet.*, 51, 1, 12–18, Jan. 2019.
33. National Research Council, *Toward Precision Medicine: Building a Knowledge Network for Biomedical Research and a New Taxonomy of Disease*, National Academies Press, Washington, DC, 2011.
34. Biesecker, A.G. and Biesecker, L., An approach to integrating genomics into health care. *Nat. Rev. Genet.*, 15, 409–414, 2014.
35. van Allen, M.A. and Wagle, E.M., Clinical applications of cancer genomics in precision oncology. *J. Clin. Oncol.*, 35, 9, 929–933, 2017.
36. Weinshilboum, R.M. and Wang, L., Pharmacogenomics: Precision medicine and drug response. *Mayo Clin. Proc.*, 92, 11, 1711–1722, 2017.
37. Rehm, H., *et al.*, ClinGen — The Clinical Genome Resource. *N. Engl. J. Med.*, 372, 23, 2235–2242, 2015.
38. Khoury, M.J., Iademarco, M., Riley, D., Precision public health for the era of precision medicine. *Am. J. Prev. Med.*, 50, 3, 398–401, 2016.
39. Alberts, B., *et al.*, *Molecular Biology of the Cell*, 6th ed., Garland Science, New York, 2014.
40. Nelson, D.L. and Cox, M.M., *Lehninger Principles of Biochemistry*, 7th ed., W. H. Freeman, 2017.
41. Franklin, H.R., X-ray diffraction studies of DNA. *Nature*, 171, 740, 1953.
42. Crick, F.H.C., The structure of the hereditary material. *Sci. Am.*, 191, 4, 54–62, 1954.

43. Watson, J.D., *The Double Helix: A Personal Account of the Discovery of the Structure of DNA*, Atheneum, 1968.
44. Kornberg, A. and Baker, T.A., *DNA Replication*, 2nd ed., W. H. Freeman, 1992.
45. Stryer, L., *Biochemistry*, 5th ed., W. H. Freeman, New York, 2002.
46. Lewin, B., *Genes XII*, Jones & Bartlett Learning, 2017.
47. Meselson, M. and Stahl, F.W., The replication of DNA in Escherichia coli. *Proc. Natl. Acad. Sci.*, 44, 7, 671–682, 1958.
48. Reece, R.H., *Analysis of Genes and Genomes*, 3rd ed., John Wiley & Sons, 2004.
49. Brown, T.A., *Genomes*, 3rd ed., Garland Science, 2006.
50. Serjeant, G.R., Sickle-cell disease. *Lancet*, 350, 9079, 725–730, 1997.
51. Ratjen, F. and Döring, G., Cystic fibrosis. *Lancet*, 361, 9358, 681–689, 2003.
52. Futuyma, D.J., *Evolution*, 4th ed., Sinauer Associates, 2023.
53. Sancar, A., Mechanisms of DNA repair by photolyase and excision nuclease. *Biochemistry*, 33, 1, 2–9, 1994.
54. Olby, R.C., *The Origins of Mendelism*, University of Chicago Press, Chicago, 1985.
55. Holland, M.J., Rediscovery of Mendel's Work. *Nature*, 127, 587–589, 1931.
56. Johannsen, W., Die Genetik und die phänotypische Variabilität. *Jena Z. Naturwiss.*, 44, 1–39, 1911.
57. Watson, J., *et al.*, *Molecular Biology of the Gene*, 7th ed., Pearson, 2013.
58. Lander, E.S., *et al.*, Initial sequencing and analysis of the human genome. *Nature*, 409, 6822, 860–921, 2001.
59. Simmons, M.J. and Crowley, J.L., *Principles of Genetics*, 5th ed., Wiley, 2009.
60. Kimmel, C.B. and Clark, M.W., *Genetics and Genomics in Medicine*, 2nd ed., Elsevier, 2016.
61. Griffiths, A., *et al.*, *Introduction to Genetic Analysis*, 11th ed., W. H. Freeman, 2015.
62. Hartl, B. and Jones, E.W., *Genetics: Analysis of Genes and Genomes*, 7th ed., Jones & Bartlett, 2018.
63. Sepulveda, J.L.H., X-linked inheritance patterns and human genetic diseases. *J. Med. Genet.*, 52, 6, 391–398, 2015.
64. Szyf, M., Epigenetics, DNA methylation, and chromatin modifying drugs. *Annu. Rev. Pharmacol. Toxicol.*, 46, 375–401, 2006.
65. Brinkman, B.R., Epigenetics: Principles and practice. *Methods Mol. Biol.*, 1050, 1–18, 2013.
66. Bird, A., DNA methylation patterns and epigenetic memory. *Genes Dev.*, 16, 1, 6–21, 2002.
67. Greally, J.B., X-chromosome inactivation and epigenetic regulation. *Epigenomics*, 1, 2, 149–158, 2009.
68. Allis, C.D. and Jenuwein, T., The molecular hallmarks of epigenetic control. *Nat. Rev. Genet.*, 17, 8, 487–500, 2016.

69. Iorio, A.E. and Croce, C., MicroRNAs in cancer. *Adv. Exp. Med. Biol.*, 774, 53–70, 2013.
70. Wallace, D.C., Mitochondrial genetics: A paradigm for aging and degenerative diseases? *Science*, 256, 5057, 628–632, 1992.
71. Schon, R.J., Maternal inheritance of mitochondrial DNA. *Annu. Rev. Genet.*, 38, 119–141, 2004.
72. Shoubridge, E.A., Mitochondrial diseases and the brain. *Neuroscientist*, 8, 2, 73–82, 2002.
73. DiMauro, J. and Schon, E.A., Mitochondrial respiratory-chain diseases. *N. Engl. J. Med.*, 348, 26, 2656–2668, 2003.
74. Venter, J.M., *et al.*, The sequence of the human genome. *Science*, 291, 5507, 1304–1351, 2001.
75. Mardis, M., Next-generation sequencing platforms. *Annu. Rev. Anal. Chem.*, 6, 287–303, 2013.
76. Shendure, N., *et al.*, DNA sequencing at 40: Past, present and future. *Nature*, 550, 345–353, 2017.
77. Glusman, A.M., The bioinformatics of next-generation sequencing. *Briefings Bioinf.*, 15, 6, 1030–1042, 2014.
78. Sherry, S.T., *et al.*, dbSNP: The NCBI database of genetic variation. *Nucleic Acids Res.*, 29, 1, 308–311, 2001.
79. Pozhitkov, L.D., Variant calling in next-generation sequencing data. *Front. Genet.*, 9, article 1054, 2018.
80. Richards, S., *et al.*, Standards and guidelines for the interpretation of sequence variants. *Genet. Med.*, 17, 5, 405–424, 2015.
81. Biesecker, M. and Green, J., Diagnostic clinical genome and exome sequencing. *N. Engl. J. Med.*, 370, 25, 2418–2425, 2014.
82. Wang, Y., Bioinformatics tools for data analysis in next-generation sequencing. *Briefings Bioinf.*, 14, 5, 667–682, 2013.
83. Ashley, K., Towards precision medicine. *Nat. Rev. Genet.*, 17, 9, 507–522, 2016.
84. Ng, S., *et al.*, Genetic diagnosis using whole exome sequencing in inherited disease. *Nat. Rev. Genet.*, 12, 10, 745–755, 2011.
85. Garraway, C. and Lander, E., Lessons from cancer genomics. *Cell*, 153, 1, 17–37, 2013.
86. Gardy, A. and Loman, N., Towards a genomics-informed, real-time, global pathogen surveillance system. *Nat. Rev. Genet.*, 19, 1, 9–20, 2018.
87. Rehm, E.M., *et al.*, Clinical genomic sequencing for precision medicine: Opportunities and challenges. *J. Am. Med. Assoc.*, 319, 13, 1314–1325, 2018.
88. Russell, R., *et al.*, Efficacy and safety of voretigene neparvovec (AAV2-hRPE65v2) in patients with RPE65-mediated inherited retinal dystrophy: A randomised, controlled, open-label, phase 3 trial. *Lancet*, 390, 10097, 849–860, 2017.
89. Hsu, H., Lander, E., Zhang, F., Development and applications of CRISPR-Cas9 for genome engineering. *Cell*, 157, 6, 1262–1278, 2014.

90. Davidson, S. and McCray, M., Current prospects for RNA interference-based therapies. *Nat. Rev. Genet.*, 12, 5, 329–340, 2011.
91. June, M., *et al.*, CAR T cell therapy for cancer. *Science*, 359, 6382, 1361–1365, 2018.
92. Ginn, L., *et al.*, Gene therapy clinical trials worldwide to 2012 – An update. *J. Gene Med.*, 15, 2, 65–77, 2013.
93. Barrangou, K. and Doudna, J., Applications of CRISPR technologies in research and beyond. *Nat. Biotechnol.*, 36, 933–941, 2018.
94. Komor, S., Badran, A., Liu, D., CRISPR-based technologies for the manipulation of eukaryotic genomes. *Cell*, 168, 1, 20–36, 2017.
95. Carroll, J., Genome editing: past, present, and future. *Yale J. Biol. Med.*, 90, 4, 653–661, 2017.
96. Sander, F. and Joung, J., CRISPR-Cas systems for editing, regulating and targeting genomes. *Nat. Biotechnol.*, 32, 4, 347–355, 2014.
97. Lanphier, E., *et al.*, Don't edit the human germ line. *Nature*, 519, 7544, 410–411, 2015.
98. Cox, A., *et al.*, Therapeutic genome editing: prospects and challenges. *Nat. Med.*, 21, 2, 121–131, 2015.
99. Pharmacogenomics, Wikipedia, last updated recently WikipediaPMC.
100. Relling, A.V. and Evans, M., Pharmacogenomics in the clinic. *Nature*, 526, 7573, 343–350, 2015.
101. Crews, M.F., *et al.*, Clinical pharmacogenetics implementation consortium guidelines for CYP2C19 and CYP2D6 genotypes and dosing of antidepressants. *Clin. Pharmacol. Ther.*, 98, 2, 127–134, 2015.
102. CYP2C9, Wikipedia, last updated recently WikipediaWikipedia+15 Wikipedia+15PMC+15.
103. Schärfe, C.P.I., *et al.*, Genetic variation in human drug-related genes. *arXiv*, 2017 arXiv.
104. Pharmacogenetic Testing and Application: Laboratory Medicine Clinical Practice Guidelines, PMC.
105. Role of warfarin pharmacogenetic testing in clinical practice PMC+15PubMed+15PMC+15.
106. Testing of VKORC1 and CYP2C9 alleles to guide warfarin dosing PMC+14PubMed+14PubMed+14.
107. Pharmacogenomic testing: Relevance in medical practice.
108. Strachan, T. and Read, A.P., *Human Molecular Genetics*, 4th ed., Garland Science, New York, 2010.
109. Collins, F.J., Cystic Fibrosis: Molecular Biology and Therapeutic Implications. *N. Engl. J. Med.*, 332, 7, 416–418, 1995.
110. Hassold, T. and Hunt, P., To err (meiotically) is human: The genesis of human aneuploidy. *Nat. Rev. Genet.*, 2, 4, 280–291, 2001.
111. Sherman, S., *et al.*, Risk of Down Syndrome by Maternal Age. *Obstet. Gynecol.*, 89, 4, 501–507, 1997.

112. Roizen, M. and Patterson, A., Down's syndrome. *Lancet*, 361, 9365, 1281–1289, 2003.
113. Guttmacher, A. and Collins, F., Welcome to the genomic era. *N. Engl. J. Med.*, 349, 10, 996–998, 2003.
114. Lichtenstein, C., *et al.*, Environmental and Heritable Factors in the Causation of Cancer. *N. Engl. J. Med.*, 343, 2, 78–85, 2000.
115. Patel, A.D. and Freudenrich, L., The Role of the Environment in Complex Disease. *Nat. Rev. Genet.*, 9, 2, 87–98, 2008.
116. Ott, J., *Analysis of human genetic linkage*, Johns Hopkins University Press, 1999.
117. Dabelea, H. and Pettitt, D., Epidemiology of type 2 diabetes in youth. *Pediatr. Diabetes*, 2, 1, 1–15, 2001.
118. Stratton, M., Campbell, P., Futreal, P., The cancer genome. *Nature*, 458, 7239, 719–724, 2009.
119. Roberts, R. and Stewart, A., Inherited risk of cardiovascular disease. *N. Engl. J. Med.*, 349, 1, 60–72, 2003.
120. Ober, J. and Vercelli, L., Gene-environment interactions in human disease: Nuisance or opportunity? *Trends Genet.*, 27, 3, 107–115, 2011.
121. Goate, A., Genetics of Alzheimer's disease. *Annu. Rev. Genet.*, 30, 1, 603–630, 1996.
122. Ginsburg, G.S. and Willard, H.F., Genomic and Personalized Medicine: Foundations and Applications. *Transl. Res.*, 154, 6, 277–287, 2009.
123. Shabani, S. and McGuire, J.M., Data governance for health data sharing: A comparative analysis of infrastructure and implementation. *Health Aff.*, 38, 7, 1161–1167, 2019.
124. Langmead, B. and Nellore, A., Cloud computing for genomic data analysis and collaboration. *Nat. Rev. Genet.*, 19, 4, 208–219, 2018.
125. Obermeyer, C. and Emanuel, E.J., Predicting the future — big data, machine learning, and clinical medicine. *N. Engl. J. Med.*, 375, 1216–1219, 2016.
126. Manolio, T.M., *et al.*, Implementing genomic medicine in the clinic: The future is here. *Genet. Med.*, 21, 3, 767–774, 2019.
127. Vassy, J., Murray, J., Green, R., Genetic testing for hereditary risk: Is the juice worth the squeeze? *JAMA*, 320, 18, 1818–1819, 2018.
128. Gymrek, M., *et al.*, Identifying personal genomes by surname inference. *Science*, 339, 6117, 321–324, 2013.
129. Wolf, S.M., *et al.*, Managing incidental findings in human subjects research: Analysis and recommendations. *J. Law Med. Ethics*, 36, 2, 219–248, 2008.
130. Hudson, B., Holohan, K., Collins, S., Keeping pace with the times—The Genetic Information Nondiscrimination Act of 2008. *N. Engl. J. Med.*, 358, 2661–2663, 2008.
131. Tabor, P.T. and Berkman, L.S., Ethical considerations in the era of genomic medicine. *J. Med. Ethics*, 44, 1, 1–6, 2018.
132. Popejoy, M. and Fullerton, S., Genomics is failing on diversity. *Nature*, 538, 161–164, 2016.

133. Sirugo, A., Williams, S.M., Tishkoff, S.A., The missing diversity in human genetic studies. *Cell*, 177, 1, 26–31, 2019.
134. Rotimi, C., *et al.*, The genomic landscape of African populations in health and disease. *Hum. Mol. Genet.*, 26, R2, R225–R236, 2017.
135. Gutman, E. and Haga, J., Regulation of genomic tests: Oversight of laboratory developed tests under CLIA and FDA. *J. Pers. Med.*, 8, 1, 8, 2018.
136. Bunnik, A., *et al.*, The new EU regulation on *in vitro* diagnostic medical devices: Implications for genetic testing. *Eur. J. Hum. Genet.*, 25, 5, 621–623, 2017.
137. Rini, D.L., *et al.*, Genetic counseling for genomic sequencing: The implications for training and practice. *J. Genet. Couns.*, 26, 3, 432–443, 2017.
138. Bernhardt, J., Biesecker, B., Mastromarino, K., Informed consent for genetic research: Current practices and future directions. *Nat. Rev. Genet.*, 6, 7, 559–565, 2005.
139. Williams, M., *et al.*, Genomic decision support tools for primary care: A scoping review. *J. Am. Board Fam. Med.*, 34, 4, 778–790, 2021.
140. Tarczy-Hornoch, A., *et al.*, A survey of informatics approaches to whole-exome and whole-genome clinical reporting in the electronic health record. *Genet. Med.*, 15, 10, 824–832, 2013.
141. Lewis, A. and Vassos, A., Prospects for using risk scores in polygenic medicine. *Genome Med.*, 12, 44, 2020.
142. Phillips, A., Deverka, L., Hooker, D., Douglas, J., Genetic test availability and spending: Where are we now? Where are we going? *Health Aff.*, 37, 5, 710–716, 2018.
143. Hasin, L., Seldin, M., Lusis, A., Multi-omics approaches to disease. *Genome Biol.*, 18, 1, 83, 2017.
144. Mega, M., *et al.*, Cytochrome P-450 polymorphisms and response to clopidogrel. *N. Engl. J. Med.*, 360, 4, 354–362, 2009.
145. Middleton, M., *et al.*, The National Lung Matrix Trial of personalized therapy in lung cancer. *Nature*, 583, 807–812, 2020.
146. Stark, R., *et al.*, A prospective evaluation of whole-exome sequencing as a first-tier molecular test in infants with suspected monogenic disorders. *Genet. Med.*, 21, 3, 713–721, 2019.

2

From Gene Discovery to Clinical Applications: The Journey of Genomic Research

Bolaji Fatai Oyeyemi[1]*, Muhammad Abiodun Sulaiman[2], Saheedat Olabisi Dauda[3], Habeebat Adekilekun Oyewusi[4], Oluwatosin Olubunmi Oladipo[3] and Habeebulahi Ajibola Adekilekun[5,6]

[1]*Cell Biology and Genetics Unit, Department of Zoology, Faculty of Life Sciences, University of Ilorin, Ilorin, Nigeria*
[2]*Department of Statistics, Faculty of Physical Sciences, University of Ilorin, Ilorin, Nigeria*
[3]*Microbiology Unit, Department of Science Technology, The Federal Polytechnic, Ado-Ekiti, Nigeria*
[4]*Department of Medical Biochemistry, Federal University of Health Sciences, Ila-Orangun, Nigeria*
[5]*Department of Pharmacology, Federal University of Health Sciences, Ila-Orangun, Nigeria*
[6]*Department of Pharmacology and Toxicology, Kampala International University Western Campus, Bushenyi, Ishaka, Uganda*

Abstract
Fast-moving genomic research has cut across our understanding of biology, mechanisms underlying diseases, and personalized medicine. This chapter gives a bird's-eye view of the transformational journey of genomics—from fundamental discoveries in genes to impactful clinical applications. From the basic number of critical breakthroughs regarding gene discovery and sequencing technologies, we trace the growth in genomic tools, including high-throughput sequencing, CRISPR gene editing, and bioinformatic-driven data analysis. The chapter emphasizes critical landmarks, like the Human Genome Project, and the consequences those have had for identifying genetic causes of diseases. This also includes a discussion on integrating multi-omics approaches in deciphering complex biological

**Corresponding author*: oyeyemi.bf@unilorin.edu.ng

Swetha Kumari Koduru, Sarad Pawar Naik Bukke, Hussaini Bello, Muhamad Mustafa Mundu and Ayush Madan (eds.) Genomics and Precision Healthcare: Innovations, Applications and Challenges, (39–66) © 2026 Scrivener Publishing LLC

networks and pathways. Strong emphasis has been placed on translating these genomic insights into clinical practice, developing novel diagnostics, targeted therapies, and personalized treatments in oncology, rare genetic disorders, and infectious diseases. This chapter summarizes current issues in genomic research, including ethical issues, data privacy, and equity in access to genomic medicine. Through case stories of clinical successes, it underlines the commitment of researchers, clinicians, and policy thinkers in collaboration to bridge the gap from gene discovery to patient care. This chapter provides a wide-ranging perspective on how medically relevant research based on the study of genomes will shape the future of medicine by reflecting backwards and speculating forwards on its transformational potential.

***Keywords*:** Genomic research, genetics, precision healthcare, genes, gene discovery, genomics

2.1 Introduction

A. Background on Genomic Research

Since the 19th century, genome investigation has changed our understanding of life significantly, starting with the work of several different scientists. Gregor Mendel, a well-known geneticist, intensively studied pea plants, and his work set the fundamental changes for classical genetics [1, 2]. Mendel's experiments established that traits are inherited as discrete units—a concept that provided the first empirical basis for studying heredity [1]. This very early data set forth a sequence of profound discoveries that provided insights into the complexity of genetic structures.

Further refinements of the field of genetics were achieved after the early 1900s, owing to the groundwork laid by classical geneticists. Using Drosophila melanogaster as a model organism, researchers like Thomas Hunt Morgan were able to prove chromosomal inheritance and genetic linkage [3]. Edward A. Sharples's revelations confirmed the established connection between genes and chromosomes, which is followed by new terms such as genes and chromosomes having linkage groups and genetic mapping describing the relative positions of these genes. At this point, biochemical studies established that DNA (deoxyribonucleic acid) was the medium for storing genetic material [4]. Influential studies, such as those by Avery, MacLeod and McCarty, decisively proved that DNA and not proteins was the biochemical basis of heredity. The success of these studies, in conjunction with Watson and Crick's discoveries of DNA's double helix, marked the beginning of the modern era of molecular biology and laid the foundations for the genomic revolution.

B. Importance of Genomics in Precision Healthcare

The application of genomic research to clinical practice has created a new frontier in precise medicine [5, 6]. In the past, medical interventions were simply based on generalized guidelines developed for a large population of patients. Nonetheless, with a deeper comprehension of genetic differences, it is becoming clearer that individual genetic makeup affects disease risk, progression, treatment and prognosis. This understanding has facilitated the shift from conventional to personalized or precision medicine, where treatment approaches are created for the specific genetic profile of every patient [5].

Genomics has transformed diagnostics by recognizing genetic indicators that show the predisposition of individuals to certain diseases and estimations [6, 7]. Sophisticated bioinformatics tools, together with high-throughput sequencing technologies, have given clinicians the ability to detect diseases early and assess how individual patients will respond to treatment [7–11]. In oncology, for example, tumor genetic profiling has led to the development of targeted therapies, which increase treatment efficacy while decreasing adverse effects. This provides better patient outcomes and enables healthcare professionals to allocate resources where they are needed and ensure proper interventions are made.

C. Overview of the Chapter's Scope

This chapter focuses on the entire scope with respect to the changes that happened after discovering a particular gene until the application of the research findings in medicine, or the so-called clinical practice, documenting the impact of genomic medicine on contemporary healthcare. The story moves from the elementary to the complicated by starting with the context and the information regarding early experiments and discoveries, which are considered breakthroughs and inform the rest of the field. Following that, the focus then shifts to other developments in the field, especially those that are considered groundbreaking within genomic research, including the recently invented high-throughput sequencing, CRISPR gene editing, and many other bioinformatic analysis techniques that have collectively expanded our ability to decode and interpret the genome.

In addition, this chapter highlights the process from the earliest pioneering discoveries of the constituent genes up to their clinical application, illustrating how these scientific feats contributed to the development of tools for working with novel diseases, targeted interventions, and personalized medicine. It also covers the incorporation of multi-omics, which fuse information from all levels of genomic, transcriptomic, proteomic, and other myriad spheres to decipher intricate biologic systems and pathways. Important ethical, legal, and social issues are also raised to show the implications of these matters and responsibilities that need to be shouldered due to the rapid changes in genomic medicine. From the comprehensive analysis of these issues, the chapter reflects on the prominent technological and conceptual advances that are and will continue to be important for developing this field and, by extension, the future of health services.

2.2 Historical Perspectives and Key Breakthroughs

A. Early Discoveries in Gene Research

1. Mendelian Genetics and Foundational Experiments

Modern-day genetics traces its origins to the remarkable efforts of Gregor Mendel in the 19th century. Mendel undertook juvenile experiments with pea plants and was able to show that traits are inherited as units (later called "genes") [1, 2]. He also established the laws of segregation and independent assortment, which provided the first quantitative explanations of heredity, thus laying the foundation for the study of genetic inheritance. Through his success, Mendel became the scientific figure who is often regarded as the father of genetics. Unreasonably overlooked during his lifetime, Mendel's experiments became known as the foundation of genetic theory, later inspiring countless scientists to research genetic mutations in various organisms. This was a new approach in biological research—predicting which traits will be inherited, which is the cornerstone of modern genetics [1].

2. Discovery of DNA as the Genetic Material

After Mendel's initial work, people tried to find the physical carrier of genetic information. Initially, they speculated that it could be proteins due to their relative complexity, but several deordenisng experiments changed the attention to DNA instead. In the 1940s, Avery, MacLeod, and McCarty's work proved beyond reasonable doubt that DNA was responsible for transferring genetic information. In their experiments, they showed that purified DNA was capable of transforming bacterial phenotypes, which changed everything that was understood about heredity. Further support came from the famous Hershey-Chase experiments, which finally demonstrated with a bacteriophage that the carrier of heredity is DNA, not protein. Watson and Crick resolved the crystallographic structure of DNA in 1953, which is regarded as one of the major milestones in molecular biology because of how it changed the landscape of genetic research in the postmodern era of molecular genetics [12].

Since the term "gene" was first used more than a century ago [13], its definition has gradually changed in several scientific fields [14–16]. According to

Laird [17], a gene is a hereditary unit composed of a piece of DNA sequence that contains genetic information that defines a biological function. Double-stranded DNA makes up the genome's molecular structure, and the base pair (bp), which consists of two nucleotides joined by hydrogen bonds across the double helix, is the smallest unit of genetic information. Only four molecules—adenine (A), cytosine (C), guanine (G), and thymine (T)—make up the intricate structure of the human genome, which has 3 billion base pairs (bps) that compose tens of thousands of genes [4]. Most genes instruct our cells to produce a particular protein required for a certain activity, and protein collections with other organic molecules carry out all of the functions required for life, such as cell signaling, the construction of bodily structures, and the treatment of illnesses.

According to Flavell, Glover [18], most human genes have a discontinuous structure, with noncoding portions, or introns, interspersed with protein-coding regions or exons. The longest known human gene, titin (TTN), has 365 exons that span 109 224 bp and encode a protein with 35991 amino acids. The average human gene has nine exons [19]. Scientists have long estimated that over 50,000 genes, including 21,000 genes, that coded for proteins [20]. However, this estimate has been repeatedly challenged as genetics and genome research have advanced.

B. Sequencing Technologies and their Evolution

1. Sanger Sequencing to Next-Generation Sequencing

The capability to interpret the genetic code underwent a transformational change after the introduction of various techniques surrounding DNA sequencing. The first valid method of identifying the nucleotide sequence of a DNA molecule was the Sanger sequencing method, which was developed in the 1970s [21]. Sanger sequencing was pivotal in the early gene discovery because, during the ionization of DNA, it was used to decode genetic information through the chain termination approach. Sanger sequencing was extremely accurate; however, it was challenging and slow pace, which made it unduly difficult to use for substantial genomic exercises.

The introduction of genomic projects in the early 2000s is when next-generation sequencing (NGS) changed the face of genomics [22]. NGS focuses on massively parallel sequencing technology that simultaneously stores million DNA pieces at once [23]. This drastically enhanced the speed and the cost of sequencing, thus allowing for the sequencing of the entire genomes, transcriptomes, and epigenomes. The shift from Sanger sequencing to NGS has not only accelerated the rate of gene discovery but genomic research as well, making it possible to understand intricate genetic structures in multiple organisms and populations [24].

2. Impact on Gene Discovery and Research Scalability

The invention of new sequencing technologies has profoundly changed the magnitude of genetic research [24]. The availability of next-generation sequencing has made genomic information easily accessible, permitting the conduct of large-scale studies that were previously deemed rest cost-used and time-consuming [22]. Detection of new genes, genetic alterations, and even novel regulatory sequences is done in record time and with great accuracy. This technological progress is high-throughput genome sequencing technology that helps understand the genetic makeup of different people, the history of mankind, and even the molecular pathology of diseases [25]. Therefore, the modern sequencing devices have been described as the game-changers that permitted genomics to evolve from a mere research area to one of the fundamental fields in medicine, enabling the rapid evolution of precision medicine as a reality rather than a dream.

C. Landmark Projects (e.g., The Human Genome Project)

1. Global Collaboration and Mapping the Human Genome

The most ambitious and impactful project in the realm of genomics is the Human Genome Project (HGP). Starting in the 1990s, the HGP intended to map and sequence the entire human genome collaboratively [26]. It was a one-of-a-kind project where resources, technological advancements and expertise were consolidated, and people worked towards a single end goal. After 13 years, the HGP was completed in 2003. This provided a comprehensive reference map of human DNA, including coding and non-coding regions. The HGP was the first to prove that international collaboration in scientific research was beneficial. It also provided a platform for large-scale genomic projects by creating standardized procedures for data collection and distribution [26].

2. Catalyzing Clinical and Research Innovations

The successful mapping of human's genetic code has impacted science on both a fundamental level and far beyond [22, 27]. The introduction of Clinical and Biomedical Informatics during the HGP has revolutionized innovation in clinics and research facilities. The readily available human genome sequence allows researchers to associate several mutations with their corresponding diseases, resulting in the intellectual development of precise diagnostics tools and individualized treatments [28]. Furthermore, the output information from the HGP enhanced the bioinformatics industry, resulting in tools that allow for more sophisticated genomic analysis and interpretation [27]. These advancements have enabled the translation of molecular breakthroughs into tangible health improvements, thereby entering the world of modern medicine. HGP continues to fuel contemporary research, giving birth to new endeavors that attempt to disentangle the human genome and convert those discoveries into applicable healthcare solutions.

2.3 Genomic Tools and Technologies

A. High-Throughput Sequencing

1. Advances in Sequencing Methods and Throughput

The field of genomics has changed due to the introduction of high-throughput sequencing due to the faster speed of data collection and an increase in the volume of data that a researcher can collect. The initial sequencing methods such as Sanger are now considered an important milestone in DNA analysis but were quite limiting in terms of their capabilities due to how DNA fragments were processed in order. The Next-generation sequencing (NGS) technologies led to a paradigm change by allowing the growth of multiple fragments to be sequenced simultaneously [29]. With the introduction of Illumina's sequencing by synthesis, Ion Torrent's semiconductor sequencing and more new 3rd-generation methods like Pacific Biosciences' single molecule real time (SMRT) sequencing and Oxford NanoPore's nanopore technology, the volume and accuracy of sequencing reached new heights. The advances have decreased the cost tremendously while leading to an increase in detection of different types of rearrangements in the sequences [22]. With the constant changes to the methods of the preparation of the library, algorithms for correction of errors, as well as the chemistry of sequencing, HTS has become key for basic science and clinical studies.

2. Contributions to Large-Scale Genomic Studies

The consequences of high-throughput sequencing (HTS) technology are most visible in its impact on genomic studies that changed the perception of diversity and disease intricacy [22, 30]. By enabling the sequencing of whole genomes, transcriptomes, and epigenomes like never before, HTS has allowed for the investigation of genetic variation among different populations on a never-before scale. The Cancer Genome Atlas (TCGA), the 1000 Genomes Project, several national biobanks, and other endeavors have been able to use these technologies to find novel mutations, resolve highly intricate networks, and identify disease loci with astounding accuracy. These investigations have improved the understanding of cancer diversity, rare genetic syndromes, and the molecular pathology of non-communicable diseases, thereby assisting in the formulation of comprehensive therapies and personalized medicine. High-throughput sequencing not only increased the rate of research innovation but also enabled scientists to do population-level studies which were previously thought to be impossible.

B. CRISPR Gene Editing

1. Mechanism and Precision of CRISPR-Based Editing

CRISPR (Clustered Regularly Interspaced Short Palindromic Repeats) has changed the game in genome editing as it provides flexibility, efficiency, and accuracy in changing a sequence of nucleotides within a DNA molecule [31, 32]. Essentially, the CRISPR-Cas9 system is based on a bacterial adaptive immune system. It has a gRNA which is a guide RNA and can be programmed to a specific target. It takes advantage of the base pairing rule and leads the Cas9 nuclease to the desired location in the genome. When the Cas9 protein binds, it makes a double strand break (DSB) within the genome at the binding site and C10 meddles with B so as to enlist the cell genes repair systems—NHEJ, non-homologus end joining, or HDR, homology directed repair—to make specific changes. The off-target effects during genomic editing are minimized to an absolute when changes are verified designed properly which is a hallmark of precision editing. Moreover, new techniques that boost gRNA's efficiency, high-fidelity Cas9, and new editing approaches (base editors, prime editors) have to some extent alleviated the constraints imposed on this technology. Today, CRISPR is an essential instrument in functional genomics in which scientists can systematically examine the function of individual genes and precisely model human diseases [31–33].

2. Therapeutic Potential and Ethical Considerations

The therapeutic potential of gene-editing using CRISPR technology is enormous, from fixing a genetic flaw to designing immune cells for cancer treatment [32]. Initial and preclinical trials have shown the effectiveness of CRISPR for a wide array of conditions, including, but not limited to, monogenic diseases, blood cancers, and viral diseases. Nevertheless, the use of this innovation in practice raises numerous ethical and safety issues. Risks of off-target changes, mosaicism in the altered tissues, and other changes to the genome that are not intended or predicted highlight the necessity for further examination and follow strict legal guidelines [34]. Moreover, changes made to a germline structure give rise to major ethical dilemmas concerning the ability to inherit, whether the future generations will have a say and the creation of augmentations that are difficult to attain and so-called "inequality augmentations." These challenges will have to be addressed to ensure the effective use of CRISPR in medicine. Finding a middle ground between these risks and the positive potential of CRISPR for therapeutic intervention is a prominent challenge in the area of bioethics [34].

C. Bioinformatic-Driven Data Analysis

1. Role of Computational Tools in Managing Genomic Data

The vast amounts of data produced by high-throughput sequencing brings about the need for effective and advanced computational methods for data analysis. Bioinformatics is a multidisciplinary field that aims to convert the primary sequencing information into valuable biological understanding through the use of IT, statistics, and biology [30]. High-performance computing and sophisticated algorithms underpin essential tasks like sequence alignment, *de novo* assembly, variant calling, and gene annotation [35]. The software tools Bowtie, BWA, and GATK are now commonplace applications in the bioinformatics field, making it possible for researchers to work with big data sets in a more precise and effective manner. Furthermore, the introduction of cloud and parallel computing has made it easier to work with datasets of torabytes in size and ensures that data storage, access, and analysis are responsive to rapid improvements in sequencing technologies [36].

2. Integration of Multi-Omics Datasets

Integration of omics data enables researchers to gather chunk information systematically [10, 11]. It has become necessary in the biological field to use an integrated approach to build a complex picture of cellular activities and diseases. Integrating these disparate datasets to build multi-dimensional models that reflect biological systems' intricacies is the prerogative of bioinformatics. Bioinformatics serves as the core discipline when combining multiple datasets. Systems biology, machine learning technologies, and network-based and other advanced computation techniques permit researchers to find links, determine causal relations, and anticipate phenotypic results from multi-omics data gathering. This holistic strategy serves a wide spectrum of biomedical problems from novel biomarker discoveries and regulatory mechanism elucidation to therapeutic target identification. The rise of new computational approaches and tools facilitate the merging of multi-omics datasets [10, 11, 37]. This would further promote a better understanding of the complex biological systems and significantly improve the speed of translating genomic findings into clinical applications, making clinical practice more effective and relevant [11, 38]. Advanced computation techniques permit finding links, determining causal relations, and anticipating phenotypic results from multi-omics data gathering. Meanwhile, this technology opens up more possibilities in novel biomarker discoveries, therapeutic target identification, and improving the efficiency of clinical practices.

2.4 Integration of Multi-Omics Approaches

A. Overview of Multi-Omics Strategies

1. Combining Genomics, Transcriptomics, Proteomics, and More

Recently, adding new omics layers, like genomic, transcriptomic, proteomic, metabolomic, and epigenomic, has become a practical approach to understanding a biological system's depth (Figure 2.1). Each omics field has its own distinct advantage; for example, genomics provides the organism's base genetic structure, transcriptomics provides the metabolic processes of genes, proteomics provides the interactions of functional proteins, metabolomics shows the flow of pathways biochemically, and epigenomics provides the changes made to the genome that does not alter the DNA sequence. The combination of these distinct data types enables researchers to map out the interplay of various molecular structures and create a scientific model of all cellular activities [11, 39]. This all-encompassing technique is useful in understanding the pathological basis of an illness since such phenotypes usually result from changes at varying levels. Building models through multi-omics data integration is more sophisticated and

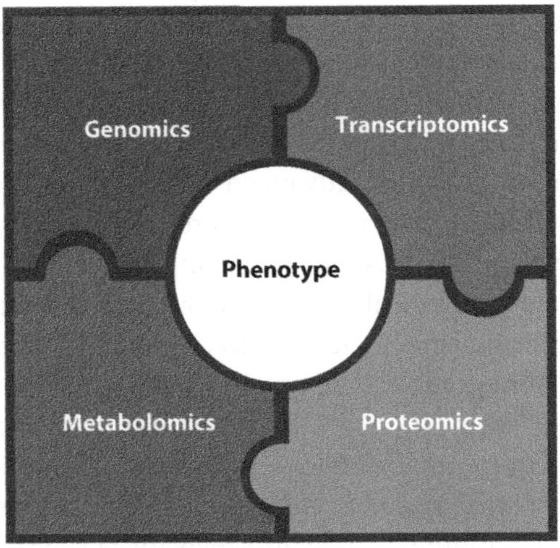

Figure 2.1 Understanding biological systems through multi-omics.

accurate than single-layer models, making it easier to identify robust disease biomarkers, effectively allowing for predictive models of disease advancement and response to drugs as well as enabling a better understanding of the molecule basis of such conditions [11, 40].

2. Data Harmonization and Analytical Frameworks

The integration of data from multiple sources is incredibly difficult because of the heterogeneity of combinations. Differences in units, magnitude, and noise differentiate datasets and therefore require normalization pipelines that take care of batching, quality control, and all other preprocessing routine tasks [11, 40]. More sophisticated methods such as multi-view machine learning, Bayesian fusion, and network-based approaches are used to merge different types of data into assembled analytical models. These frameworks enable the mapping and linkage of omics layers, which in turn helps uncover cross-omics relationships and correlations and trace their causal pathways. Analytic tools such as integrative clustering, matrix factorization, and other dimensionality-reducing tools like PCA or t-SNE enable more complex biologically meaningful pattern identification within the given dataset [11, 40, 41]. As a result of these advanced crafts, it is becoming easier and easier to sift through, "understand," and/or make use of complex biological datasets within the fundamental or clinical environments.

B. Deciphering Complex Biological Networks and Pathways

1. Systems Biology and Network Modeling

The fusion of multi-omics data has led to the development of systems biology methods aimed at modelling complex networks of cellular functions [42, 43]. Network modeling is the process of creating networks in which nodes are molecular entities such as genes, proteins, or metabolites, while edges represent functional interactions or regulatory relations among them. With the use of network theory and systems biology, it is feasible to depict the dynamic aspects of biological systems and to predict the consequences of a change in one component of the network on the rest of the system. Tools like Bayesian network inference, co-expression network analysis and other graph theoretical approaches are important for the detection of central hubs and modules within these networks. These hubs are often critical regulatory genes or proteins that are pathed to disease causing dysregulation. Network models with multi-omics data not only capture the essence of cellular mechanisms, but also enhance prediction of emergent properties that are otherwise blind to the investigation of separate omic layers [42].

2. Uncovering Regulatory Mechanisms and Disease Pathways

A primary goal for an interdisciplinary approach is figuring out how different regulatory integrate cells work cause diseases. Combining change of genes with their expression, proteins and metabolites activities offers a possibility of pathological processes elucidation [44]. For example, combining transcriptomic and epigenostic data makes it possible to define transcription factors and epigenetic notch positioners for principal oncogenic signaling in cancer. Likewise, proteomic and metabolomic data intertwining makes it possible to get a primary understanding of signaling and metabolic changes in main metabolic and neurodegenerative diseases. Genes, as targets, utilize GSEA technologies and pathway analysis by network topology ganged enable specific biological phenomena that result from the genetic changes or environmental changes. Integrative deal with these studies deepens the understanding the etiology of the disease and equally broadens the window for effective therapeutic interventions and approaches to personalized medicine.

2.5 Clinical Applications of Genomic Research

A. Translating Genomic Insights into Clinical Practice

1. Bridging the Gap between Research and Patient Care

One of the clinical innovations is the incorporation of genomic information into everyday clinical practice. It had been opined that anthropometry traits can be used to predict genetic information [45–47]. This integrates scientific discovery and clinical practice and translates into actionable processes for diagnosing, prognosing, and treating patients [48, 49]. Integrating molecular biology, clinical practice, bioinformatics, and even regulation creates collaborative networks that ensure the rigorous evaluation and practical application of genomic research and development [50, 51]. These efforts include cancer treatment decision-making molecular tumor boards that incorporate genomic risk factors of cancer into cancer treatment decisions. In addition, these innovations improve public health by creating extensive databases and cutting-edge clinical decision support systems to manage genomic risk information and implement personalized medicine. Even though tremendous progress has been made, one of the glaring issues still to be addressed is equity in distributing genomic research benefits among the patient populations.

2. Implementation Challenges and Success Factors

Genomic medicine's promise is more significant than what has currently been achieved. However, its clinical implementation leaves a lot to be desired. Primary challenges include the absence of universal protocols for genomic testing, the integration of sophisticated genomic information into electronic health record systems, and the capability of the healthcare personnel to utilize genomic knowledge in practice. Furthermore, challenges with privacy of data, regulatory obstacles, and reimbursement issues are serious impediments to adoption. Factors that promote successful implementation include collaborative engagement, continuous bioinformatics investment, and the creation of scalable affordable technologies. These elements allow primary programs and the initial clinical studies to be implemented seamlessly. But, once these factors are tended to, the gained genomic knowledge is certain to improve diagnosis, optimally intervene with therapeutics, and above all, increase patient's overall health.

B. Development of Novel Diagnostics

1. Genomic Biomarkers and Non-Invasive Testing

The progress made in genomic research has made it possible to track markers that demonstrate critical indications of a disease's presence, progression, and body's response to treatment [49, 52]. These markers can be determined using various molecular methods, which supports early diagnosis and valuable prognostic details [48]. Non-invasive testing like saliva and liquid biopsies use 16S rRNA microbiome and circulating tumor DNA (ctDNA) or cell-free DNA (cfDNA) obtained from blood to detect genetic changes linked to cancer [49, 53]. Such methodologies not only help in minimizing the use of traditional invasive tissue biopsies but also enable constant tracking of tumor changes over time. Apart from oncology, genomic biomarkers are being studied in other disease areas, such as cardiovascular disorders and neurodegenerative diseases, thus increasing the scope of genomic diagnostics in many clinical fields.

2. Companion Diagnostics for Targeted Therapies

Companion diagnostics have gained special importance with the advent of targeted therapies that focus on specific genetic mutations and alterations. They are crucial for patient stratification and treatment selection for ensured therapeutic efficacy [54, 55]. In Molecular Oncology, determination of eligibility for targeted therapeutic agents, such as tyrosine kinase inhibitors or monoclonal antibodies, is performed through companion diagnostic tests to find actionable mutations like EGFR, ALK, or BRAF. The creation and marketing of companion diagnostics always must have research, clinical, and regulatory synergies so that all of these bodies can work within the frameworks of the legal and scientific regulations of accuracy, reliability, and clinical relevance that these tests have to achieve.

C. Targeted Therapies and Personalized Treatments

1. Oncology

a. Genetic Profiling and Targeted Cancer Treatments

The field of oncology has witnessed a monumental change in the management of cancer with the development of genetic profiling of tumors [56]. The discovery of somatic mutations, copy number variants, and gene-fusions with oncogenic potential through comprehensive genomic profiling are now considered hallmarks of cancer. These discoveries make it possible to construct therapies that block critical molecular pathways associated with tumor growth. Examples of such therapies are the EGFR, ALK, or BRAF targeting inhibitors which have shown remarkable positive survival rates in patients with numerous types of cancer [56]. Furthermore, the discovery of changes during treatment for cancer that may contribute to the resistance of cancer treatment is also possible, which makes it easier to adapt treatment methods during the process. It is now possible to develop treatment strategies that involve a combination of targeting RNAs and proteins based on genetic information and make sure therapy does not contribute to the problems associated with cancer treatment.

2. Rare Genetic Disorders

a. Gene Therapy and Personalized Interventions

Proposed treatment methods, especially as they relate to genomic technologies, have become a major advancement in diagnosing and treating various unique genetic disorders. Nowadays, rare single-gene mutation conditions are considered highly identifiable by whole exome or whole genome sequencing. This accurate genetic diagnosis has enabled the development of approaches to gene therapy aimed at correcting the underlying defect. Affected individuals are targeted by applying advanced techniques like viral vector gene delivery systems, RNA interference, or genome editing using CRISPR-Cas systems [57]. By focusing on the precise genetic mutation and the disease mechanisms involved, these approaches go beyond symptom management to a true cure for previously incurable diseases. Early trials of gene therapy in spinal muscular atrophy and some forms of inherited retinal dystrophies highlight the effectiveness of genomic medicine in rare diseases.

3. Infectious Diseases

a. Genomic Surveillance and Pathogen Tracking

Tracking pathogens and surveillance of diseases has improved significantly due to genomic research, especially for contagious diseases. Sequencing pathogen genomes during outbreaks enables public health authorities to monitor the progress of infection with the outbreak within a specific region [58]. Public health policies are adjusted needing the spread of infection and determine when to declare an intervention as an example quarantine and vaccination. Host and pathogen genomic data integration also provides information on immune responses and pathogen-host interactions and aids in developing adequate disease prevention measures. Unceasing observation of the pathogen genomes is important to new risks, threats, and healthier measures.

2.6 Ethical, Legal, and Social Considerations

A. Ethical Issues in Genomic Research

1. Informed Consent and Incidental Findings

The emergence of the new field of genomics research prompted changes to be made with regard to its ethical rules, especially concerning informed consent. In large genomic studies, informed consent is difficult to obtain because there are incidental findings, such as genetic data that may put the subject at risk for other diseases that were not focused on during the research. Informed consent implies that participants understand that there are several new medically useful, but irrelevant genetic changes that can be uncovered and their consequences. The complexity of such research requires thoughtful consideration of the consent process used to obtain consent. It is necessary to provide not only vivid, but also easy-to-understand information at the beginning, while communication must also be done in relation to the new findings. This is critical knowing that genomic research has far-reaching effects on individuals and their privacy. In informed consent, participants must know the risks and the consequences of their personal information being exposed, which touches on more ethical autonomy and trust.

2. Gene Editing and Implications for Future Generations

New accurate gene-editing technologies, like the CRISPR-Cas systems, have encouraged the treatment of diseases and defective genes [31, 33, 59]. However, when applied to germline cells, these technologies modify cells that are inheritable. This modification impacts not only the specific individual but also future generations. This gives rise to consent problems as future people have no say on their genetic structure, having been altered already. Also, the potential for undetermined effects—including off-target effects and concerns about human enhancement—presents complex ethical dilemmas that require restraint [60]. The ability to edit the human germline interferes with social customs and ethical boundaries, bringing about the need for tight regulation and global agreements to guarantee the responsible and fair use of gene editing technologies.

B. Data Privacy and Security

1. Protecting Sensitive Genomic Information

Genomic data is personal because it reveals sensitive details about an individual's genetic structure, family history, and the likelihood of certain health issues. Therefore, sensitive information of humans should be carefully managed and protected from breach. Protective strategies include complex data encoding, data protection measures, and anonymization procedures that reduce the chance for re-identification [61]. Furthermore, stringent access restrictions and comprehensive documentation of all actions taken must be in place to mitigate unauthorized usage and access to genomic information. Protecting the privacy and safety of genomic data is a purely technical issue and at the same time, a deeply ethical one because breaches can cause discrimination or even stigmatization and other forms of harm.

2. Regulatory Frameworks and Technological Safeguards

Because genomic information is extraordinarily sensitive, comprehensive regulations to safeguard individual privacy have continued to evolve. In Europe, regulations like the GDPR enforced complex frameworks for data protection that are now being replicated elsewhere in the world. These frameworks set clear policies regarding the collection, storage, and sharing of genomic data, requiring everyone involved to meet the minimum level of data security. In parallel, technological safeguards such as blockchain for genomic data sharing and advanced encryption methods are being implemented in data management systems. These regulatory and technological approaches provide a comprehensive solution to data breaches and build public trust that real benefits can be obtained from genomic research without infringing on the rights of individuals [61].

C. Equity in Access to Genomic Medicine

1. Addressing Socioeconomic and Geographic Disparities

The unequal distribution of genomic medicine is vast across various populations [62]. Access to holistic genomic diagnostics and personalized therapies is severely limited due to socioeconomic and geographical disparities [27]. Marginalized and underserved communities indeed face barriers such as an absence of representative genomic technologies. Adjusting such barriers poses the challenge of improving access, including providing affordable genomic testing, public health tackling underprivileged areas, and reducing healthcare inequality through policies. In order for genomic medicine to be efficient on a broader context, these gaps must be closed.

2. Promoting Inclusivity in Genomic Research
Inclusive participation of different populations is necessary and central to effective genomic research, especially with its mare penetrative focus [62]. It is to be noted that precision medicine cannot be fully actualized without broadening the scope of genomic research to include people from different ethnic and racial groups, as many genomic studies tend to be over-focused on people of European descent. Deliberate global collaborative efforts that prioritize diversity, culturally sensitive engagement and proactive recruitment strategies are also required. This will promote inclusivity that not only enhances the scientific validity of genome studies but also ensures that personalized healthcare approaches apply to all segments of the global population. Efforts to foster diversity in research must be supported by policy interventions, funding initiatives, and community outreach programs aimed at building trust and participation among underrepresented groups.

2.7 Case Studies and Clinical Success Stories

A. Real-World Examples of Genomic Applications
1. Impact on Oncology, Rare Disorders, and Infectious Diseases
The effects of genomic research are well demonstrated in oncology and several other medical fields. For example, cancer treatment and early diagnosis have been transformed through genomic profiling of tumours [49, 63]. Clinicians can now select targeted therapies based on other driver mutations, gene fusions, and differing aberrations on molecular levels [7, 64]. For instance, the detection of actionable mutations in the EGFR, ALK, and BRAF genes have enabled the design of new inhibitors that outperform traditional chemotherapeutic agents in improving the clinical outcomes of patients while significantly lowering toxicity levels [7, 64].

With regard to rare genetic conditions, genomic sequencing has made it possible to achieve early diagnosis and correct them after years of clinical guesswork. Moreover, the developing whole exome and whole genome sequencing technologies have genetically solved many cryptic disorders and helped to formulate tailored strategies including gene therapy and precision drugs. These target therapies markedly provide disease symptom relief, but in some cases, they could change the course of the disease entirely [64].

Moreover, infectious diseases profited from real-time pathogen surveillance, tracking how these diseases evolve. During epidemics, the widespread sequencing of viral and bacterial genomes has allowed public health officials to track the movement of infections, monitor the changes occurring in the pathogens, and make appropriate changes in treatment within the outbreaks. These distinct possibilities illustrate how genomic technologies are transforming diagnostics and treatment options and the public health response to numerous diseases.

2. Success Stories from Translational Research

Translational research has many success stories where cross-discipline effort results in positive and innovative outcomes [65]. It is applied where the biology of cancer is studied in laboratories to apply it in hospitals for patients. In cancer management, the use of liquid biopsies is an excellent example. With ctDNA that oncologists can collect from blood draws, tumors can be monitored, and treatments assessed noninvasively [53]. The corrective therapeutic approaches that can be put into practice are already enhancing treatment outcomes and prognostic precision.

In this domain, the terminology used is complex, but its application brings powerful innovation to rare genetic disorders. It is opined that comprehensive genomic analyses will pave the way for targeted treatments. Patients suffering from spinal muscular atrophy have seen unfathomable increases in their overall motor function and longevity through gene therapy trials. Such feats signify revolutionary progress in medical fields previously thought unsolvable.

In epidemic management, the progressive detection of changes in pathogens, notably *via* genomic sequencing, directly correlates with creating vaccines and antiviral medications. This modern approach is integral when dealing with a public health crisis, as it can aid in decreasing fatalities and the prevalence of diseases, accordingly.

B. Collaborative Efforts among Researchers, Clinicians, and Policy Makers

1. Multidisciplinary Consortia and Public-Private Partnerships

The achievement of genomic uses in clinical environments is greatly due to the effective cooperation among stakeholders. The translation of genetic findings into their practical uses has greatly benefitted from the efforts of consortia which include, at minimum, basic scientists, clinicians, bioinformaticians, and data analysts [66]. Such efforts are further aided through public-private collaborations which integrate the innovative aspects of academic institutions with the resources and market reach of industries. Some of the best-known such networks are The Cancer Genome Atlas (TCGA) and the International Cancer Genome Consortium (ICGC), which have constructed extensive databases which are invaluable for both scientific and clinical domains. These international consortia not only stimulate research in cancer genomics but also promote the development and sharing of standardized methods and procedures among the institutions, thus facilitating the implementation of cancer genomic research results into practice.

2. Establishing Standards and Best Practices

To guarantee that genomic inventions are integrated into patient care proficiently, and safely, there is a need to put in place stringent standards and best practices. Efforts by researchers, clinicians, and policy makers have produced a framework for genomic testing, sharing of data, and ethical supervision [67]. These standards are being developed and are valuable for regulatory agencies and professional bodies to set because they encompass important aspects of assay validation, quality control, and the interpretation of genomic data. Such measures not only improve the accuracy of genomic tests, but also foster public confidence in genomic medicine [67]. Moreover, standardization helps comparison and data merging between institutions, which makes it easy to apply genomic knowledge in clinical practice. Therefore, these collaborative frameworks are important in translating advanced scientific research into routine clinical care so as to make progress in the field of precision medicine.

2.8 Future Directions and Challenges

A. Emerging Innovations in Genomic Technologies

1. Next-Generation and Single-Cell Sequencing Advances

The frontiers of genomic study are persistently being reshaped as the technologies around sequencing continue to progress. From their inception, next-generation sequencing (NGS) platforms have always strived to achieve higher throughput levels without compromising on accuracy, and to decrease costs such that whole genome, transcriptome, and epigenome analyses are subsequently made easier to perform. Simultaneously, the development of single cell sequencing has provided a powerful new method for resolving the heterogeneity of complex tissues [68]. Scientists are now capable of recognizing rare cell populations, tracing their developmental lineages, as well as dissecting their responses at astonishingly high resolutions. Gaining this level of detail will improve our understanding of the pathogenesis of many diseases, with cancer being an important one given that intratumoral heterogeneity is an increasingly recognized phenomenon in cancer biology [68]. These advanced techniques in sequencing are likely to stimulate further innovations in basic and translational research, leading to more accurate diagnosis and treatment in the future.

2. CRISPR Advancements and AI-Driven Genomics

Gene editing has become pivotal in the modern world, with multitasking capabilities being accomplished using tools like CRISPR, base editing and prime editing. Unlike traditional systems, these advanced systems allow for alterations to be made without double strand breaks. This shift enables increased safety and optimization with the use of these systems, creating pathways for tackling and treating genetic disorders. Moreover, the invention of artificial intelligence has enabled the reformation of systems through optimization of previously used methods [69–71]. AI facilitates the swift analysis of complex genomic data allowing for the prediction of types of genes, identification of diseases, and enhancing methods for editing genes like CRISPR [69, 71]. Altogether, these opportunities will redefine our approach towards various diseases and allow wide-scale integration onto patients using modern techniques.

B. Challenges in Implementation and Integration

1. Data Management and Computational Demands

The rapid increase of genomic data comes along with a plethora of challenges with regard to data tracking and analysis. High-throughput sequencing and multi-omics approaches are a major source of big data that goes beyond available storage, processing, and interpretation capabilities. Ensuring data of good quality and integrating various data types is also highly important, alongside the construction of scalable bioinformatics tools which are essential for analyzing big data of such high dimensions. These complex data streams can only be achieved with the enhancement of modern computational algorithms combined with powerful resources to provide high performance, which makes surpassing these hurdles incredibly crucial for effectively utilizing genomic research in both scientific and clinical domains.

2. Ethical, Regulatory, and Workforce Training Issues

The advent of modern genomic tools poses substantial ethical, regulatory, and training-related concerns. Ethical issues such as informed consent, privacy of data, and the broader consequences of gene editing, particularly in germline cells, require stringent protective measures to enable responsible research and clinical practices. Regulatory authorities face the challenge of coming up with rules that encourage innovation but restrict harm, which is made more difficult by the rapid advancement of technology. Moreover, the need to educate a new breed of biologically and computationally skilled genomic professionals is more urgent than ever. Closing this gap is crucial for the effective incorporation of genomic technologies into everyday healthcare practices, as well as for promoting cross-disciplinary partnerships that will lead to groundbreaking new discoveries.

C. Speculations on the Future Impact of Genomics on Medicine

1. Shifts toward Proactive and Personalized Care

The advancement of medicine is more projected towards a proactive, bespoke approach optimized by genetic understanding. The ability to predict health risks and track disease through genetic profiling is set to improve, shifting healthcare from a reactive approach of waiting for disease onset to proactive prevention and intervention. With a caring system put into place, journaling a patient's health and ensuring there is a reduction in adverse drug reactions as well as enhancing the treatment efficiency is remarkably achievable. This change has the potential to greatly alter the healthcare system, as risks can easily be marked, followed by quick and precise diagnosis, and individualized treatment that attends to the demands of each patient.

2. Integration with Other Omics and Digital Health Innovations

The omics revolution is taking shape through the intersection of genomics with proteomics, metabolomics, epigenomics, and advancements in digital health technologies. This will allow the development of health profiles that consider the dynamic interrelationships of genes, molecules, and environment. Digital tools such as wearable devices and telemedicine will enable real-time, personalized patient monitoring which can transform healthcare into a more proactive industry. The integration of multi-omics and digital health will enable new levels of disease prevention, diagnosis, treatment, and precision medicine. This will significantly change the paradigm of health systems.

2.9 Conclusion

A. Recap of Key Milestones and Developments

1. Summary of Historical Breakthroughs and Technological Advancements

In the past hundred years, genomic research has transformed from Mendelian genetics experiments to today's high-throughput technology. Many twin- and cross-breeding experiments set the basic principles of heredity which resulted in identifying DNA as the genetic material and determining its double-helical structure. Sanger sequencing brought a revolution in the field allowing for the first direct reading of the genetic code, which although limited, set the stage for future advancements. The emergence of next-generation sequencing technologies revolutionized the passage of reading and coding DNA, leading to lower costs and feasibility of conducting large-scale genomic studies. With the radical projects, especially with the Human Genome Project, the world was provided with comprehensive maps of the human genome, radically increasing the pace of research and clinical practice innovation. This increased the understanding of biological complexity and made the use of genomics in precision medicine possible, forming the basis for unparalleled biomedical research today.

B. The Transformational Potential of Genomic Research in Healthcare

1. Future Prospects for Precision Medicine and Patient Care

The direction of genomic studies suggests that precision medicine will become an essential part of healthcare in the coming years. With the progress in human genome sequencing and interpretation, genomic information is being used to customize diagnosis, prognosis, and treatment plan for each patient. This evolution from a blanket to a more individualized approach is expected to improve treatment success rates and patient health outcomes while reducing harm. Merging genomic knowledge with the latest digital health technologies and multi-omics is expected to enable the monitoring of a patient's state of health in real time, making it possible to have a more proactive approach. Such a comprehensive systems approach will allow for the earlier diagnosis of diseases, better risk stratification, and the formulation of targeted interventions tailored to the patient's specific genetic and molecular make-up. The advancement of these technologies will undoubtably have the greatest impact on population health worldwide, enabling the widespread adoption of precision medicine.

AI disclosure statement: The author acknowledges the use of AI-based tools for minor language editing or grammatical correction. However, all intellectual contributions, analyses, and conclusions presented in this chapter are entirely the author's own.

References

1. Berry, A. and Browne, J., Mendel and Darwin. *Proc. Natl. Acad. Sci.*, 119, 30, e2122144119, 2022.
2. Panchal, S., Fundamentals of genetics, in: *Genetics Fundamentals Notes*, pp. 3–51, Springer, 2022.
3. Roberts, D.B., Drosophila melanogaster: The model organism. *Entomol. Exp. Appl.*, 121, 2, 93–103, 2006.
4. Alberts, B., Heald, R., Johnson, A., Morgan, D., Raff, M., Roberts, K., *et al.*, *Molecular Biology of the Cell: Seventh International Student Edition with Registration Card*, WW Norton & Company, 2022.
5. Elhaie, M., Koozari, A., Abedi, I., Beyond the One-Size-Fits-All Approach: The Role of Imaging Biomarkers and Genomics in Precision Oncology: A Comprehensive Review. *Front. Biomed. Technol.*, 12, 2025.
6. Crimi, M., Digital genetics and genomics, in: *The Digital Doctor*, pp. 309–316, Elsevier, 2025.
7. Oldfield, L.E. and Pugh, T.J., Cancer Genomics, Biomarkers and Precision Medicine, in: *Cancer Systems and Control for Health Professionals*, pp. 243–252, 2025.
8. Zhang, S., Liu, K., Liu, Y., Hu, X., Gu, X., The role and application of bioinformatics techniques and tools in drug discovery. *Front. Pharmacol.*, 16, 1547131, 2025.
9. Vinarski, B., Ramanujam, A., Paz, R., Khader, A.H.S.A., Artificial intelligence in personalized medicine: Application of genomics to influence therapy decisions, in: *Artificial Intelligence in Urologic Malignancies*, pp. 77–113, Elsevier, 2025.
10. Papadaki, E., Kakkos, I., Vlamos, P., Petropoulou, O., Miloulis, S.T., Palamas, S., *et al.*, Recent Web Platforms for Multi-Omics Integration Unlocking Biological Complexity. *Appl. Sci. (2076-3417)*, 15, 1, 2025.
11. Ewald, J.D., Zhou, G., Lu, Y., Kolic, J., Ellis, C., Johnson, J.D., *et al.*, Web-based multi-omics integration using the Analyst software suite. *Nat. Protoc.*, 19, 5, 1467–1497, 2024.
12. Watson, J.D. and Crick, F.H., The structure of DNA, in: *Cold Spring Harbor Symposia on Quantitative Biology*, pp. 123–131, Cold Spring Harbor Laboratory Press, 1953.

13. Miller, H., Konkel, D., Leder, P., An intervening sequence of the mouse β-globin major gene shares extensive homology only with β-globin genes. *Nature*, 275, 5682, 772–774, 1978.
14. Zeeshan, S., Xiong, R., Liang, B.T., Ahmed, Z., 100 years of evolving gene–disease complexities and scientific debutants. *Briefings Bioinf.*, 21, 3, 885–905, 2020.
15. Friedmann, T., A brief history of gene therapy. *Nat. Genet.*, 2, 2, 93–98, 1992.
16. Maglott, D., Ostell, J., Pruitt, K.D., Tatusova, T., Entrez Gene: Gene-centered information at NCBI. *Nucleic Acids Res.*, 33, suppl_1, D54–D58, 2005.
17. Laird, C.D., Chromatid structure: relationship between DNA content and nucleotide sequence diversity. *Chromosoma*, 32, 4, 378–406, 1971.
18. Flavell, R., Glover, D., Jeffreys, A., Discontinuous genes. *Trends Biochem. Sci*, 3, 4, 241–244, 1978.
19. LeWinter, M.M. and Granzier, H.L., Titin is a major human disease gene. *Circulation*, 127, 8, 938–944, 2013.
20. Lander, E.S., Initial impact of the sequencing of the human genome. *Nature*, 470, 7333, 187–197, 2011.
21. Sanger, F., Determination of nucleotide sequences in DNA. *Science*, 214, 4526, 1205–1210, 1981.
22. Abdi, G., Tarighat, M.A., Jain, M., Tendulkar, R., Tendulkar, M., Barwant, M., Revolutionizing genomics: exploring the potential of next-generation sequencing, in: *Advances in Bioinformatics*, pp. 1–33, Springer, 2024.
23. Ramanathan, B., Jadhav, S.A., Kumar, D.K., Ramsait, S.P., Kumaramanickavel, G., Genomic tools and techniques—An introduction, in: *Genomics, Populations, and Society*, pp. 43–65, Elsevier, 2025.
24. Mersha, T.B., From Mendel to multi-omics: shifting paradigms. *Eur. J. Hum. Genet.*, 32, 2, 139–142, 2024.
25. Lu, Y., Li, M., Gao, Z., Ma, H., Chong, Y., Hong, J., *et al.*, Advances in Whole Genome Sequencing: Methods, Tools, and Applications in Population Genomics. *Int. J. Mol. Sci.*, 26, 1, 372, 2025.
26. Wang, T., Antonacci-Fulton, L., Howe, K., Lawson, H.A., Lucas, J.K., Phillippy, A.M., *et al.*, The Human Pangenome Project: a global resource to map genomic diversity. *Nature*, 604, 7906, 437–446, 2022.
27. Jenkins, S.M., Palmquist, R., Shayota, B.J., Solorzano, C.M., Bonkowsky, J.L., Estabrooks, P., *et al.*, Breaking barriers: Fostering equitable access to pediatric genomics through innovative care models and technologies. *Pediatr. Res.*, 1–8, 2025.
28. Peters, C., 1997: "Your genes, your choices" and public education about the ethical, legal and social issues of the Human Genome Project. *Public Underst. Sci.*, 2025. 09636625241300390.
29. Dongare, D.B., Nishad, S.S., Mastoli, S.Y., Saraf, S.A., Srivastava, N., Dey, A., High-throughput sequencing: A breakthrough in molecular diagnosis for precision medicine. *Funct. Integr. Genomics*, 25, 1, 1–23, 2025.

30. Dhanapandi, V., Mahendran, G., Nivetha, S., Sonali, M., Susmitha, D., Artificial Intelligence in Personalized Medicine: A Survey on FuncPredict: AI Driven Genetic Variant Analysis, in: *2024 3rd International Conference on Automation, Computing and Renewable Systems (ICACRS)*, IEEE, pp. 999–1005, 2024.
31. Pacesa, M., Pelea, O., Jinek, M., Past, present, and future of CRISPR genome editing technologies. *Cell*, 187, 5, 1076–1100, 2024.
32. Tiwari, S.K., Wong, W.J., Moreira, M., Pasqualini, C., Ginhoux, F., Induced pluripotent stem cell-derived macrophages as a platform for modelling human disease. *Nat. Rev. Immunol.*, 25, 2, 108–124, 2025.
33. Azeez, S.S., Hamad, R.S., Hamad, B.K., Shekha, M.S., Bergsten, P., Advances in CRISPR-Cas technology and its applications: Revolutionising precision medicine. *Front. Genome Ed.*, 6, 1509924, 2024.
34. Lange, V. and Kappel, K., CRISPR Gene-Therapy: A Critical Review of Ethical Concerns and a Proposal for Public Decision-Making. *Can. J. Bioeth.*, 5, 2, 78–87, 2022.
35. Pham, T.-T., Le Bui, N., Trinh, C.-D., Nguyen, D.T., Chu, D.-T., Computational Tools for Genomics Data Analysis, in: *Advances in Genomics: Methods and Applications*, pp. 41–54, Springer, 2024.
36. Schadt, E.E., Linderman, M.D., Sorenson, J., Lee, L., Nolan, G.P., Computational solutions to large-scale data management and analysis. *Nat. Rev. Genet.*, 11, 9, 647–657, 2010.
37. Oyewusi, H.A., Wahab, R.A., Akinyede, K.A., Albadrani, G.M., Al-Ghadi, M.Q., Abdel-Daim, M.M., *et al.*, Bioinformatics analysis and molecular dynamics simulations of azoreductases (AzrBmH2) from Bacillus megaterium H2 for the decolorization of commercial dyes. *Environ. Sci. Eur.*, 36, 1, 31, 2024.
38. Adekilekun, H.A., Oyewusi, H.A., Wahab, R.A., Huyop, F., Albadrani, G.M., Al-Ghadi, M.Q., *et al.*, A computational discovery of hexokinase 2 inhibitors from Newbouldia laevis for Hepatocellular carcinoma (HCC) treatment. *S. Afr. J. Bot.*, 169, 12–26, 2024.
39. Oyewusi, H.A., Wahab, R.A., Huyop, F., Whole genome strategies and bioremediation insight into dehalogenase-producing bacteria. *Mol. Biol. Rep.*, 48, 3, 2687–2701, 2021.
40. Pang, Z., Zhou, G., Ewald, J., Chang, L., Hacariz, O., Basu, N., *et al.*, Using MetaboAnalyst 5.0 for LC–HRMS spectra processing, multi-omics integration and covariate adjustment of global metabolomics data. *Nat. Protoc.*, 17, 8, 1735–1761, 2022.
41. Han, Z., Oyeyemi, B.F., Zenobio, J.E., Salawu, O.A., Adeleye, A.S., Perfluorooctanoic acid dominates the molecular-level effects of a mixture of equal masses of perfluorooctanoic acid and perfluorooctane sulfonic acid in earthworm. *J. Hazard. Mater.*, 131718, 2023.

42. Li, F., Song, X., Fan, W., Pei, L., Liu, J., Zhao, R., *et al.*, SPathDB: A comprehensive database of spatial pathway activity atlas. *Nucleic Acids Res.*, 53, D1, D1205–D1214, 2025.
43. Oyewusi, H.A., Adedamola Akinyede, K., Wahab, R.A., Susanti, E., Syed Yaacob, S.N., Huyop, F., Biological and molecular approaches of the degradation or decolorization potential of the hypersaline Lake Tuz Bacillus megaterium H2 isolate. *J. Biomol. Struct. Dyn.*, 42, 12, 6228–6244, 2024.
44. Schwarzerova, J., Mate, E.G., Idkowiak, J., Olesova, D., Kvasnicka, A., Dobesova, D., *et al.*, MInfer: Bridging MetaboAnalyst and Jacobian analysis for metabolomic networks. *Comput. Methods Programs Biomed.*, 108672, 2025.
45. Oyeyemi, B.F., Iyiola, O.A., Oyeyemi, A.W., Oricha, K.A., Anifowoshe, A.T., Alamukii, N.A., Sexual dimorphism in ratio of second and fourth digits and its relationship with metabolic syndrome indices and cardiovascular risk factors. *J. Res. Med. Sci.: The Official Journal of Isfahan University of Medical Sciences*, 19, 3, 234–239, 2014.
46. Oyeyemi, B.F., Adebayo, J.O., Anifowoshe, A.T., Iyiola, O.A., Relationship between Ratio of Second and Fourth Digit and Obesity Traits among Different Ethnic Groups in Ilorin, North Central Nigeria. *Not. Sci. Biol.*, 8, 4, 396–400, 2016.
47. Oyeyemi, F.B., Adebayo, J.O., Oyeyemi, A.W., Adeola, I.C., Ayanwole, E.T., Jegede, F.O., *et al.*, Second to Fourth Digit Ratio (2D:4D) as a Predictor of Adult Circulating Sex Hormones and Overweight/Obesity in Ado-Ekiti Nigeria. *Biomed. Environ. Sci.*, 31, 7, 539–544, 2018.
48. Oyeyemi, B.F., Ologunde, C.A., Olaoye, A.B., Alamukii, N.A., FTO Gene Associates and Interacts with Obesity Risk, Physical Activity, Energy Intake, and Time Spent Sitting: Pilot Study in a Nigerian Population. *J. Obes.*, 20172017, doi: 10.1155/2017/3245270.
49. Oyeyemi, B.F., Kaur, U.S., Paramraj, A., Tandon, R., Kumar, A., Bhavesh, N.S., Microbiome analysis of saliva from oral squamous cell carcinoma (OSCC) patients and tobacco abusers with potential biomarkers for oral cancer screening. *Heliyon*, 9, 11, e21773, 2023.
50. Yadav, P., Oyeyeymi, B.F., Jamling, T.C., Kumar, A., Bhavesh, N.S., Multiomics approach for precision wellness, in: *Epigenetics and Metabolomics*, pp. 147–180, Elsevier, 2021.
51. Clark, A.J. and Lillard Jr., J.W., A comprehensive review of bioinformatics tools for genomic biomarker discovery driving precision oncology. *Genes (Basel)*, 15, 8, 1036, 2024.
52. Ojo, G.J., Onile, O.S., Momoh, A.O., Oyeyemi, B.F., Omoboyede, V., Fadahunsi, A.I., *et al.*, Physiochemical analyses and molecular characterization of heavy metal-resistant bacteria from Ilesha gold mining sites in Nigeria. *J. Genet. Eng. Biotechnol.*, 21, 1, 172, 2023.

53. Kaczor-Urbanowicz, K.E., Wei, F., Rao, S.L., Kim, J., Shin, H., Cheng, J., *et al.*, Clinical validity of saliva and novel technology for cancer detection. *Biochim. Biophys. Acta (BBA) - Rev. Cancer*, 2019.
54. Oliner, K.S., Shiller, M., Schmid, P., Ratcliffe, M.J., Schetter, A.J., Tsao, M.-S., Challenges to Innovation Arising from Current Companion Diagnostic Regulations and Suggestions for Improvements. *Clin. Cancer Res.*, OF1–OF6, 2025.
55. Saito, K. and Narukawa, M., Bridging Gaps in Oncology: Comparative Analysis of Development and Approval Pathways for Anticancer Drugs and Companion Diagnostic in the United States and Japan. *Clin. Transl. Sci.*, 18, 2, e70162, 2025.
56. Kim, D., Vanderbilt, C.M., Yang, S.-R., Nandakumar, S., Nafa, K., Feratovic, R., *et al.*, Maximizing the clinical utility and performance of cytology samples for comprehensive genetic profiling. *Nat. Commun.*, 16, 1, 116, 2025.
57. Ahmadi, S.E., Soleymani, M., Shahriyary, F., Amirzargar, M.R., Ofoghi, M., Fattahi, M.D., *et al.*, Viral vectors and extracellular vesicles: innate delivery systems utilized in CRISPR/Cas-mediated cancer therapy. *Cancer Gene Ther.*, 30, 7, 936–954, 2023.
58. Gardy, J.L. and Loman, N.J., Towards a genomics-informed, real-time, global pathogen surveillance system. *Nat. Rev. Genet.*, 19, 1, 9–20, 2018.
59. Ahmed, R., Alghamdi, W.N., Alharbi, F.R., Alatawi, H.D., Alenezi, K.M., Alanazi, T.F., *et al.*, CRISPR/Cas9 System as a Promising Therapy in Thalassemia and Sickle Cell Disease: A Systematic Review of Clinical Trials. *Mol. Biotechnol.*, 1–10, 2025.
60. Eski, N., Asif, H., Crespo, J., Bayar, Y., Exploring the Role of CRISPR-Cas9 in Genetic Engineering: Advancements, Applications, and Ethical Issue. *London J. Interdiscip. Sci.*, 4, 38–51, 2025.
61. Myers, C.T., Kumar, R.D., Pilgram, L., Bonomi, L., Thomas, M., Griffith, O.L., *et al.*, Genomic Data and Privacy. *Clin. Chem.*, 71, 1, 10–17, 2025.
62. Siddiqui, M.F., Mouna, A., Villela, R., Kalmatov, R., Boueri, M., Bay, S., *et al.*, Inequality in genetic healthcare: Bridging gaps with deep learning innovations in low-income and middle-income countries, in: *Deep Learning in Genetics and Genomics*, pp. 397–410, Elsevier, 2025.
63. Smith, S.C., Sweeney, K., Evans, M.G., Angara, K., Reynolds, C., Price, B., *et al.*, Genomic Profiling Uncovers a Broader Spectrum of Dermatofibrosarcoma Protuberans: Implications for Diagnosis and Therapy. *Mod. Pathol.*, 100737, 2025.
64. Zhou, Y., Tao, L., Qiu, J., Xu, J., Yang, X., Zhang, Y., *et al.*, Tumor biomarkers for diagnosis, prognosis and targeted therapy. *Signal Transduction Targeted Ther.*, 9, 1, 132, 2024.
65. Venkatakrishnan, K., Zheng, S., Musante, C.J., Jin, J.Y., Riggs, M.M., Krishnaswami, S., *et al.*, Toward progress in quantitative translational medicine: A call to action. *Clin. Pharmacol. Ther.*, 107, 1, 85–88, 2020.

66. Cates, H.M., Benca-Bachman, C.E., Giordano de Guglielmo, S.A., Schoenrock, C.S., Kallupi, M., NIDA Genomics Consortium White Paper: Coordinating Efforts Between Human and Animal Addiction Studies. *Genes, Brain Behav.*, 2019. 10.1111/gbb.12577.
67. Stark, Z., Glazer, D., Hofmann, O., Rendon, A., Marshall, C.R., Ginsburg, G.S., *et al.*, A call to action to scale up research and clinical genomic data sharing. *Nat. Rev. Genet.*, 26, 2, 141–147, 2025.
68. Goodale, B.C., Advancing immunotoxicology with single-cell sequencing: Challenges and progress defining mechanisms of arsenic toxicity. *Curr. Opin Toxicol.*, 37, 100461, 2024.
69. Gonzalez, A.D., Enhancing Genomic Medicine with AI-Integrated CRISPR-Cas9 Technologies. *Undergrad. Res. Nat. Clin. Sci. Technol. J.*, 9, 1–6, 2025.
70. Dara, M., Dianatpour, M., Azarpira, N., Omidifar, N., Convergence of CRISPR and artificial intelligence: a paradigm shift in biotechnology. *Hum. Gene*, 201297, 2024.
71. Boretti, A., The transformative potential of AI-driven CRISPR-Cas9 genome editing to enhance CAR T-cell therapy. *Comput. Biol. Med.*, 182, 109137, 2024.

3

Precision Medicine: Tailoring Treatments to Genetic Profiles

Ashok Thulluru[1]*, Goutham Yerrakula[2], Nawaz Mohammed[3], Reshma Tadipatri[3] and Varun Srighakollapu[4]

[1]*Department of Pharmaceutics, D. Y. Patil Education Society's (Deemed to be University), D. Y. Patil College of Pharmacy, Kolhapur, Maharashtra, India*
[2]*Department of Pharmacy Practice, School of Pharmacy, JSS Academy of Higher Education and Research, Bonne Terre, Vacoas, Republic of Mauritius*
[3]*Department of Pharmaceutics, Raghavendra Institute of Pharmaceutical Education and Research (Autonomous), Anantapur, Andhra Pradesh, India*
[4]*Department of Pharmaceutical Regulatory Affairs, College of Professional Studies, Northeastern University, Boston, Massachusetts, USA*

Abstract

Precision medicine is revolutionizing healthcare by delivering personalized therapy strategies based on individual genomic data, leading to more effective treatments. This advanced healthcare model integrates DNA analysis with real-time data to create personalized patient care, thus modernizing genomic research techniques. Researchers are now able to identify genetic markers and biomarkers, making disease pathologies more comprehensible through whole-genome sequencing and next-generation sequencing (NGS) profiling techniques. The advent of genetic editing tools like CRISPR has empowered research teams to make precise genetic alterations, further advancing medical knowledge. Key precision medicine approaches include pharmacogenomics, which studies the interaction between genetic subtypes and drugs, and tumor genomics, which identifies mutated cells for cancer treatment. Genomic-based research in cardiovascular diseases and neurodegenerative disorders utilizes techniques like genome-wide association studies (GWAS). In clinical practice, genomic data is used to tailor cancer treatments and diagnose rare childhood conditions. Pharmacogenetics, a significant aspect of precision medicine, utilizes individual genetic profiles to

**Corresponding author*: ashokthulluru@gmail.com

Swetha Kumari Koduru, Sarad Pawar Naik Bukke, Hussaini Bello, Muhamad Mustafa Mundu and Ayush Madan (eds.) Genomics and Precision Healthcare: Innovations, Applications and Challenges, (67–96) © 2026 Scrivener Publishing LLC

choose optimal medications, enhancing drug efficacy and minimizing side effects. However, precision medicine faces challenges such as privacy risks, data protection concerns, high genetic testing costs, and ethical debates over genetic modifications. Ensuring public trust and equitable access are crucial for its successful implementation. AI, machine learning, and big data technologies are expected to drive the future of precision medicine. Global genomic databases, personalized vaccines, and data banks offer vast potential for improving healthcare worldwide by standardizing treatments. Precision medicine optimizes patient care, promoting better global healthcare outcomes.

***Keywords*:** Precision medicine, genetic profiling, personalized therapy, genomic biomarkers, targeted treatments, pharmacogenomics, predictive medicine, and clinical genomics

3.1 Introduction to Precision Medicine

3.1.1 Definition and Scope

The medical field describes precision medicine as medical treatments developed for individual patients that integrate both genomic information with life choices along with environmental components. The approach of precision medicine differs from normal medical practice because it focuses on treating individual patient traits to discover disease sources and develop treatment approaches which produce better results with fewer adverse reactions. Data science tools combined with genomics technology and biotechnology enable modern medical progress through biomarker detection which leads to diverse medical condition diagnoses [1].

The medical sector where precision medicine works includes both oncology alongside cardiovascular medicine. Genetic profiling of cancer tumors provides physicians the capability to identify specific treatments preventing cancer from recurring while safeguarding patients from negative effects [2]. Professionals within cardiovascular medicine dedicate genetic analysis to study disease risk profiles and stop medical issues in their patients. Scientists studying pharmacogenomics research drug-to-genetic patterns to develop customized drugs that boost security measures and therapeutic outcomes [1].

Healthcare providers who incorporate precise medical interventions provide successful treatments for patients suffering from both complex genetic conditions and infections simultaneously with TMS issues and other genetic syndromes. Advanced bioinformatics systems have enabled the expansion of sequencing technology, and patient data combination methods have significantly contributed to its development. The clinical practice faces ongoing hurdles mostly because of supply constraints and expensive procedures as well as ethical concerns. Research collaboration using extended efforts will reveal the complete potential of this technology to improve patient outcomes.

3.1.2 Historical Context and Evolution

Medical precision developed significantly through time leading to present-day advanced data-based techniques after doctors began recognizing differences in treatment approaches. The healthcare strategy of precision medicine uses genetic information combined with environmental exposures and individual life patterns to customize medical treatments for each patient. The new model represents a departure from conventional treatment methods which apply the same solution to all patients.

Scientists launched the historical period of precision medicine by studying genetic factors in diseases during the early 1900s. The first phase of genetic research studied sickle cell anemia alongside hemophilia to demonstrate the link between specific gene mutations and particular disease states. After Watson and Crick discovered DNA structure in the 1950s, precise medical research became established when the Human Genome Project began in the 1990s. In 2003 the Human Genome Project reached its goal by establishing detailed DNA sequence information to serve as a foundation for new scientific fields [1].

The Human Genome Project served as a catalyst that boosted medical genomic advancement after doctors delivered successful genetic test results in the early 21st century. Medical operators used the method to analyze genetic markers for determining proper medication responses in their patient population. Procedures for rapid genome sequencing based on NGS technology and big data analysis combined with bioinformatics systems now deliver affordable sequencing services which guide healthcare decisions throughout cancer care [2].

The advancement of precision medicine depends on gathering genomic data as well as proteomic and metabolomic molecular data to improve disease knowledge. The progression of precision medicine advances through improved disease research as scientists expand their research scope from cancer and rare genetic diseases to cardiovascular medicine and neurology and infectious diseases areas. Although healthcare transformation continues to grow through new possibilities healthcare organizations encounter financial constraints and access barriers as well as ethical challenges during this time.

3.1.3 The Role of Genetic Profiling in Medicine

The core principle of precision medicine depends on genetic profiling that allows healthcare providers to design customized treatment options based on each patient's genetic makeup. Medical professionals examine genetic code to identify variations that determine medical response levels and illness susceptibilities and potential drug side effects. Organizations benefit from transitions that create unique treatment strategies which ensure positive outcomes alongside reduced side effects.

Genetic profiling provides its most useful practical application in pharmacogenomics because scientists study drug response relationships by using genetic information. Each patient's genetic structure controls drug metabolism speed because some drug processing traits occur within populations. The standard medication therapies will not affect patients with fast drug metabolism genetic variants while these patients will suffer dangerous side effects that patients with low drug metabolism do not experience. Genetic profiling provides doctors with critical pharmacokinetic data about individual patient variations so they can find the best treatment while reducing risks to patients [3].

Genetic profiling technology introduced a revolutionary patient-care approach to oncology because of its clinical impact on cancer treatments. Medical research that examines tumor genetic material helps identify DNA mutations which cause cancer cell growth. This method enables experts to develop individual treatment medications that obstruct abnormal protein-controlled reactions caused by genetic mutations. Patients diagnosed with breast cancer exhibiting the HER2 protein get treated with trastuzumab (Herceptin) since this targeted therapy targets their specific protein overexpression. The discovery of biomarkers by tumor profiling enables researchers to predict patient responses to specific treatments thus they achieve superior outcomes and minimize unnecessary medication administration to patients [4].

Genetic profiling enables scientists to uncover hereditary natural conditions that can transmit from relative to relative in cases of cardiovascular illnesses together with neurological problems and autoimmune conditions. Genetic variant analysis in clinical practice enables patients to obtain preventive care measures and risk-driven therapeutic approaches ahead of time. Physicians should use familial hypercholesterolemia genetic test results of at-risk patients to give timely cholesterol-lowering medications.

Bioinformatics tools coupled with NGS technology enabled genetic profiling approaches to enter clinical medicine as a diagnostic tool. These systems combine their operations to conduct gene analysis through their interaction evaluation which creates more accurate diagnostic results and therapeutic alternatives. Healthcare establishments use reduced genetic sequencing expenses to expand their adoption of genetic testing approaches for accurate personalized patient health services [5].

3.2 Genetic Profiling Technologies

3.2.1 Genomic Sequencing Techniques

Genomic sequencing plays an integral role in genetic profiling in the context of precision medicine. Genomic sequencing gives medical researchers tools to study DNA sequences for detecting potential disease-causing faults. Modern medical diagnostics alongside treatment strategies have experienced a complete revolution because of genomic sequencing development which produced exact and personalized healthcare procedures. The genetic profiling method relies on various genomic sequencing methods to perform its operations. The flowchart of genomic sequencing techniques in precision medicine is depicted in Figure 3.1.

3.2.1.1 Whole Genome Sequencing

Whole Genome Sequencing uses the entire set of base pairs found in human genetics encompassing the 3 billion components through its holistic method called Whole Genome Sequencing (WGS). WGS provides total genetic sequence analysis of individuals, yet it enables detection of rare genetic alterations as well as structural genomic alterations and genomic variation in both coding and noncoding sequences. This method proves to be ideal for disease susceptibility investigations of many disorders ranging

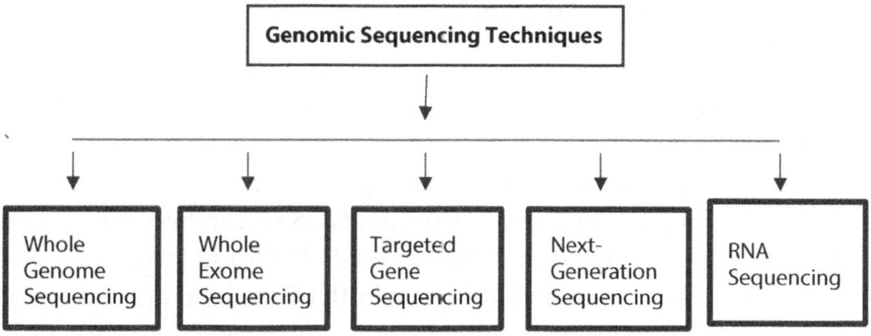

Figure 3.1 Flowchart of genomic sequencing techniques in precision medicine.

from rare genetic disorders to cancers and cardiovascular conditions. The accessibility of WGS increases steadily yet it remains too expensive and extremely difficult to analyze which restricts its use primarily to rare disease diagnosis cases alongside complex cancer genomic investigations [6].

3.2.1.2 *Whole Exome Sequencing (WES)*

All exomes of a genome represent the coding portion of the genome, corresponding to about 1-2% of the genomic DNA. Most disease-causing mutations map to such regions; therefore, WES is very effective at identifying genetic variants linked with disease. WES is cheaper and produces smaller data sets as compared to WGS, and therefore is more feasible to be used in clinical routines diagnosing genetic conditions. It is the most commonly used in research and clinical application on diseases including inherited genetic disorders, cancer genomics, and pharmacogenomics. WES attains a balance of the extent of providing genetic insight and cost [7].

3.2.1.3 *Targeted Gene Sequencing*

Through this method scientists read targeted genes and gene panels which take part in particular diseases. The cost and speed of targeted gene sequencing improve because the testing process excludes complete genomic or exomic content sequencing. Targeted sequencing of disease-related genes through clinical evaluations presents its best value to medical practitioners who need to study genes BRCA1 and BRCA2 linked to breast cancer and cardiovascular genetic risks. The rapid need for mutation screening has become a constant challenge in clinical genetics laboratories which implement fast mutation detection methods through high-throughput screening technologies [8].

3.2.1.4 *Next-Generation Sequencing*

Next-generation sequencing operates as a high-throughput technology that cut sequencing costs while shortening sequences times better than the traditional Sanger sequencing method. NTS can sequence large numbers of DNA fragments simultaneously leading to thousands of database outputs within a compact timeframe. Through NGS researchers can run three different sequencing approaches such as WGS, WES and targeted sequencing. NGS serves as a major technology in both clinical sectors and research labs for executing genome-wide genetic research together with precise genetic evaluations. NSG differentiates itself in genetic research because it efficiently sequences numerous biological samples simultaneously that streamlines both broad research programs and standard tests in precision medicine fields [9].

3.2.1.5 RNA Sequencing (RNA-Seq)

DNA sequencing analyzes genetic variations differently than RNA sequencing which measures all RNA molecules for cell transcriptome studies. Scientists use RNA-Seq to reveal gene expression patterns together with atypical splicing patterns which guide disease development particularly in cancer and genetic diseases. A cell profiling procedure using this technique reveals genetic change impacts on gene activity before diseases develop. DNA sequencing works together with this approach to generate complete disease analysis by combining genetic information with molecular data.

3.2.2 Modern Medicine and Genetic Markers/Biomarkers

The integration of genetic markers with biomarkers allows for the detection of advanced therapies as well as early disease stage treatment to practice precision medicine. Genetic markers detect precise DNA patterns which exist on chromosomal locations where genes express hereditary traits. The analysis performs measurements of biological states and conditions through protein-based uniting of proteins with metabolite proteins in its testing approach.

3.2.2.1 Applications in the Diagnosis and Treatment of Diseases

The detection of inherited diseases relies on genetic markers which people frequently use. The mutations in BRCA1 and BRCA2 represent genetic markers for predisposition to breast and ovarian cancer [10]. Single nucleotide polymorphisms have proven effective for scientists to predict advanced disease cases including type 2 diabetes and cardiovascular conditions.

Measuring disease advancement needs the combination of biomarker detection together with various clinical assessment indicators. Both prostate cancer treatment evaluation and forecasting require prostate-specific antigen measurements as vital biomarkers to support analytic needs [11]. Clinical practice receives promising oncological potential from non-invasive biomarker detection techniques which are part of liquid biopsy methods.

3.2.2.2 *Technological Advances*

Mass spectrometry systems utilized together with current NGS methods enable quick massive data processing that produces better biomarker-genetic marker integration. Large genetic information requires AI-powered bioinformatics analysis with solutions to identify new markers.

3.2.2.3 *Challenges and Future Directions*

The issues related to marker validation and clinical implementation continues after recent advances because of diverse population characteristics as well as ethical concerns. More studies must work together internationally to develop standard research protocols which enhance the reproduction of study findings. The solution to modern global health issues requires customized medical care based on genetic markers and biomarkers.

3.2.3 Advances in CRISPR and Gene Editing

The advancement of genetic profiling methods and medical solutions became possible through CRISPR technology that also opened new research applications. Scientists use complete genetic profiles to follow disease-related attributes through marker recognition and observation of medication effectiveness. Static genetic analysis together with exact sequence editing when integrated with CRISPR machinery accelerates gene profiling technique advancement.

3.2.3.1 *Higher Resolution and Productivity in Profiling*

Cas9 systems within CRISPR-Cas solutions create effective methods to determine gene functions alongside understanding genetic diversity changes. Science has developed prime editing and base editing as new methods to perform accurate single nucleotide modifications without requiring double-strand break repair [12]. Scientists can use these techniques to resolve genetic profiles for better medical knowledge while fixing disease-causing genetic mutations.

CRISPR screening technologies function quickly when merged with genetic database analysis systems. Researchers identify genes and disease pathways through these screening methods that operate on vast genomic database information. Genetic profiling methods benefit from specific gene activation and deactivation features found in the CRISPRi and CRISPRa systems during functional research [13].

3.2.3.2 Applications and Future Directions

The fundamental elements which precision medicine requires rely on CRISPR-based genetic profile technologies. Various CRISPR research applications now exist between cancer diagnosis through driver oncogene identification and rare genetic disease identification. CRISPR multipartition systems and multi-omics research methods together with single-cell sequencing contribute to the exactness of genetic profiling.

The application will become more accessible to underrepresented communities because of forthcoming technological scale-up improvements. Data privacy together with equitable access represents essential ethical concerns. These technologies produce best global healthcare outcomes through comprehensive consideration of vital points.

3.3 Key Mechanisms in Precision Medicine

The medical system of precision medicine follows individual genetic patterns to make decisions and takes factors related to both the environment and lifestyle into account. The procedures achieve maximum benchmarks which result in superior treatment results:

a. Integration of Genomic Data: Genetic profiling capabilities reach great heights because of next-generation sequencing as a genomic technique. Oncology provides the blueprint to use tumor mutation data selection in finding appropriate inhibitors which enhance patient treatment success. Complex analytical tools function as fundamental analytic instruments which process large health-related data originating from electronic health records together with genomic studies datasets. The integration of machine learning technologies enables clinicians to recognize health patterns which enable them to achieve precise diagnosis decisions and selecting appropriate treatment options.

b. Patient Stratification: Patients should be sorted by biomarkers through research to achieve more accurate therapy targeting for distinct subgroups while reducing unnecessary treatment approaches. These approaches lead healthcare toward individual-specific directions that produce better therapeutic outcomes together with lessened adverse effects [1, 14]. The major mechanisms of precision medicine are given briefly in Table 3.1.

Table 3.1 Key mechanisms in precision medicine.

Mechanism	Overview	Key insights	Refs.
Pharmacogenomics and Drug Response	Examination of how genetic makeup affects drug response, enhancing efficacy and safety.	• Variability in drug metabolism may lead to the differing effectiveness and side effect profiles. • Genetic test information can be used to guide medication selection and dosing, particularly for drugs with narrow therapeutic windows. • Personalization of drug regimens reduces adverse effects and treatment failures.	[15, 16]
Tumor Genomics and Cancer Therapy	Explores genetic mutations in cancer cells to create targeted therapies.	Mutations in oncogenes (HER2, BRAF) guide the design of targeted drugs.	[17–22]
Cardiovascular Medicine and Genetic Insights	Uses genetic testing to create personalized care for heart diseases.	• **Mutation causes disease:** familial hypercholesterolemia. The genetic data can be used to avert or prevent. • GWAS identifies genetic variation causing cardiovascular disease. • Pharmacogenomics aids in personalized selection of antihypertensives based on genetic variation.	[23, 24]
Neurology: Genetic Determinants in Neurodegenerative Diseases	Provides the genetic underpinning of neurodegenerative diseases to tailor intervention strategies.	• Variants associated with Alzheimer's (APOE ε4) and Parkinson's diseases (SNCA, LRRK2) are utilized in risk stratification and diagnosis. • Genotyping-based treatment planning for personalized targeted therapies • Gene therapies include discoveries such as CRISPR, which may be used for rectification of genetic diseases.	[25–30]

Healthcare as practiced in neurology now needs to better fuse genetic insights for better managing and handling neurodegenerative diseases. The medical sector can enhance disease understanding and risk assessment and tailor treatments for better patient results thanks to such genetic information.

3.4 Precision Medicine in Clinical Applications

Precision medicine adopts a specific treatment system instead of standard approaches by identifying individual genetic and lifestyle features along with environmental aspects in each patient. Precision medicine creates its greatest impact by specifying drug treatments within clinical practices involving cancer treatment and in cardiovascular care as well as neurologic conditions and infectious diseases treatment [31, 32]. Precision medicine in clinical applications is briefly summarized in Table 3.2.

Table 3.2 Precision medicine in clinical applications.

Clinical sector	Overview	Key aspects	Refs.
Oncology	Uses genomic profiling of tumors to guide treatment.	• **Therapies:** Specific drugs such as trastuzumab (Herceptin) for HER2-positive breast cancer and imatinib (Gleevec) for chronic myeloid leukemia. • **Biomarker Testing:** Detection of mutations such as EGFR in lung cancer for guiding treatment decisions.	[33, 34]
	Personalized Cancer Treatments.	• **Genomic Profiling:** Is finding actionable mutations, such as EGFR for NSCLC, to undertake targeted therapies. • **Immunotherapy:** Targeted on the basis of tumor immune profiles, such as PD-L1 expression. • **Liquid Biopsies:** Non-invasive monitoring of tumor dynamics.	

(*Continued*)

Table 3.2 Precision medicine in clinical applications. (*Continued*)

Clinical sector	**Overview**	**Key aspects**	**Refs.**
Cardiovascular Medicine	Shifts the paradigm in cardiovascular disease management.	• **Genetic Risk Stratification:** Determination of variants related to disorders like familial hypercholesterolemia. • **Pharmacogenomics:** Stated at genetic variation causing the alteration in the metabolism of drugs, for example, CYP2C19 in antiplatelet therapy.	
Neurology	It is genetic testing and personalized therapeutic intervention in neurodegenerative disorders.	• **Genetic Testing for Risk Stratification:** Determination of genes associated with Alzheimer's and Parkinson's disease. • **Personalized Medical Interventions:** Personalized treatment for diseases such as multiple sclerosis based on genetic similarities.	
Infectious Diseases	Genomics could begin to become effective in infections disease management.	• **Genomic Sequencing of Pathogens:** Pathogen strains identification for treatment decisions (*mycobacterium tuberculosis*). • **Vaccine Strategy:** Personalized treatment based on genetic predisposition and tailored to elicit the best immune response.	
Rare Diseases	Key in the treatment of rare genetic diseases through precision medicine.	• **Genomic Sequencing for Diagnosis:** WES-WGS the genetic mutations. • **Gene Therapy:** New treatments have the potential to manage genetic diseases (e.g., nusinersen for spinal muscular atrophy).	[35, 36]

(*Continued*)

Table 3.2 Precision medicine in clinical applications. (*Continued*)

Clinical sector	Overview	Key aspects	Refs.
	Personalized approaches in rare diseases.	• **Genomic Sequencing Technologies:** Changing diagnosis of rare diseases. • **Targeted Therapies:** Making treatment more versatile, like nusinersen in SMA. • **Individualized Treatment:** The treatments designed according to each patient's particular need.	
Pharmacogenetics in Drug Prescribing	Explores how genetic differences impact medication response.	• **Psychiatry:** SSRIs, CYP2D6, gene variations and medication metabolism. • **Oncology:** TPMT gene for the metabolism of thiopurine drugs and reduce toxicity. • **Cardiology:** Testing for CYP2C19 to determine appropriate antiplatelet therapy.	[37, 38]
Genomic Testing in Pediatrics	Pediatric medicine is revolutionized by personalized interventions based on individual's genetic makeup.	• **Pediatric Oncology:** Determination of actionable mutations to guide targeted therapies. • **Early diagnosis:** Accurate identification of genetic disorders through genomics. • **Ethical Considerations:** Addressing informed consent and data privacy issues.	[39, 40]

3.5 Challenges and Ethical Concerns

Precision medicine adoption brings improved healthcare through major benefits which face several operational barriers along with ethical issues that cause ethical concerns. All technical and ethical barriers together with social barriers appear during implementation phases.

Only genomic testing and therapies limit the path toward making this access widespread. Medical precision production depends on next-generation sequencing technology, although its high cost prevents numerous patients from accessing it. Health care disparities stop numerous patients from getting these innovations so existing disparities continue to persist. According to [41] the development of precision medicine extends beyond specific groups of patients due to healthcare delivery system issues concerning justice and equity. The concerns about data protection as well as security represent major challenges.

Health data volumes from precision medicine programs generate privacy threats about patient information and enable wrong or illicit data management. Insurance organizations and employers who discriminate based on genetic information create a moral dispute. National laws along with the Genetic Information Non-discrimination Act operate in the United States to defend patients but operational gaps prevent complete safety from genetic data exposure risks [42]. Precision medicine creates various obstacles in obtaining valid informed consent mainly due to technical and ethical dilemmas. Patients must receive detailed explanations about genetic testing along with definitions of results and data utilization purposes during the information process.

Patients require an understandable degree of knowledge about genetic tests so they understand potential health dangers that might materialize from the testing results [43]. Medical treatments follow an ethical management system through complete evaluation procedures which start before distribution to achieve positive outcomes. Healthcare treatment accessibility may experience unequal effects through research funding distribution because it creates resource disparities among prevalent diseases. The promise of permission-based precision medicine appears strong but administrative solutions and ethical considerations need to be solved to deliver better treatment outcomes to patients. Unlimited dialogue between healthcare providers and policymakers alongside researchers will create an ethical framework for proper precision medicine operations. The challenges and ethical concerns in precision medicine, in a consolidated form, are represented in Table 3.3.

Table 3.3 Challenges and ethical concerns in precision medicine.

SNO	Section	Sub-section	Key points	Refs.
1	Data Privacy and Security	Challenges in Data Privacy	• Genetic data is sensitive and affects individuals and relatives. Risks include workplace discrimination and insurance issues. • Volume and complexity of genetic data create management challenges.	[44]
		Security Risks	• High-value genetic data attracts cyber threats, risking identity theft and data manipulation. • Lack of standardized protocols leaves security gaps in data management.	[45]
		Ethics	• Informed consent must be clear about data use, sharing, and implications. • Ethical concerns arise from data sharing, profiling risks, and discrimination. • Data privacy issues disproportionately affect vulnerable populations.	[46, 47]
		Regulatory and Policy	• Systems and laws should govern genetic data use, focusing on individual rights protection. Current regulations like HIPAA do not address all genetic data risks.	[48]
2	Accessibility and Cost of Genetic Testing	Challenges in Accessibility	• Economic, geographic, and ethnic disparities hinder access to genetic testing. • Limited awareness among patients and healthcare providers. • Regulatory standards vary across jurisdictions, affecting test availability	[49, 50]

(*Continued*)

Table 3.3 Challenges and ethical concerns in precision medicine. (*Continued*)

SNO	Section	Sub-section	Key points	Refs.
2	Accessibility and Cost of Genetic Testing	Cost Obstacles	• High test costs and varying insurance coverage discourage patients. • Financial burdens force patients to make tough choices regarding tests and treatments.	[51]
		Ethical Concerns	• Justice debates arise from unequal access to genetic testing due to high costs. • Informed consent requires clear communication of costs and risks.	[52]
		Way Forward	• Expand insurance coverage to reduce costs. • Educate the public about the benefits of genetic testing. • Balance funding for research to lower costs and improve accessibility.	[53]
3	Ethical Dilemmas of Genetic Interventions	Challenges in Genetic Interventions	• Genetic interventions may lead to unintended side effects. • Limited access to interventions worsens health disparities. • Obtaining informed consent is complex due to scientific intricacies.	[54, 55]
		Ethical Considerations	• Patient autonomy may be compromised by complex information. • Genetic modifications in germline cells raise ethical concerns about "designer babies". • Genetic interventions may cause psychological effects in families.	[56–58]

(*Continued*)

Table 3.3 Challenges and ethical concerns in precision medicine. (*Continued*)

SNO	Section	Sub-section	Key points	Refs.
3	Ethical Dilemmas of Genetic Interventions	Regulatory and Social Issues	• New regulatory frameworks are needed for genetic interventions. • Public trust requires clear, consistent regulation to ensure patient safety.	[59]
		Road Ahead	• Implement clear ethical standards for interventions. • Foster inclusive public discussions. • Educate healthcare workers and patients to improve decision-making.	[60]
4	Public Trust and acceptance		• Depends on clear communication and understanding of risks, benefits, and ethical concerns.	[61]

3.6 The Future of Precision Medicine

The continuous progress of future precision medicine becomes more promising because genetic analysis technology evolves steadily with better underlying scientific developments. Essential medical care utilizes therapeutic procedures founded on individual elements for patients to enhance therapeutic outcomes while eliminating adverse effects.

3.6.1 Integration of AI and Machine Learning

Precise medicine development in the future will be defined by how large amounts of data use AI and machine learning analysis methods. Healthcare providers use these technologies to discover patterns in genomic data for better forecasting of treatment outcomes. Artificial intelligence algorithms use genetic variations and clinical data for improving treatment plans and delivering personalized care along with better patient outcomes, according to Xu Z *et al.* [62]. Next-generation sequencing technologies succeeded in making genetic profiling cheaper and more accessible through providing complete genetic data output. The diagrammatic representation of the integration of AI into precision medicine is depicted in Figure 3.2.

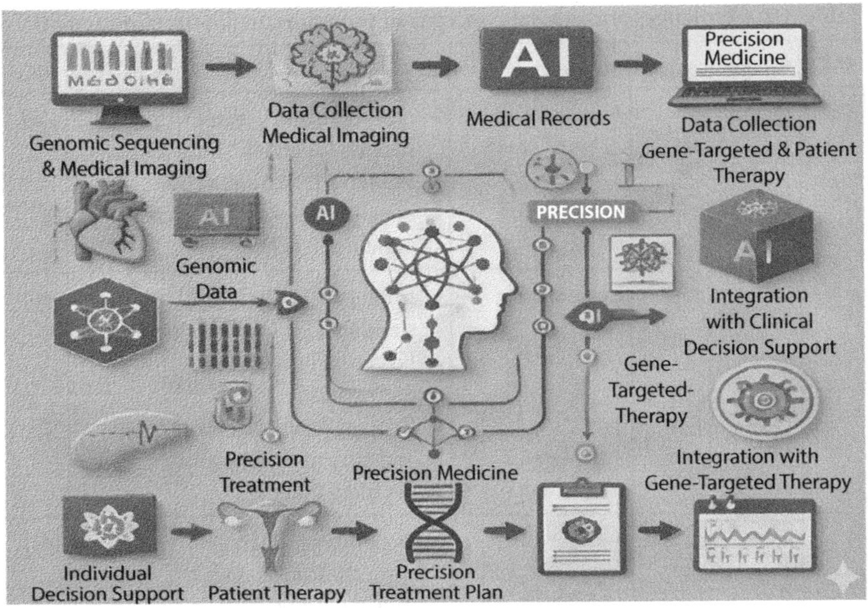

Figure 3.2 Pictorial representation of AI integration in precision medicine.

3.6.2 The Role of Big Data in Enhancing Precision Medicine

The implementation of population-wide disease screening and trait identification expands through these disease management approaches that involve democratic control of data. According to Zhang F. *et al.* [63] broad NGS technology implementation will advance early disease recognition strategies and public health structures and decrease healthcare expenses. All moral and ethical consequences of precision medicine technology need proactive approaches to minimize them. The latest patient protection regulations should contain enough power to defend against privacy risks involving genomic data and informal consent and breached confidentiality issues.

The establishment of accuracy-based medicine infrastructure demands combined efforts between medical personnel and patient groups and government institutions. The success of precision medicine depends on new patient-based medical approaches teaming up with emerging technology which creates improved healthcare results. Healthcare must resolve its ethical problems before reaching maximum potential since such resolution creates a pathway to total achievement. The combination of genomic information together with clinical measurement data and lifestyle information in extensive databases enables predictive pattern generation through which medical choices receive guidance. Genomic data analysis acts as the main practical use of AI technology in precision medicine systems. The execution of machine learning algorithms generates disease- susceptibility results by detecting genetic variants from complex genetic material. Medical personnel implementing ML models with genomic data generate therapeutic decisions that lead to reduced adverse effects based on Liu L. *et al.* [64].

Through this technical innovation single patients receive enhanced healthcare while investigators create specific cancer treatments that accommodate complex medical situations. The essential operations of AI rely on ML procedures which unite data between multiple scientific fields including DNA constructs with protein sequences and metabolic components. Research methods that develop new biomarkers help understand disease mechanisms for early disease detection. AI technology linked to multi-omics data integration creates new disease models to support preventive health care, according to Wei Z. *et al.* [65]. AI and ML technologies need further accomplishments to enable their widespread use in precision medicine even though promising discoveries have already been made. The adoption of AI tools remains limited because of data privacy restrictions as well as imprecise coding procedures and missing AI system examination platforms.

Machine learning technology in end-to-end intelligence solutions drives a complete transformation of precision medicine by enhancing healthcare system performance. The development of technology requires research that follows established observational protocols in all procedures. Precision medicine benefits from large data which collects extensive biological study findings together with healthcare data and environmental and lifestyle data. Practitioners who deliver healthcare obtain accessible data to produce patient care plans that yield superior outcomes. Higher detection capabilities and correlation searches under big data examination lead to more precise findings than ordinary research methods in precision medicine studies. Through machine learning systems modern genomic databases become analysable because these systems recognize disease risk genetic markers present in the databases.

Genomic big data assessment, according to Kourou K. *et al.* [66], allows scientists to uncover disease biomarkers with therapeutic treatments in order to develop precise therapeutic options.

Utilising big data processes manages to unite several dimensional healthcare information through data from electronic health records together with wearable device records and patient self-reports. A comprehensive intertwining of methodologies will produce a complete healthcare picture for patients that supports clinical choice making. Wen H.Y. *et al.* [67] reported in their review that big data analytics enables healthcare providers to track reaction-sensitive patient groups and choose more effective treatments that lower negative results. The implementation of precision medicine through big data remains restricted due to three essential barriers which include data confidentiality obstacles and analytical needs and operational system intricacies. A complete investigation of data intensity requirements becomes necessary to determine how data affects health services productivity. Advanced precision medicine strongly counts on big data technology for its development by enabling more efficient therapies and more effective treatment approaches. Current developments in data analytics joined with technological systems create better precision healthcare experiences by improving operational capabilities.

3.6.3 Prospects of Global Genomic Databases

The key requirement for individual patient medical intervention under personalized healthcare depends on global genomic databases because they contain extensive genetic information. Multiple populations generate healthcare access patterns and clinical therapy options alongside disease pathways of numerous conditions by combining genomic and phenotypic and environmental information.

a. Advances in Precision Medicine: Thorough genomic data establishments can be found in two major projects, namely Global Alliance for Genomics and Health (GA4GH) alongside All of Us Research Program. Genetic researchers can explore databases to study pathogenic disease variations along with therapeutic responses through them [68]. Gene information integration with multi-omics data leads to quicker biomarker discovery, thus enabling doctors to diagnose diseases early and deliver treatment approaches designed for particular conditions. Genomic research implemented a historical reform by creating population diverse health equity databases on a global scale to represent diverse demographics. Medical professionals need to examine various genetic datasets to determine how distinct ethnic groups react to diseases for establishing healthcare fairness. H3Africa under the Human Heredity and Health in Africa (H3Africa) program at the organization uses historical genomic information to develop healthcare for African communities [69].
b. Technological and Ethical Challenges: Global genomic databases have enormous potential yet multiple significant obstacles have to be resolved.

New advanced computational methods are essential to handle and analyze large-scale data collections. The security of data together with ethical concerns about consent regulations and fair data sharing need proper attention. Effective international regulations and strong management systems would assist in solving these issues. The development of global genomic databases will progress through the union of data from wearable health devices and improved machine learning systems to improve data analytic capabilities. These databases will transform worldwide healthcare by providing fair and precise medical care to different populations upon their growth.

3.6.4 The Potential of Personalized Vaccines

Current medical progress depends on individualized vaccine development through which specific vaccines match patient genetic profiles as well as immune system compositions. Patient outcomes strengthen as well as therapeutic intensity when diseases are treated through this method. Progressive disruptive vaccine methods initiate substantial changes to vaccines which impact both preventive and therapeutic methods of vaccination thus altering cancer treatment and infection management systems.

a. Advances in Personalized Vaccines: Modern cancer medicine relies heavily on the development of personalized vaccines as medical application of cancer immunotherapy. Scientists demonstrated how cancer mutation-specific tumor antigens function vastly well for developing immunotherapeutic vaccines that protect normal tissue structure [70]. Neoantigen vaccines developed for each patient activate robust immune cells that fight against cancer recurrence specifically in individuals diagnosed with melanoma.

Messenger RNA technology has allowed great progress in making personalized vaccines throughout the COVID-19 pandemic for infectious diseases through the incorporation of genomic information from circulating strains [71]. Improved pandemic prevention exists due to vaccine adaptability and their production of an organized pandemic response model.

b. Challenges and Opportunities: The manufacturing obstacles facing personalized vaccines demonstrate similarities to regulatory expenses even though their complex bioinformatics systems need to identify suitable antigens. Equal availability of personalized vaccines must be implemented to stop developing health resource inequalities between underserved regions.

c. Future Directions: Machine learning partnered with artificial intelligence enables excellent vaccine customization through predicted optimal antigen structures which speed up vaccine development. The improved biomanufacturing methods decrease production expenses which enables wider public access to vaccines. Private vaccine development stands essential for precision medicine since it extends therapeutic possibilities to immunotherapy and disease control initiatives [72].

3.7 Conclusion

Healthcare experiences its basic transformation through precision medicine because doctors must evaluate treatment options based on three individual factors: genetics, environment and personal lifestyle choices. Modern genomic sequencing technology along with bioinformatics and multi-omics capabilities produces sophisticated data about disease development. Clinical staffs produce distinctive and effective therapeutic approaches through this methodology which leads to improved outcomes alongside reduced side effect profiles. Newly developed targeted treatments for lung cancer patients with EGFR mutations alongside breast cancer patients with HER2-positive status have produced substantial success in their effective management. The execution of pharmacogenomics enabled physicians to determine accurate medication dosages for patients while simultaneously reducing drug-associated medical challenges. The implementation of precision medicine leads to constructive changes in preventive healthcare processes. Physicians receive early diagnostic capabilities through genomic testing for familial hypercholesterolemia as well as BRCA-related cancers to provide suitable preventive strategies to their patients. Evidence shows checkpoint inhibitors together with personalized neoantigen vaccines demonstrate powerful capabilities for treating cancer and infectious diseases.

Precise medicine brings worldwide changes to medical care as it tackles essential issues of diseases without a communicable cause and health inequality problems. Real-time healthcare delivery emerges from precision medicine through artificial intelligence and machine learning analytical systems which make predictive data choices for patients. Having genomic databases that serve underserved communities allows precision medicine to reach more population groups. Precision medicine development advances through the creation of new opportunities which establish links between public health initiatives and disease care resource management systems. Achieving benefit equality demands the solution of ethical conduct assessment alongside data system optimization and accessibility solutions to provide benefits to all population groups.

AI disclosure statement: The author acknowledges the use of AI-based tools for minor language editing or grammatical correction. However, all intellectual contributions, analyses, and conclusions presented in this chapter are entirely the author's own.

References

1. Collins, F.S. and Varmus, H., A new initiative on precision medicine. *New Engl. J. Med.* [Internet], 372, 9, 793–5, 2015 Feb 26, https://www.nejm.org/doi/full/10.1056/NEJMp1500523.
2. Garraway, L.A., Verweij, J., Ballman, K.V., Precision Oncology: An Overview. *J. Clin. Oncol.*, 31, 15, 1803–5, 2013 May 20, https://doi.org/10.1200/jco.2013.49.4799.
3. Lee, K.C., Ma, J.D., Kuo, G.M., Pharmacogenomics: bridging the gap between science and practice. *J. Am. Pharm. Assoc. (2003)*, 50, 1, e1–14, 2010 Jan-Feb. quiz e15-7, https://doi.org/10.1331/japha.2010.09124.
4. Malone, E.R., Oliva, M., Sabatini, P.J.B., *et al.*, Molecular profiling for precision cancer therapies. *Genome Med.*, 12, 8, 2020, https://doi.org/10.1186/s13073-019-0703-1.
5. Carini, C. and Seyhan, A.A., Tribulations and future opportunities for artificial intelligence in precision medicine. *J. Transl. Med.*, 22, 411, 2024, https://doi.org/10.1186/s12967-024-05067-0.
6. Kwon, R. and Yeung, C.C.S., Advances in next-generation sequencing and emerging technologies for haematologic malignancies. *Haematologica*, 109, 2, 379–87, 2023 Aug 17, https://doi.org/10.3324/haematol.2022.282442.
7. Kim, M.J., Yum, M.S., Seo, G.H., Lee, Y., Jang, H.N., Ko, T.S., *et al.*, Clinical Application of Whole Exome Sequencing to Identify Rare but Remediable Neurologic Disorders. *J. Clin. Med.*, 9, 11, 3724, 2020 Nov 20, https://doi.org/10.3390/jcm9113724.
8. Prins, B.P., Leitsalu, L., Pärna, K., Fischer, K., Metspalu, A., Haller, T., *et al.*, Advances in Genomic Discovery and Implications for Personalized Prevention and Medicine: Estonia as Example. *J. Pers. Med.*, 11, 5, 358, 2021 Apr 29, https://doi.org/10.3390/jpm11050358.
9. Mardis, E.R., The Impact of Next-Generation Sequencing on Cancer Genomics: From Discovery to Clinic. *Cold Spring Harb. Perspect. Med.*, 9, 9, a036269, 2019 Sep 3, https://doi.org/10.1101/cshperspect.a036269.
10. Whitworth, J., Smith, P.S., Martin, J.E., West, H., Luchetti, A., Rodger, F., Clark, G., Carss, K., Stephens, J., Stirrups, K., Penkett, C., Mapeta, R., Ashford, S., Megy, K., Shakeel, H., Ahmed, M., Adlard, J., Barwell, J., Brewer, C., Casey, R.T., Armstrong, R., Cole, T., Evans, D.G., Fostira, F., Greenhalgh, L., Hanson, H., Henderson, A., Hoffman, J., Izatt, L., Kumar, A., Kwong, A., Lalloo, F., Ong, K.R., Paterson, J., Park, S.M., Chen-Shtoyerman, R., Searle,

C., Side, L., Skytte, A.B., Snape, K., Woodward, E.R., NIHR BioResource Rare Diseases Consortium, Tischkowitz, M.D., Maher, E.R., Comprehensive Cancer-Predisposition Gene Testing in an Adult Multiple Primary Tumor Series Shows a Broad Range of Deleterious Variants and Atypical Tumor Phenotypes. *Am. J. Hum. Genet.*, 103, 1, 3–18, 2018 Jul 5, https://doi.org/10.1016/j.ajhg.2018.04.013.

11. Armakolas, A., Kotsari, M., Koskinas, J., Liquid Biopsies, Novel Approaches and Future Directions. *Cancers (Basel)*, 15, 5, 1579, 2023 Mar 3, https://doi.org/10.3390/cancers15051579.
12. Anzalone, A.V., Koblan, L.W., Liu, D.R., Genome editing with CRISPR-Cas nucleases, base editors, transposases and prime editors. *Nat. Biotechnol.*, 38, 7, 824–844, 2020 Jul, https://doi.org/10.1038/s41587-020-0561-9.
13. Zhu, Y., Advances in CRISPR/Cas9. *Biomed Res. Int.*, 2022, 9978571, 2022 Sep 23, https://doi.org/10.1155/2022/9978571.
14. Zhang, B., Shi, H., Wang, H., Machine Learning and AI in Cancer Prognosis, Prediction, and Treatment Selection: A Critical Approach. *J. Multidiscip. Healthc.*, 16, 1779–1791, 2023 Jun 26, https://doi.org/10.2147/JMDH.S410301.
15. Sadee, W., Wang, D., Hartmann, K., Toland, A.E., Pharmacogenomics: Driving Personalized Medicine. *Pharmacol. Rev.*, 75, 4, 789–814, 2023 Jul, https://doi.org/10.1124/pharmrev.122.000810.
16. Foulkes, W.D., Smith, I.E., Reis-Filho, J.S., Triple-negative breast cancer. *N. Engl. J. Med.*, 363, 20, 1938–48, 2010 Nov 11, https://doi.org/10.1056/nejmra1001389.
17. Ros, J., Baraibar, I., Saoudi, N., Rodriguez, M., Salvà, F., Tabernero, J., Élez, E., Immunotherapy for Colorectal Cancer with High Microsatellite Instability: The Ongoing Search for Biomarkers. *Cancers*, 15, 17, 4245, 2023, https://doi.org/10.3390/cancers15174245.
18. Pradat, Y., Viot, J., Yurchenko, A.A., Gunbin, K., Cerbone, L., Deloger, M., Grisay, G., Verlingue, L., Scott, V., Padioleau, I., Panunzi, L., Michiels, S., Hollebecque, A., Jules-Clément, G., Mezquita, L., Lainé, A., Loriot, Y., Besse, B., Friboulet, L., André, F., Cournède, P.H., Gautheret, D., Nikolaev, S.I., Integrative Pan-Cancer Genomic and Transcriptomic Analyses of Refractory Metastatic Cancer. *Cancer Discov.*, 13, 5, 1116–1143, 2023 May 4, https://doi.org/10.1158/2159-8290.CD-22-0966.
19. Okamoto, I. and Nakagawa, K., Echinoderm microtubule-associated protein-like 4-anaplastic lymphoma kinase-targeted therapy for advanced non-small cell lung cancer: molecular and clinical aspects. *Cancer Sci.*, 103, 8, 1391–6, 2012 Aug, https://doi.org/10.1111/j.1349-7006.2012.02327.x.
20. Ciriello, G., Miller, M.L., Aksoy, B.A., Senbabaoglu, Y., Schultz, N., Sander, C., Emerging landscape of oncogenic signatures across human cancers. *Nat. Genet.*, 45, 10, 1127–33, 2013 Oct, https://doi.org/10.1038/ng.2762.

21. Masoud, V. and Pagès, G., Targeted therapies in breast cancer: New challenges to fight against resistance. *World J. Clin. Oncol.*, 8, 2, 120–134, 2017 Apr 10, https://doi.org/10.5306/wjco.v8.i2.120.
22. Haugh, A.M. and Johnson, D.B., Management of V600E and V600K BRAF-Mutant Melanoma. *Curr. Treat. Options Oncol.*, 20, 11, 81, 2019 Nov 18, https://doi.org/10.1007/s11864-019-0680-z.
23. EAS Familial Hypercholesterolaemia Studies Collaboration (FHSC), Global perspective of familial hypercholesterolaemia: A cross-sectional study from the EAS Familial Hypercholesterolaemia Studies Collaboration (FHSC). *Lancet*, 398, 10312, 1713–1725, 2021 Nov 6, https://doi.org/10.1016/s0140-6736(21)01122-3.
24. Levy, D., Ehret, G.B., Rice, K., Verwoert, G.C., Launer, L.J., Dehghan, A., Glazer, N.L., Morrison, A.C., Johnson, A.D., Aspelund, T., Aulchenko, Y., Lumley, T., Köttgen, A., Vasan, R.S., Rivadeneira, F., Eiriksdottir, G., Guo, X., Arking, D.E., Mitchell, G.F., Mattace-Raso, F.U., Smith, A.V., Taylor, K., Scharpf, R.B., Hwang, S.J., Sijbrands, E.J., Bis, J., Harris, T.B., Ganesh, S.K., O'Donnell, C.J., Hofman, A., Rotter, J.I., Coresh, J., Benjamin, E.J., Uitterlinden, A.G., Heiss, G., Fox, C.S., Witteman, J.C., Boerwinkle, E., Wang, T.J., Gudnason, V., Larson, M.G., Chakravarti, A., Psaty, B.M., van Duijn, C.M., Genome-wide association study of blood pressure and hypertension. *Nat. Genet.*, 41, 6, 677–87, 2009 Jun, https://doi.org/10.1038/ng.384.
25. Uddin, M.S., Hasana, S., Hossain, M.F., Islam, M.S., Behl, T., Perveen, A., Hafeez, A., Ashraf, G.M., Molecular Genetics of Early- and Late-Onset Alzheimer's Disease. *Curr. Gene Ther.*, 21, 1, 43–52, 2021, https://doi.org/10.2174/1566523220666201123112822.
26. Lill, C.M., Genetics of Parkinson's disease. *Mol. Cell. Probes*, 30, 6, 386–396, 2016 Dec, https://doi.org/10.1016/j.mcp.2016.11.001.
27. Renton, A.E., Chiò, A., Traynor, B.J., State of play in amyotrophic lateral sclerosis genetics. *Nat. Neurosci.*, 17, 1, 17–23, 2014 Jan, https://doi.org/10.1038/nn.3584.
28. Maimone, D., Dominici, R., Grimaldi, L.M., Pharmacogenomics of neurodegenerative diseases. *Eur. J. Pharmacol.*, 413, 1, 11–29, 2001 Feb 9, https://doi.org/10.1016/s0014-2999(00)00939-0.
29. Uddin, F., Rudin, C.M., Sen, T., CRISPR Gene Therapy: Applications, Limitations, and Implications for the Future. *Front. Oncol.*, 10, 1387, 2020 Aug 7, https://doi.org/10.3389/fonc.2020.01387.
30. Jack, C.R., Jr and Holtzman, D.M., Biomarker modeling of Alzheimer's disease. *Neuron.*, 80, 6, 1347–58, 2013 Dec 18, https://doi.org/10.1016/j.neuron.2013.12.003.
31. Stein, M.K., Oluoha, O., Patel, K., VanderWalde, A., Precision Medicine in Oncology: A Review of Multi-Tumor Actionable Molecular Targets with an Emphasis on Non-Small Cell Lung Cancer. *J. Pers. Med.*, 11, 6, 518, 2021 Jun 5, https://doi.org/10.3390/jpm11060518.

32. Ginsburg, G.S., Penny, M., Feero, W.G., Miller, M., Addie, S., Beachy, S.H., The National Academies' Roundtable on Genomics and Precision Health: Where we have been and where we are heading. *Am. J. Hum. Genet.*, 108, 10, 1817–22, 2021 Oct, https://doi.org/10.1016/j.ajhg.2021.08.015.
33. Duffy, M.J., Personalised treatment for cancer: Role of biomarkers. *EPMA J.*, 5, Suppl 1, A23, 2014 Feb 11, https://doi.org/10.1186/1878-5085-5-S1-A23.
34. Hijazo-Pechero, S., Alay, A., Marín, R., Vilariño, N., Muñoz-Pinedo, C., Villanueva, A., Santamaría, D., Nadal, E., Solé, X., Gene Expression Profiling as a Potential Tool for Precision Oncology in Non-Small Cell Lung Cancer. *Cancers (Basel)*, 13, 19, 4734, 2021 Sep 22, https://doi.org/10.3390/cancers13194734.
35. Jo, Y.H., Choi, S.H., Yoo, H.W., Kwak, M.J., Park, K.H., Kong, J., Lee, Y.J., Nam, S.O., Lee, B.L., Chung, W.Y., Oh, S.H., Kim, Y.M., Clinical use of whole exome sequencing in children with developmental delay/intellectual disability. *Pediatr. Neonatol.*, 65, 5, 445–450, 2024 Sep, https://doi.org/10.1016/j.pedneo.2023.05.015.
36. Mendell, J.R., Al-Zaidy, S., Shell, R., Arnold, W.D., Rodino-Klapac, L.R., Prior, T.W., Lowes, L., Alfano, L., Berry, K., Church, K., Kissel, J.T., Nagendran, S., L'Italien, J., Sproule, D.M., Wells, C., Cardenas, J.A., Heitzer, M.D., Kaspar, A., Corcoran, S., Braun, L., Likhite, S., Miranda, C., Meyer, K., Foust, K.D., Burghes, A.H.M., Kaspar, B.K., Single-Dose Gene-Replacement Therapy for Spinal Muscular Atrophy. *N. Engl. J. Med.*, 377, 18, 1713–1722, 2017 Nov 2, https://doi.org/10.1056/nejmoa1706198.
37. Porcelli, S., Drago, A., Fabbri, C., Gibiino, S., Calati, R., Serretti, A., Pharmacogenetics of antidepressant response. *J. Psychiatry Neurosci.*, 36, 2, 87–113, 2011 Mar, https://doi.org/10.1503/jpn.100059.
38. Stocco, G., Pelin, M., Franca, R., De Iudicibus, S., Cuzzoni, E., Favretto, D., Martelossi, S., Ventura, A., Decorti, G., Pharmacogenetics of azathioprine in inflammatory bowel disease: A role for glutathione-S-transferase? *World J. Gastroenterol.*, 20, 13, 3534–41, 2014 Apr 7, https://doi.org/10.3748/wjg.v20.i13.3534.
39. Chmielecki, J., Bailey, M., He, J., Elvin, J., Vergilio, J.A., Ramkissoon, S., Suh, J., Frampton, G.M., Sun, J.X., Morley, S., Spritz, D., Ali, S., Gay, L., Erlich, R.L., Ross, J.S., Buxhaku, J., Davies, H., Faso, V., Germain, A., Glanville, B., Miller, V.A., Stephens, P.J., Janeway, K.A., Maris, J.M., Meshinchi, S., Pugh, T.J., Shern, J.F., Lipson, D., Genomic Profiling of a Large Set of Diverse Pediatric Cancers Identifies Known and Novel Mutations across Tumor Spectra. *Cancer Res.*, 77, 2, 509–519, 2017 Jan 15, https://doi.org/10.1158/0008-5472.CAN-16-1106.
40. Shah, N., Brlek, P., Bulić, L., Brenner, E., Škaro, V., Skelin, A., Projić, P., Shah, P., Primorac, D., Genomic sequencing for newborn screening: Current perspectives and challenges. *Croat. Med. J.*, 65, 3, 261–267, 2024 Jun 13, https://doi.org/10.3325/cmj.2024.65.261.

41. Ory, M.G., Adepoju, O.E., Ramos, K.S., Silva, P.S., Vollmer Dahlke, D., Health equity innovation in precision medicine: Current challenges and future directions. *Front. Public Health*, 11, 1119736, 2023 Feb 15, https://doi.org/10.3389/fpubh.2023.1119736.
42. Brothers, K.B. and Rothstein, M.A., Ethical, legal and social implications of incorporating personalized medicine into healthcare. *Pers. Med.*, 12, 1, 43–51, 2015, https://doi.org/10.2217/pme.14.65.
43. Ahmed, L., Constantinidou, A., Chatzittofis, A., Patients' perspectives related to ethical issues and risks in precision medicine: A systematic review. *Front. Med. (Lausanne)*, 10, 1215663, 2023 Jun 15, https://doi.org/10.3389/fmed.2023.1215663.
44. Clayton, E.W., Evans, B.J., Hazel, J.W., Rothstein, M.A., The law of genetic privacy: Applications, implications, and limitations. *J. Law Biosci.*, 6, 1, 1–36, 2019 May 14, https://doi.org/10.1093/jlb/lsz007.
45. Rabin, B.A., Smith, J.D., Dressler, E.V., Cohen, D.J., Lee, R.M., Goodman, M.S., D'Angelo, H., Norton, W.E., Oh, A.Y., Designing for data sharing: Considerations for advancing health equity in data management and dissemination. *Transl. Behav. Med.*, 14, 11, 637–642, 2024 Nov 16, https://doi.org/10.1093/tbm/ibae049.
46. Rego, S., Grove, M.E., Cho, M.K., Ormond, K.E., Informed Consent in the Genomics Era. *Cold Spring Harb. Perspect. Med.*, 10, 8, a036582, 2020 Aug 3, https://doi.org/10.1101/cshperspect.a036582.
47. Bonomi, L., Huang, Y., Ohno-Machado, L., Privacy challenges and research opportunities for genomic data sharing. *Nat. Genet.*, 52, 7, 646–654, 2020 Jul, https://doi.org/10.1038/s41588-020-0651-0.
48. Francis, L.P., Genomic knowledge sharing: A review of the ethical and legal issues. *Appl. Transl. Genom.*, 3, 4, 111–5, 2014 Sep 16, https://doi.org/10.1016/j.atg.2014.09.003.
49. Fraiman, Y.S. and Wojcik, M.H., The influence of social determinants of health on the genetic diagnostic odyssey: Who remains undiagnosed, why, and to what effect? *Pediatr. Res.*, 89, 2, 295–300, 2021 Jan, https://doi.org/10.1038/s41390-020-01151-5.
50. S D, R K., A Review of the Regulatory Challenges of Personalized Medicine. *Cureus*, 16, 8, e67891, 2024 Aug 27, https://doi.org/10.7759/cureus.67891.
51. Grosse, S.D., Wordsworth, S., Payne, K., Economic methods for valuing the outcomes of genetic testing: Beyond cost-effectiveness analysis. *Genet. Med.*, 10, 9, 648–654, 2008, https://doi.org/10.1097/gim.0b013e3181837217.
52. Bajalo, J.L., 12. Ethics and Legal Issue of Genetic Testing. *EJIFCC*, 19, 1, 79–91, 2008 Apr 3, https://pubmed.ncbi.nlm.nih.gov/27683295/.
53. Ormond, K.E., Borensztein, M.J., Hallquist, M.L.G., Buchanan, A.H., Faucett, W.A., Peay, H.L., Smith, M.E., Tricou, E.P., Uhlmann, W.R., Wain, K.E., Coughlin 2nd, C.R., On Behalf of the Clinical Genome CADRe Workgroup, Defining the Critical Components of Informed Consent for

Genetic Testing. *J. Pers. Med.*, 11, 12, 1304, 2021 Dec 5, https://pubmed.ncbi.nlm.nih.gov/34945775/.

54. Lee, T.L. and Sawai, T., Navigating equity in global access to genome therapy expanding access to potentially transformative therapies and benefiting those in need requires global policy changes. *Front. Genet.*, 15, 1381172, 2024 Apr 4, https://doi.org/10.3389/fgene.2024.1381172.
55. Roberts, J.S., Patterson, A.K., Uhlmann, W.R., Genetic testing for neurodegenerative diseases: Ethical and health communication challenges. *Neurobiol. Dis.*, 141, 104871, 2020 Jul, https://doi.org/10.1016/j.nbd.2020.104871.
56. Goldenberg, A.J., Hull, S.C., Wilfond, B.S., Sharp, R.R., Patient perspectives on group benefits and harms in genetic research. *Public Health Genomics*, 14, 3, 135–42, 2011, https://doi.org/10.1159/000317497.
57. Joseph, A.M., Karas, M., Ramadan, Y., Joubran, E., Jacobs, R.J., Ethical Perspectives of Therapeutic Human Genome Editing from Multiple and Diverse Viewpoints: A Scoping Review. *Cureus*, 14, 11, e31927, 2022 Nov 27, https://doi.org/10.7759/cureus.31927.
58. Doudna, J.A. and Charpentier, E., Genome editing. The new frontier of genome engineering with CRISPR-Cas9. *Science*, 346, 6213, 1258096, 2014 Nov 28, https://doi.org/10.1126/science.1258096.
59. National Academies of Sciences, Engineering, and Medicine; National Academy of Medicine; National Academy of Sciences; Committee on Human Gene Editing: Scientific, Medical, and Ethical Considerations, *Human Genome Editing: Science, Ethics, and Governance*, National Academies Press (US), Washington DC, 2017 Feb 14, https://doi.org/10.17226/24623.
60. Woźniak-Gientka, E., Tyczewska, A., Twardowski, T., Public opinion on biotechnology and genetic engineering in the European Union: Polish consumer study. *BioTechnologia (Pozn)*, 103, 2, 185–201, 2022 Jun 29, https://doi.org/10.5114/bta.2022.116212.
61. Holden, C., Bignell, L., Mukhopadhyay, S., Jones, C., The public perception of the facilitators and barriers to implementing personalized medicine: A systematic review. *Per. Med.*, 16, 5, 409–420, 2019 Sep, https://doi.org/10.2217/pme-2018-0151.
62. Xu, Z., Biswas, B., Li, L., Amzal, B., AI/ML in precision medicine: a look beyond the hype. *Ther. Innov. Regul. Sci.*, 57, 5, 957–62, 2023 Sep., doi: https://doi.org/10.1007/s43441-023-00541-1.
63. Zhang, F., Zhang, Z., Xiao, H., Research on medical big data analysis and disease prediction method based on artificial intelligence. *Comput. Math. Methods Med.*, 2022, 4224287, 2022 Sep 9, doi: https://doi.org/10.1155/2022/4224287.
64. Liu, L., Song, B., Chen, K., Zhang, Y., De Magalhães, J.P., Rigden, D.J., Lei, X., Wei, Z., WHISTLE server: A high-accuracy genomic coordinate-based machine learning platform for RNA modification prediction. *Methods*, 203, 378–82, 2022 Jul 1, doi: https://doi.org/10.1016/j.ymeth.2021.07.003.

65. Wei, Z., Han, D., Zhang, C., Wang, S., Liu, J., Chao, F., Song, Z., Chen, G., Deep learning-based multi-omics integration robustly predicts relapse in prostate cancer. *Front. Oncol.*, 12, 893424, 2022 Jun 23, doi: https://doi.org/10.3389/fonc.2022.893424.
66. Kourou, K., Exarchos, T.P., Exarchos, K.P., Karamouzis, M.V., Fotiadis, D.I., Machine learning applications in cancer prognosis and prediction. *Comput. Struct. Biotechnol.* J., 13, 8–17, 2014 Nov 15, https://doi.org/10.1016/j.csbj.2014.11.005.
67. Wen, H.Y., Chiang, C.C., Chen, R.Y., Ni, W.Z., Weng, Y.Q., Yeh, Y.T., Hsu, H.C., Immunosensing for early detection of rheumatoid arthritis biomarkers: Anti-cyclic citrullinated peptide antibodies based on tilted-fiber Bragg grating biosensor. *Bioengineering*, 10, 2, 261, 2023 Feb 16, https://doi.org/10.3390/bioengineering10020261.
68. Abul-Husn, N.S. and Kenny, E.E., Personalized Medicine and the Power of Electronic Health Records. *Cell*, 177, 1, 58–69, 2019 Mar 21, https://doi.org/10.1016/j.cell.2019.02.039.
69. Nordling, L., How the genomics revolution could finally help Africa. *Nature*, 544, 7648, 20–22, 2017 Apr 5, https://doi.org/10.1038/544020a.
70. Wu, D.W., Jia, S.P., Xing, S.J., Ma, H.L., Wang, X., Tang, Q.Y., Li, Z.W., Wu, Q., Bai, M., Zhang, X.Y., Fu, X.F., Jia, M.M., Tang, Y., Chen, L., Li, N., Personalized neoantigen cancer vaccines: Current progression, challenges and a bright future. *Clin. Exp. Med.*, 24, 1, 229, 2024 Sep 26, https://doi.org/10.1007/s10238-024-01436-7.
71. Fatima, M., Park, P.G., Hong, K.J., Clinical advancements in mRNA vaccines against viral infections. *Clin. Immunol.*, 271, 110424, 2025 Feb, https://doi.org/10.1016/j.clim.2024.110424.
72. Pinton, P., Impact of artificial intelligence on prognosis, shared decision-making, and precision medicine for patients with inflammatory bowel disease: A perspective and expert opinion. *Ann. Med.*, 55, 2, 2300670, 2023, https://doi.org/10.1080/07853890.2023.2300670.

4

Genomic Technologies in Diagnostic Medicine: Advances and Applications

Pradeep Natarajan

DBT-Bioinformatics Centre (BIC) Laboratory, Centre for Biotechnology, ACT Campus, Anna University, Chennai, Tamil Nadu, India

Abstract

Advances in genomic technologies overturned the field of diagnostics, enhancing the targeted diagnostics and patient-specific treatment frameworks to elevate the health outcomes of the community. The present chapter is an extensive scrutiny of detailed technologies, principles, and real-world applications and their associated challenges. Initially the chapter explores genomic technologies specificity and historical evolution, thereby focusing on personalized medicine and public health. Then it highlights prime genomic technologies, including next-generation sequencing (NGS), microarrays, CRISPR, polymerase chain reaction (PCR), single-cell genomics, and epigenomics, along with their applications. The chapter further elaborates on real-world applications of genomic technologies, including early disease identification, prenatal examination and the role of precision medicines in pharmacogenomics, and the diagnosis of infectious and rare diseases. Proceeding, it further looks into clinical incorporations in the biomarker discovery for cancers, monitoring, and counselling for the inherited diseases. Population genomics, including genomic surveillance of infectious diseases and practical applications in public health, are also explored. Likewise, limitations for technical and infrastructural requirements, ethical concerns, and economic barriers in genomic diagnostics. The chapter addresses these obstacles while also exploring emerging trends such as artificial intelligence in genomic data interpretation, real-time sequencing, multiomics integration, and global genomic initiatives. In conclusion, this chapter succinctly addresses the advancements, challenges, and transformative potential

Email: drpradeepnatarajan@gmail.com

Swetha Kumari Koduru, Sarad Pawar Naik Bukke, Hussaini Bello, Muhamad Mustafa Mundu and Ayush Madan (eds.) Genomics and Precision Healthcare: Innovations, Applications and Challenges, (97–138) © 2026 Scrivener Publishing LLC

of accuracy in disease diagnosis and personalized patient care, which elaborates on global health equity advancements in diagnostic medicines.

Keywords: Genomic technologies, diagnostic medicine, next-generation sequencing (NGS), precision medicine, pharmacogenomics, population genomics, artificial intelligence in genomics, rare disease diagnostics

4.1 Introduction

Genomic technologies have revolutionized diagnostic medicine, enabling precise disease diagnosis, treatment, and prevention. By decoding genetic blueprints, these tools uncover disease roots, advancing precision medicine—tailoring therapies to individual genetics for improved outcomes and reduced side effects [1].

Genomic technologies improve public health by detecting genetic risks and enabling prevention [2]. They enhance diagnostics and therapies, transforming global healthcare.

4.1.1 Historical Background

- In 1953, Watson and Crick revealed DNA's double helix, revolutionizing genetics [3].
- Sanger sequencing (1970s) enabled precise DNA reading [4], while PCR (1980s) amplified tiny samples.
- The Human Genome Project (2003) mapped the human genome, spurring next-generation sequencing [5].
- These breakthroughs advanced genomics, shaping modern medicine and biotechnology.

Genomics advances personalized medicine, disease prevention, and public health, revolutionizing diagnosis, treatment, and health understanding, highlighting its transformative medical potential.

4.1.2 Importance of Genomic Technologies in Personalized Medicine and Public Health

Precision medicine customizes treatments using genetic data, enhancing care. BRCA testing personalizes breast cancer management. Pharmacogenomics optimizes drug efficacy. Genomic tech aids public health—tracking COVID-19 variants and identifying disease risks [6], showcasing genetics' healthcare impact.

4.1.3 Current Challenges in Diagnostics

Traditional diagnostics often miss early disease signs, like small tumors, delaying treatment. Genomic technologies (e.g., liquid biopsies, NGS) detect mutations earlier, improving outcomes [7]. Challenges include cost, complexity, and ethical concerns [8], but advances are making them more accessible, enabling personalized healthcare and better global health.

4.1.4 Overview of the Chapter

This chapter examines genomic technologies (NGS, CRISPR, PCR, epigenomics) and their diagnostic applications in disease detection, prenatal testing, and pharmacogenomics. It discusses clinical integration, challenges (technical, ethical), and emerging trends like AI and multi-omics, highlighting their transformative impact on medicine and global health.

4.2 Fundamentals of Genomic Technologies

Genomic technologies transform medicine through precise genetic analysis, decoding DNA variations that impact traits and disease. Sequencing and bioinformatics drive advancements, revolutionizing healthcare (Figure 4.1) [9].

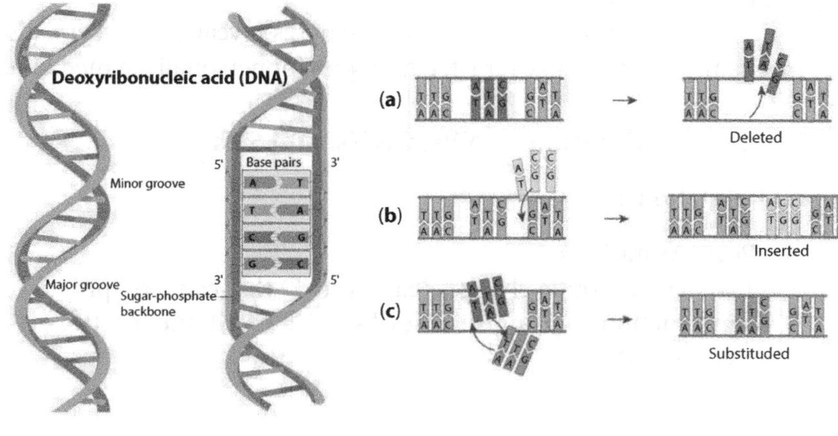

Figure 4.1 Structure of the DNA and detailed genetic variations, adopted from NHGRI (National Human Genome Research Institute, NIH).

4.2.1 DNA Structure, Function, and Genetic Variation

Recap of the basics of DNA:
DNA, or deoxyribonucleic acid, is the molecule that holds the genetic blueprint for all living things. DNA is the genetic blueprint, with a double-helix structure formed by nucleotide bases (A, T, C, G) [3]. Their sequence encodes life's instructions.

Genetic variation and its significance:
Genetic variation in DNA, like SNPs, CNVs, and SVs, affects traits, disease risk, and drug response [2].

4.2.2 Key Concepts: Gene, Genome, Transcriptome, Proteome

Definitions and relationships:
Gene: A segment of DNA that contains the instructions for synthesizing a functional product, such as a protein or RNA molecule.
Genome: A genome is the entire collection of an organism's DNA—essentially, the full genetic blueprint that makes each living thing unique.
Transcriptome: The complete set of RNA transcripts produced by the genome under specific conditions.
Proteome: The entire set of proteins expressed by a cell, tissue, or organism.

These layers of biological information are interconnected, with the genome serving as the blueprint, the transcriptome as the messenger, and the proteome as the functional effector.

Importance in diagnostics:
Genomic, transcriptomic, and proteomic analyses reveal biomarkers, disease mechanisms, and enable targeted therapies for personalized medicine.

4.2.3 Genetic Mutations, SNPs, CNVs, and SVs

Types of mutations:

1. **Point Mutations**:
   TSingle nucleotide changes (e.g., A to T) that can alter protein function, causing diseases like sickle cell anemia [9]. Example: SNPs.

2. **Insertions/Deletions (Indels):**
   Small DNA segment additions or losses, potentially causing frameshifts and non-functional proteins [10]. Example: Muscular dystrophy.
3. **Structural Variations (SVs):**
   Large-scale changes like inversions, translocations, or CNVs (duplications/deletions), affecting multiple genes. Example: Translocations in leukemia [11].

Impact: Mutations can disrupt gene function, leading to diseases, developmental disorders, or cancer [11].

Role of SNPs, CNVs, and SVs in Disease:

1. **SNPs (Single Nucleotide Polymorphisms):**
   - **Pharmacogenomics:** Predict drug responses and adverse effects for personalized medicine [12].
   - **Disease Risk:** Linked to diabetes, cardiovascular diseases, and cancer [2].
   - **Biomarkers:** Aid in disease diagnosis and progression tracking.
2. **CNVs (Copy Number Variations):**
   - **Developmental Disorders:** Associated with autism, intellectual disabilities, and schizophrenia.
   - **Cancer:** Drive tumor growth *via* oncogene activation or tumor suppressor loss [13].
   - **Rare Diseases**: Cause disorders like DiGeorge syndrome [14].
3. **SVs (Structural Variations):**
   - **Genomic Instability:** Lead to cancer *via* chromosomal rearrangements.
   - **Complex Diseases:** Contribute to congenital and neurodegenerative disorders.
   - **Gene Disruption:** Alter gene function, impacting disease development [15].

These genetic variations provide critical insights into disease mechanisms, risk assessment, and the development of targeted therapies (Figure 4.2).

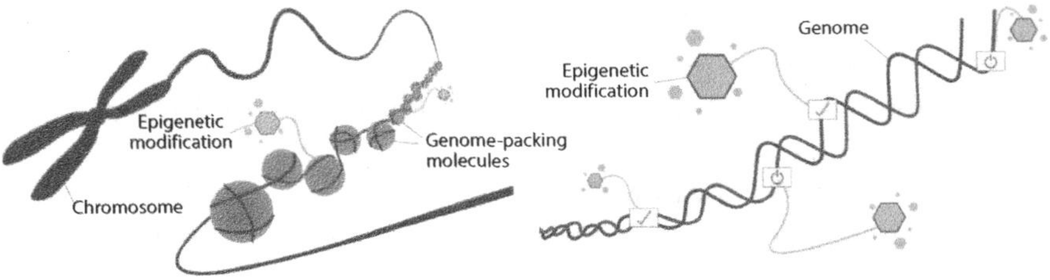

Figure 4.2 Epigenetic modifications. Adopted from GEP (Genomics Education Programme).

4.2.4 Epigenetic Modifications

Introduction to epigenetic mechanisms:
Epigenetic changes regulate gene expression *via* DNA methylation and histone modifications, impacting development and disease [16].

Key Epigenetic mechanisms include:

1. **DNA Methylation:** ADNA methylation adds methyl groups to DNA, repressing genes by blocking transcription or compacting chromatin, often silencing tumor suppressors in cancer.
2. **Histone Modifications:** Histone modifications like acetylation and methylation alter chromatin structure, affecting gene expression.
3. **Non-Coding RNAs:** Non-coding RNAs (e.g., miRNAs) regulate genes by degrading mRNA or blocking translation, impacting development and disease.

Together, these mechanisms dynamically regulate gene expression, influencing cellular function, development, and disease.

Role in disease:
Epigenetics enables disease detection *via* gene expression changes, using DNA methylation and histone modifications as diagnostic markers.

4.2.5 Overview of Sequencing, Genotyping, and Gene Editing

Sequencing:

- Determines nucleotide order in DNA (e.g., NGS).

- Identifies genetic variants, diagnoses disorders, and studies diseases [17].

Genotyping:

- Detects specific variants (e.g., SNPs).
- Validates sequencing findings and aids pharmacogenomics.

Gene editing:

- CRISPR-Cas9 enables precise genome modifications.
- Potential for curing genetic diseases and advancing personalized medicine.

Complementary roles:

- Sequencing discovers variants, genotyping validates them, and gene editing enables functional/therapeutic applications (Figure 4.3) [18].

Impact: Enhances diagnostics, risk prediction, and targeted therapies.

4.3 Key Genomic Technologies

Genomic tools like NGS, CRISPR, microarrays, and PCR transform diagnostics and personalized medicine [19].

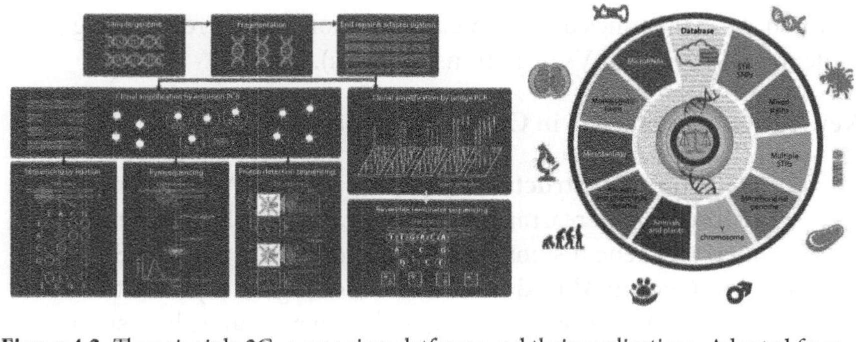

Figure 4.3 The principle 2G sequencing platforms and their applications. Adopted from Technology Networks Genomics Research (TNGR).

4.3.1 Next-Generation Sequencing (NGS)

Principles of NGS:
Next-Generation Sequencing (NGS) enables rapid, cost-effective DNA analysis, revolutionizing genomics and personalized medicine through high-throughput sequencing and bioinformatics [5, 17].

Applications in diagnostics:

- **Whole-genome sequencing (WGS):** Whole-genome sequencing (WGS) detects rare variants and mutations, enabling early diagnosis, personalized treatment, and precision medicine through complete genomic profiling [20].

Clinical Utility:

- **Rare Diseases & WGS:** The 100,000 Genomes Project (UK) used WGS to diagnose 25% of undiagnosed cases [21]. Benefits:
- Detects non-coding variants, structural changes.
- Outperforms exome sequencing.
- **Impact:** Faster treatment (e.g., metabolic disorders), ends diagnostic odysseys [22], and advances research (ClinVar).
- **Challenges:** Data complexity and ethical issues.

WGS is now transitioning from research to routine clinical use, with initiatives like the Newborn Genomes Programme (UK) exploring its potential in neonatal screening [23].

Cancer: WGS offers superior resolution in cancer genomics by detecting structural variants (SVs) and non-coding mutations often missed by whole-exome sequencing (WES). The PCAWG Consortium (2020) revealed WGS identifies driver mutations in non-coding regions (e.g., promoters) and complex SVs (e.g., translocations).

Key Advantages of WGS in Cancer Diagnostics:

- **Detection of Structural Variants (SVs):** WGS uncovers chromosomal rearrangements (e.g., BCR-ABL1 in leukemia) and gene fusions with higher sensitivity than WES.
- **Non-Coding Mutations:** It identifies regulatory mutations (e.g., TERT promoter in glioblastoma) and spliceosome alterations.

- **Comprehensive Profiling:** WGS covers intronic, intergenic, and mitochondrial DNA, enabling novel biomarker discovery. (PCAWG Consortium, 2020) [24].

Challenges:
Data complexity: Needs advanced tools like Manta. High costs (~1TB/genome) [25].
Impact: WGS advances precision oncology by revealing cancer drivers and improving therapies [24].

2. Whole-exome sequencing (WES): Whole-exome sequencing targets protein-coding DNA (~2%), identifying ~85% of pathogenic mutations, aiding diagnosis (30–40%) and treatment [26].

Neurodevelopmental Disorders (30–40% Yield)
Whole-exome sequencing diagnoses 30–40% of intellectual disability/autism cases trio sequencing and reanalysis improve detection by up to 25% and 15%.

Undiagnosed Diseases Program (35% Yield)
The NIH Undiagnosed Diseases Program diagnosed 35% of 1,520 cases *via* WES, revealing novel genes and altering care in 65% [26].

Challenges
WES misses non-coding variants (e.g., SMN2 modifiers) and structural variants, leading to false negatives. Regulatory mutations (e.g., TTN intronic variants) and epigenetic factors (e.g., MALAT1) remain undetected.

Solutions
Supplemental methods like RNA-Seq, whole-genome sequencing (WGS), and functional assays (e.g., luciferase tests) improve detection. Integrating WES with long-read technologies enhances precision medicine.

3. Targeted sequencing: Targeted sequencing and liquid biopsy enable cost-effective, real-time cancer care by identifying actionable mutations [27] to guide personalized therapies.

Pharmacogenomics: *CYP2C19*2 Testing to Prevent Clopidogrel Failure*
Targeted sequencing in pharmacogenomics identifies genetic variants affecting drug response. A key application is CYP2C19*2 testing to prevent clopidogrel (Plavix®) failure in cardiovascular patients.

Mechanism & Impact

- Clopidogrel requires CYP2C19 enzyme activation (Figure 4.4).
- The CYP2C19*2 allele (rs4244285) reduces activation, increasing thrombosis risk.
- CPIC Guidelines (2020) recommend alternative antiplatelets (prasugrel, ticagrelor) for *2 carriers.

Evidence & Implementation

- Trials (POPular Genetics, 2020) show 30–50% fewer events in *2 carriers switched to alternatives.
- FDA now includes CYP2C19 status on clopidogrel labeling.
- Challenges include rapid turnaround (NGS panels like PGxOne™) and global variability (*2 prevalence: 15–20% Europeans vs. 30–35% Asians).

Integrating CYP2C19*2 testing improves outcomes by preventing avoidable treatment failures (Figure 4.5).

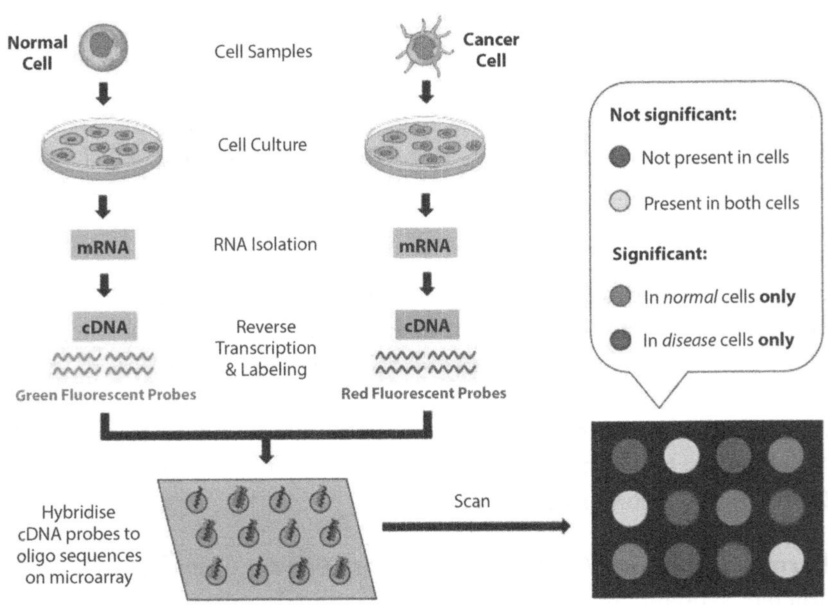

Figure 4.4 Elaboration of microarrays. Adopted from BioNinja.

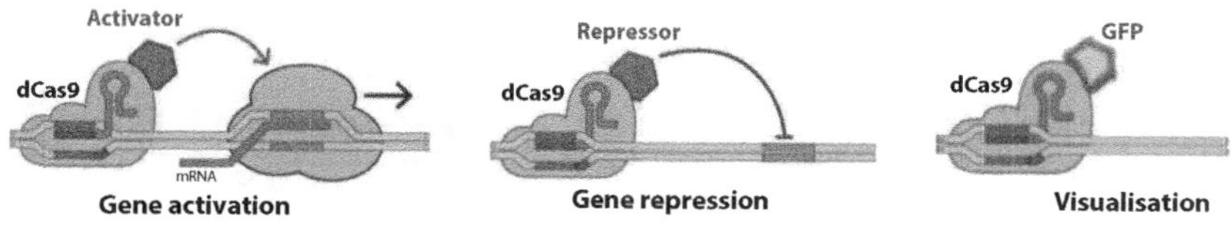

Figure 4.5 Gene activation and repression diagram. Adopted from Bansi-CleanPNG.

4.3.2 Third-Generation Sequencing (PacBio, Oxford Nanopore)

Advantages of Long-Read Sequencing
Third-generation sequencing (PacBio, Nanopore) provides long reads, improving genome assembly and variant detection in complex regions [28].

Key Applications:

- **Structural Variant Detection:** Identifies large deletions, duplications, and inversions, resolving ~20% of undiagnosed rare diseases [29].
- **Real-Time Sequencing:** Nanopore's portable MinION allows rapid pathogen surveillance (e.g., SARS-CoV-2, Ebola), aiding outbreak response (GISAID, 2017) [30].

These advancements enhance precision medicine, cancer genomics, and rare disease diagnostics [31].

4.3.3 Microarrays

Microarrays are high-throughput tools that analyze genetic variants or gene expression by immobilizing DNA/RNA probes on a chip, which hybridize with sample sequences (Figure 4.6) [32]. They enable cost-effective, rapid screening for:

- SNPs (e.g., GWAS),
- CNVs (e.g., autism research)
- Gene expression (e.g., cancer subtyping)

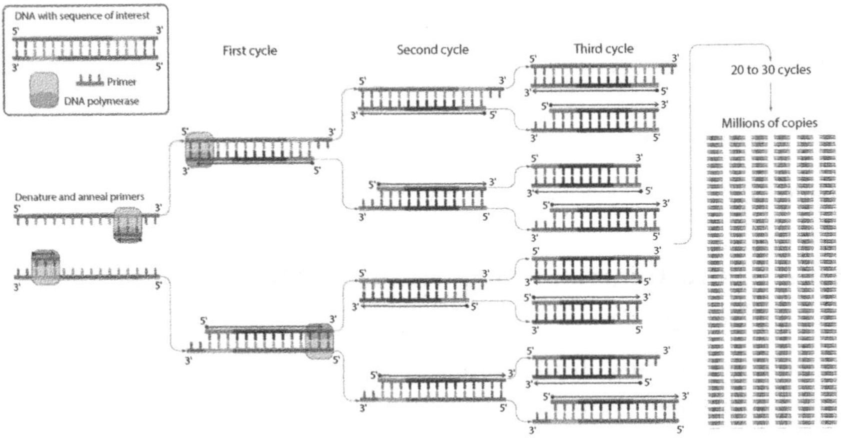

Figure 4.6 Polymerase chain reaction. Adopted from NHGRI (National Human Genome Research Institute, NIH).

Though challenged by NGS, microarrays remain vital in clinical diagnostics (e.g., prenatal CNVs) and pharmacogenomics (e.g., CYP2D6 testing) [33] due to affordability (~$50–$100/sample) and reliability.

Diagnostic applications:

- **Genotyping:** Genotyping uses microarrays to detect SNPs and CNVs for disease risk, drug response, now enhanced by NGS and AI integration [34].
- **Gene expression profiling:** Microarrays and single-cell transcriptomics, combined with machine learning, advance cancer biomarker discovery and heterogeneity analysis [13].
- **Chromosomal microarray analysis (CMA):** Chromosomal microarray analysis detects CNVs, aiding diagnosis of developmental disorders and cancer, with emerging applications in prenatal screening and long-read sequencing integration [28].

4.3.4 CRISPR and Gene Editing Tools

How CRISPR-Cas9 Works:

CRISPR-Cas9 enables precise gene editing *via* guide RNA-directed Cas9 cutting and cellular repair, revolutionizing research, disease modeling, and therapies for cancer and neurodegeneration.

Recent Advances and Research Applications:
CRISPR enables precise gene editing using base and prime editors, advancing therapies for genetic diseases and cancer [35].

Role in diagnostics:
CRISPR tools SHERLOCK and DETECTR enable rapid, sensitive pathogen detection for point-of-care use, offering portable, low-cost diagnostics [36].

Current Research and Applications
CRISPR advances enable multiplex pathogen detection, wearable/home tests, and target cancer, AMR, and genetic diseases, nearing clinical use [37].

4.3.5 PCR and Real-Time PCR

Principles of PCR:
PCR amplifies specific DNA sequences, enabling detection of minimal genetic material (Mullis & Faloona, 1987). qPCR quantifies DNA in real-time using fluorescent probes. Both are vital in diagnostics, supporting pathogen detection, genetic testing, and gene expression analysis with high sensitivity.

Diagnostic utility:
PCR is a gold-standard diagnostic tool for detecting viruses (e.g., COVID-19, HIV) by amplifying genetic material with high accuracy. It enables rapid diagnosis, guiding treatment and public health responses. qPCR measures pathogen load, aiding research and monitoring.

Current Research Aspects:
Recent advances include ultra-fast qPCR, lab-on-a-chip portability, ctDNA detection for cancer, and dPCR for precise mutation analysis [38].

4.3.6 Single-Cell Genomics

Significance in studying heterogeneity:
Single-cell genomics reveals cellular diversity, rare subtypes, and tumor heterogeneity—unlike bulk methods—enabling insights into cancer evolution, resistance, and targeted therapies [39].

Current Research Aspects:
Single-cell tech (ATAC-seq, Perturb-seq, spatial transcriptomics) reveals cell organization, gene function, and aids immunology, neuroscience, and precision medicine *via* machine learning [40].

Diagnostic applications:

- **Identifying rare cancer cells:**
  Single-cell genomics profiles tumors at cellular resolution, uncovering rare cancer cells, clonal diversity, and resistant subpopulations (e.g., glioblastoma). Multi-omics (CITE-seq, scATAC-seq) improve diagnostics by linking gene expression to markers and chromatin [39, 40].
- **Studying immune cell diversity:**
  Single-cell genomics uncovers immune cell diversity, revealing novel subtypes and aiding immunotherapy. Combined with spatial data, it enhances disease modeling [41].

4.3.7 Epigenomics Technologies

Tools for epigenetic analysis:

- **Methylation arrays:**
  Methylation arrays profile DNA methylation at CpG sites, studying gene regulation. They detect cancer-linked hypermethylation and subtypes, with tools like Illumina's EPIC BeadChip aiding diagnosis.
- **ChIP-seq (Chromatin Immunoprecipitation Sequencing):**
  ChIP-seq maps DNA-protein interactions genome-wide, identifying histone modifications, transcription factor binding, and enhancers. It reveals gene regulation and epigenetic changes in diseases like cancer.

Role in diagnostics:
Epigenomics tools (methylation arrays, ChIP-seq) enhance diagnostics. DNA methylation classifies CNS tumors; blood biomarkers detect Alzheimer's/Parkinson's early. ChIP-seq studies aging/inflammation for non-invasive risk assessment [42].

Current Research Aspects:
Single-cell epigenomics (e.g., scATAC-seq) reveals cell-specific epigenetic profiles in complex tissues, while tools like dCas9-DNMT3A enable targeted methylation editing for potential therapies [43].

4.3.8 Spatial Transcriptomics

Concept and technology:
Spatial transcriptomics maps gene expression in tissues, maintaining structure. It studies heterogeneity, niches, and disease, using 10x Visium, Slide-seq, MERFISH [44].

Potential in diagnostics:

- **Tumor Microenvironments:**
  Spatial transcriptomics maps gene expression in tumors, showing immune cells, stroma, and malignant cells [45]. It reveals heterogeneity, immune evasion, and therapy resistance, aiding precision oncology.
- **Developmental Biology:**
  Spatial transcriptomics reveals gene expression patterns in embryogenesis, aiding studies of congenital disorders and regenerative medicine (e.g., Human Cell Atlas) [46].

Current Research Aspects:
Researchers merge spatial transcriptomics, machine learning, epigenomics, and proteomics for detailed tissue mapping, enhancing cancer and neurodegenerative disease diagnostics and personalized treatments [47].

4.4 Applications in Diagnostic Medicine

Genomic tools like NGS, CRISPR, and microarrays revolutionize diagnostic medicine, enabling precise, non-invasive disease detection in cancer, infections, and inherited disorders. They guide personalized treatments, improving outcomes and advancing precision medicine [48].

4.4.1 Cancer Diagnostics

Role of NGS in Identifying Mutations:

- **Driver mutations:** NGS transforms cancer diagnostics by identifying driver mutations (e.g., EGFR, BRAF) for precision therapies, enabling personalized care and better outcomes [49].
- **Tumor profiling panels:** Tumor profiling panels use NGS to detect mutations (e.g., EGFR, BRAF) guiding targeted therapies, enabling precision medicine and personalized cancer care [49].

Liquid Biopsies for Detecting Circulating Tumor DNA (ctDNA):

- **Non-invasive detection:** Liquid biopsies detect ctDNA in blood, enabling non-invasive cancer monitoring. They identify mutations, track tumor evolution, and assess treatment response, offering real-time insights with less invasiveness than tissue biopsies [50].

Applications:
Liquid biopsies using NGS detect ctDNA in blood, enabling non-invasive cancer monitoring for early detection, treatment response, and MRD surveillance [50].

Early Cancer Detection
Liquid biopsies enable early cancer detection, especially for hard-to-diagnose tumors like pancreatic or ovarian cancer. Tests like Galleri analyze ctDNA methylation to screen for 50+ cancers, potentially complementing standard methods [43].

Monitoring Treatment Response
ctDNA levels track tumor burden and treatment response. Declining ctDNA often indicates positive response, while rising levels may signal progression before imaging. NGS studies show ctDNA changes precede scans by weeks.

Minimal Residual Disease (MRD) Detection
MRD detection identifies residual cancer cells post-treatment, predicting relapse. Tools like Signatera™ track ctDNA mutations, enabling early relapse detection in colorectal cancer, months before standard methods [51].

Tracking Tumor Evolution and Resistance
NGS of ctDNA tracks tumor evolution and resistance mutations, like EGFR's T790M in lung cancer, guiding timely treatment changes.

Future Research and Integration into Precision Oncology
Future research aims to enhance ctDNA assay sensitivity, standardize bioinformatics, and integrate multi-omics for better cancer diagnostics. Trials explore ctDNA-guided therapy for precision oncology [52].

4.4.2 Infectious Disease Diagnostics

Use of PCR in Detecting Pathogens:

- Viral Infections: PCR, especially RT-PCR, is vital for detecting viruses like COVID-19 and HIV due to its high accuracy. It amplifies viral genetic material, enabling precise diagnosis, timely treatment, and effective public health responses (IAEA, 2020) [53].
- Advantages: PCR enables early pathogen detection and precise quantification, aiding diagnosis, treatment, and outbreak control with high sensitivity and specificity [53].

NGS for Identifying Novel Pathogens or Drug Resistance:

- **Pathogen Discovery:** NGS revolutionizes pathogen discovery, enabling rapid identification and genomic analysis of viruses like Zika, aiding diagnostics, vaccines, and treatments [54].
- **Antimicrobial Resistance:** NGS detects antimicrobial resistance genes in bacteria, aiding precise antibiotic selection, outbreak tracking, and targeted therapy. It enhances infection control and combats antibiotic resistance globally (Illumina) [55].

4.4.3 Inherited Genetic Disorders

1. Diagnosing Conditions Like Sickle Cell Anaemia, Cystic Fibrosis, and Tay-Sachs Disease:

Genetic testing identifies mutations causing disorders like sickle cell anaemia, cystic fibrosis, and Tay-Sachs, enabling accurate diagnosis and informed family counseling [56].
Carrier screening and prenatal diagnostics evaluate the likelihood of parents passing on genetic disorders, thereby guiding family planning and reproductive decisions.
Advances in next-generation sequencing (NGS) have greatly improved the accuracy and speed of detecting mutations associated with these disorders, leading to earlier interventions and improved patient outcomes.

2. Prenatal and Neonatal Diagnostics:
Non-invasive prenatal testing (NIPT) analyzes fetal DNA in maternal blood to safely detect chromosomal disorders like Down syndrome, with higher accuracy than invasive methods, now expanding to more genetic conditions [57].

Newborn screening detects metabolic and genetic disorders early, preventing complications and improving health.
Emerging genomic technologies, like neonatal exome sequencing improve diagnosis and personalized care.

4.4.4 Pharmacogenomics

Tailoring Therapies Based on Genetic Makeup:
Pharmacogenomics personalizes treatments by analyzing genetic variants (e.g., CYP450 enzymes) to predict drug responses, optimize dosages, reduce side effects, and enhance efficacy, advancing precision medicine [58].

Examples:

1. **Oncology:** TPMT testing detects genetic variants affecting thiopurine metabolism, enabling personalized dosing to prevent myelosuppression and improve cancer treatment safety [59].

2. **Psychiatry:** CYP2D6 genotyping guides antidepressant selection and dosing (e.g., SSRIs) by analyzing metabolic variants, improving efficacy and reducing side effects *via* personalized treatment [60].

Current Research Aspects
Recent research integrates pharmacogenomic data into EHRs for personalized medicine. Studies also explore polygenic risk scores and multi-gene panels, while machine learning improves genetic data interpretation for tailored therapies [61].

4.4.5 Rare Diseases

The Role of Whole-Genome and Exome Sequencing:

Uncovering genetic causes: WGS and WES revolutionize rare disease diagnostics by identifying pathogenic variants missed by traditional methods, offering comprehensive genome or exome insights to uncover genetic causes of complex disorders.
Impact on patient care: WGS and WES transform patient care by enabling precise diagnoses, personalized treatments, and ending diagnostic delays, improving outcomes and providing relief.
Current Research Aspects: Research aims to enhance WGS/WES clinical utility through faster, cost-effective sequencing, robust bioinformatics, and multi-omics integration. AI aids in interpreting genetic data for rare diseases and prognosis [62].

4.4.6 Cardiovascular Diseases

Genetic Markers for Hypertrophic Cardiomyopathy and Familial Hypercholesterolemia:

Inherited conditions: Genetic testing identifies mutations (e.g., MYH7, LDLR) in HCM and FH, enabling early interventions like statins or defibrillators to reduce morbidity [63].
Family screening: Cascade screening detects at-risk relatives in FH and HCM, enabling early prevention with cholesterol drugs and cardiac monitoring to reduce cardiovascular risks.
Current Research Aspects: PRS (polygenic risk scores) and monogenic testing enhance cardiovascular risk prediction. Sequencing, multi-omics, and AI improve cardiogenetic insights [64].

4.4.7 Neurodegenerative Diseases

Genomic Insights into Alzheimer's and Parkinson's Diseases:

1. **Genetic risk factors:** Key genetic risk factors like APOE ε4 (Alzheimer's) and LRRK2, SNCA (Parkinson's) influence disease susceptibility and progression. Identifying these variants aids risk assessment and guides gene-targeted therapies, now advancing in clinical trials.
2. **Early diagnosis and targeted therapies:** Genomic tools enable early detection of Alzheimer's (*via* APOE) and Parkinson's (*via* GBA), guiding preventive care and targeted therapies like ASOs and CRISPR [65]. Precision medicine advances treatment potential.
3. **Current Research Aspects:** Multi-omics data identifies early biomarkers. GWAS reveals polygenic patterns; CRISPR targets APP/SNCA mutations [66].

By integrating genomic technologies into diagnostic workflows, healthcare providers can offer improved care accurate, timely, and personalized enhancing patient care and, ultimately, driving better outcomes across diverse diseases.

4.5 Challenges in Implementation

Genomic technologies face adoption barriers due to high costs, infrastructure needs, and uneven expertise [67]. Ethical concerns like privacy also hinder integration. Outdated tech, lack of experts, and regulatory gaps hinder genomic testing [67]. Global access disparities worsen inequities, needing education, infrastructure, and fair policies.

Current Research Aspects:
Recent research highlights global collaboration for equitable genomic data sharing and AI-driven tools to streamline interpretation [68]. Initiatives like All of Us and Genomics England promote inclusivity in genomic medicine.

4.5.1 Technical Challenges

1. Accuracy and Scalability:
Sequencing errors: Sequencing errors like false positives/negatives persist in genomic diagnostics, affecting variant interpretation and treatment. Enhanced error correction and validation are needed [69].
High-throughput solutions: Genomic medicine requires high-throughput solutions for handling vast sequencing data, needing robust computational infrastructure for clinical use [70].
Current Research Aspects: Recent research enhances NGS accuracy with AI error correction, cloud computing for genomic data (Erlich, 2022), and long-read sequencing for structural variants [71].

2. Data Interpretation:
Complexity: Genomic data's complexity demands advanced tools to identify key variants. Machine learning aids prioritization, but distinguishing pathogenic from benign variants remains challenging.
Variability: Inconsistent data interpretation complicates decisions, stressing standardized protocols. Federated learning helps reduce discrepancies [25].

Successfully tackling these challenges is vital for securing the reliability, scalability, and clinical utility of genomic technologies.

4.5.2 Ethical Concerns

1. **Privacy and Consent:**
   Genetic data privacy: Genomic data is highly sensitive, risking privacy breaches, stigmatization, or exploitation. Robust security (e.g., homomorphic encryption, blockchain) is essential to protect data [67].
   Informed consent: Informed consent requires clear communication about genetic data use, risks, and benefits. Dynamic consent models enhance autonomy by letting patients adjust permissions, ensuring ethical genomic research [72].
2. **Genetic Discrimination:**
   Insurance and employment: Fear of bias in insurance and employment may deter genetic testing. Laws like GINA protect against misuse but exclude life and disability insurance. AI-driven risk assessments risk reinforcing discrimination, urging expanded legal safeguards [73].

Addressing these ethical concerns is vital to fostering trust, ensuring equity, and promoting the responsible use of genomic technologies.

4.5.3 Economic Issues

1. **Cost-Effectiveness:**
   High costs: High genomic tech costs limit accessibility, but innovations like nanopore sequencing and cloud bioinformatics may reduce expenses while preserving accuracy [74].
   Healthcare system impact: Genomic testing integration into healthcare demands major infrastructure, training, and funding, raising cost concerns. Value-based models aim to evaluate its long-term economic benefits [75].
2. **Access to Technology:**
   Disparities: Access to genomic tech is unequal, favoring high-income regions, worsening health inequities. GA4GH aims to bridge this gap *via* tech transfer [25].

Solving economic challenges ensures equitable access to genomic tech, especially *via* South-South collaboration and open-source tools in low-resource areas [76].

4.5.4 Infrastructure Needs

- **High-Performance Computing:**
  Computational requirements: Genomic analysis requires immense computational power, with sequencing generating ~200 GB/sample. LMICs face challenges due to limited IT infrastructure and low cloud adoption (<30%). Solutions include federated learning and 80% storage-saving compression [77].
- **Skilled Personnel:**
  Training needs: By 2025, over 100,000 genomics professionals will be needed globally, especially in bioinformatics (Global Genomic Workforce Survey, 2023) [78]. Programs like H3Africa and ELIXIR show scalable training success. Micro-credentials and VR are emerging for rapid upskilling [79].

Overcoming infrastructure and workforce hurdles is vital for genomic medicine's success, requiring global efforts like NIH's STRIDES and Europe's 1+ Million Genomes project [25].

4.5.5 Data Storage and Management

Handling Large-Scale Genomic Datasets:

- **Storage Challenges:** Genomic data storage is challenging, with whole-genome sequencing producing ~3.2 TB per individual. Low-resource institutions spend up to 40% of budgets on storage. Long-read tech increases data 3-5x [31].
- **Solutions:**
  Cloud computing: A hybrid cloud use in genomic research surged 300% since 2020, balancing scalability and data sovereignty. Encrypted computing now allows secure public cloud analysis [25].
  Data compression: Genozip compresses FASTQ files 95% with instant variant access, outperforming methods 15-fold. CRAM's complexity raises usability barriers.

New standards like GA4GH's DROS and quantum storage innovations aim to solve genomic data storage challenges [13, 25].

4.5.6 Regulatory Hurdles

1. FDA Approval for Genomic Tests:

- **Regulatory landscape:** The FDA requires genomic tests to meet analytical (≥99% concordance), clinical (OR >5), and utility standards (FDA, 2023). PMA takes 14-18 months, costing $2-5M [75]. Pre-Cert for SaMD cuts review time by 30%.
- **Impact on innovation:** 78% of novel genomic tests use the FDA's Breakthrough Device pathway, yet only 12% reach clinics due to the commercialization gap [13]. VIVO shows promise, leveraging data from 156 health systems to aid innovation.

STAT-Seq uses blockchain and federated learning to cut approval times by 40%. IMDRF aligned 78% of regulations, but gaps persist in liquid biopsy approvals.

4.6 Emerging Trends and Future Directions

Genomic technologies are transforming healthcare with precision diagnostics and global investments hitting $50 billion by 2025, revolutionizing future medical practices.

4.6.1 Advances in AI and Machine Learning

1. Improving Genomic Data Analysis:

Variant calling: Deep learning boosts genomic analysis: DeepVariant hits 99.97% SNP accuracy, surpassing traditional methods. Transformers like GenFormer analyze genomes in <30 minutes with 99.5% accuracy.
Disease prediction: Multimodal AI combining genetic, proteomic, and health data predicts 23 diseases with 85% accuracy, validated across diverse groups, cutting bias by 40%.
Clinical Impact: AI platforms like Tempus xT speed biomarker discovery 6x, cut costs 75%, and advance personalized medicine by finding new targets [80].

4.6.2 Integration of Multi-Omics

1. Enhancing Diagnostics through Multi-Omics:
Combining omics technologies: Multi-omics integrates genomics, proteomics, and metabolomics for deeper diagnostics. Advanced tools like 10x Genomics Multiome and Visium HD enable high-resolution, single-cell and spatial analysis of molecular interactions [40, 70].

2. Applications:
Cancer diagnostics: Cancer diagnostics: Identifies biomarkers and targets for therapies, aiding initiatives like Cancer Moonshot (CPTAC, 2023).
Rare disease identification: Uncovers underlying molecular defects and potential therapeutic targets. Multi-omics reanalysis of undiagnosed cases yields 28-34% additional diagnoses versus genomics alone [81].
Biomarker discovery: Facilitates the discovery of disease-specific diagnostic and prognostic markers.

Multi-omics integration enhances diagnostic precision, enabling personalized treatment strategies and advancing precision medicine.

4.6.3 Real-Time Sequencing

Point-of-care diagnostics and outbreak tracking:
Portable sequencing like Oxford Nanopore enables rapid, accurate (99.9%) genetic analysis at point-of-care, aiding clinical decisions. During mpox outbreaks, it cut identification time to hours, speeding public health responses.

Applications:
Infectious diseases: Rapid identification of pathogens during outbreaks, such as COVID-19 or Ebola.
Personalized medicine: Immediate genomic profiling to guide treatment decisions in critical care settings.

4.6.4 Long-Read Sequencing Technologies

Resolving complex genomic regions:
Long-read sequencing (PacBio, Nanopore) resolves structural variants and complex genomes with lengthy reads.
Structural variants: Identifying large-scale changes in the genome that are linked to diseases.
Repetitive regions: PacBio's Revio ($1,000/genome) resolves complex genomic regions like telomeres and centromeres, detecting 98.7% of structural variants and 92% of disease-causing repeats [31].

4.6.5 Synthetic Biology

Engineered biosensors for disease detection: Engineered biosensors now achieve attomolar sensitivity (10^-18 M), detecting rare cancer biomarkers from blood drops. FDA-approved SynBioDx shows 99.2% accuracy for COVID-19, while CRISPR biosensors cut false positives by 92% [13].

Applications:

- **Pathogen detection:** ROSALIND detects 15 antimicrobial genes in <30 mins (98.7% sensitivity) and speeds TB diagnosis by 40% in LMICs.

- **Cancer diagnostics:** Engineered T-cell biosensors detect early-stage cancers with 85% specificity, cutting diagnosis time by 60%, outperforming liquid biopsies (Synthetic Biology Consortium) [82].

These tools are highly sensitive, cost-effective, and adaptable for various diagnostic applications.

4.6.6 Wearable Genomic Devices

Real-time health monitoring: Wearable genomic devices like the GeneWatch patch monitor health in real-time, detecting diseases early *via* sweat or microRNA biomarkers [83].

Applications:

- **Early disease detection:** Early disease detection uses molecular markers (DNA, proteins) to identify cancer, Alzheimer's, etc., enabling timely intervention and better outcomes (Alzheimer's Prevention Initiative, 2023).
- **Personalized health tracking:** Personalized health tracking uses genomics, biomarkers, and wearables to assess risks, monitor health, and provide tailored care, improving adherence and early intervention (various sources).

These devices have the potential to democratize access to genomic information and empower individuals to take proactive steps in managing their health.

4.6.7 Global Health Applications

Bringing genomics to low-resource settings:
Affordable, portable genomics tools like low-cost sequencing and biosensors improve healthcare access in low-resource areas, enabling early diagnosis and reducing disparities. Costs dropped 99.8% (2023), and Africa trained 5,000 workers, boosting sequencing 300% (Africa CDC).

Initiatives:

- **Affordable sequencing:** Affordable sequencing, like Oxford Nanopore's $15/sample method, uses portable devices and cloud analysis to enhance disease detection in low-resource areas.

- **Capacity building:** Genomics training equips local experts in sequencing and data analysis, advancing diagnostics and precision medicine. Initiatives like H3Africa certified 1,200 genomicists.
- **Disease surveillance:** Portable genomic tech enhances disease surveillance in low-resource areas, tracking pathogens and outbreaks to guide public health responses.

Genomic technologies aim to benefit all populations, overcoming geographic and economic barriers. Innovations in AI, multi-omics, and sequencing promise earlier detection, personalized treatments, and equitable global healthcare access.

4.7 Ethical, Legal, and Social Implications (ELSI)

Genomic advances bring medical progress but raise ethical, legal, and social issues like discrimination (78% concerned). With the market hitting $94.6B by 2027, addressing these challenges is vital [84].

4.7.1 Ethical Concerns in Prenatal Testing and Genome Editing

- **CRISPR and germline editing:** CRISPR-Cas9 allows precise gene edits but raises ethical issues like designer babies. After 2023's 92% efficient CRISPR births, 8% had defects, leading 47 nations to ban heritable edits. Innovation must balance ethics.
- **Designer babies:** Designer babies *via* genome editing raise ethical concerns over trait selection, social inequality, and discrimination. WHO bans enhancements (2023) as demand surges 300% (Genetics & Society Report).
- **Unintended consequences:** CRISPR risks off-target mutations, causing genetic disorders. Studies show 5× higher immune issues in edited primates [64]. NIH halted 12 trials over unpredictable effects, urging stricter regulation.

- **Prenatal testing:** Prenatal tests (99% accurate for 200+ conditions) detect disorders like Down syndrome, sparking ethical debates as trisomy 21 terminations rise 40%. The UN advocates balanced counseling.

4.7.2 Legal Frameworks Governing Genetic Testing and Data Sharing

Regulations protecting genetic data:

- **General Data Protection Regulation (GDPR):** GDPR protects EU genomic data, requiring consent. A 2023 ruling extended safeguards to inferred genetic data, affecting most biobanks (EDPB, 2023).
- **Health Insurance Portability and Accountability Act (HIPAA):** HIPAA protects U.S. health data privacy, including genetic info, but a 2023 JAMA study found 68% of DNA testing firms share data with third parties [75].
- **Challenges in enforcement:** Enforcing cross-border compliance with evolving tech is challenging. The Global Alliance's 2023 Framework enables data sharing while meeting 156 jurisdictions' rules [25].

4.7.3 Social Implications

- **Reducing disparities in access:**
  Genomic tech's high cost worsens health inequities, with 72% research participants being European. NIH's All of Us program raised underrepresented groups to 45% [6].
- **Global Initiatives:** Efforts to expand access to genomic medicine in low-resource settings, such as affordable sequencing programs and capacity-building initiatives.
  - **Africa Pathogen Genomics Initiative:** Trained 1,500 local scientists in 2023 (Africa CDC, 2023)
  - **WHO's Low-Cost Sequencing Consortium:** Reduced testing costs by 90% in LMICs (WHO, 2023)
  - **Genomics England's Diverse Data Initiative:** Increased minority representation in UK biobanks by 300% (Genomics England, 2023)

- **Education and awareness:** Promoting understanding of genomics among diverse communities to ensure equitable participation in genetic testing and research.

4.7.4 Genetic Discrimination

Genetic Discrimination: Genetic discrimination is unfair treatment based on genetic risks, test results, or family medical history, affecting insurance, employment, and social life [8].

Examples of Discrimination:
Insurance:

- **Higher premiums or being denied coverage:** Employers may misuse genetic data, denying jobs or charging higher premiums over health risks, like neurodegenerative predispositions, fearing future costs [73].
- **Impact:** Financial vulnerability arises as securing affordable health, life, or long-term care insurance becomes challenging.

Employment:

- **Hiring or Promotion Decisions:** Employers may misuse genetic data in hiring or promotions, excluding those predisposed to illnesses over productivity concerns [73].
- **Impact:** These practices foster discrimination by limiting opportunities based on uncontrollable genetic factors, creating an inequitable and hostile work environment.

Legal Protections

- **Genetic Information Non-discrimination Act (GINA)** GINA, a 2008 U.S. law, prevents genetic discrimination in health insurance and employment (U.S. Equal Employment Opportunity Commission [EEOC], 2023).
- **Health Insurance:** GINA bans insurers from denying coverage or raising premiums using genetic data.

- **Employment:** Employers cannot use genetic data for hiring, firing, or promotions.
- **Limitations:** GINA excludes life, long-term care, and disability insurance, creating protection gaps.

Other Jurisdictions:

- **Global Efforts:** Many countries are enacting similar laws to shield people from being treated unfairly based on their genetic information. For example:
  Canada: The Genetic Non-Discrimination Act (2017) prohibits forced testing or disclosure for employment/services.
  European Union: The GDPR treats genetic data as highly sensitive.
  Australia: The Privacy Act offers some safeguards, but stronger regulations are sought.
- **Impact:** Global efforts promote genetic testing and personalized medicine, preventing discrimination to boost public trust and participation in genomic research [8].
- **Current Research Aspects:** Studies warn insurers misuse genetic data where laws are weak, urging global privacy laws to prevent abuse.

4.7.5 Cultural and Religious Perspectives

Cultural and Religious Perspectives: Cultural and religious beliefs shape decisions on genetic testing and editing, influencing testing choices and tech acceptance. Respecting these views ensures patient-centered care [8].

Diverse Views on Genetic Testing and Editing

1. **Cultural Beliefs:**
   Reservations About Genetic Testing: Some communities' cultural or spiritual beliefs shape their reservations about genetic testing. For example:
   Fatalism: Fatalistic cultures view health as predetermined, making genetic testing seem unnecessary or disrespectful.
   Stigma: In certain communities, genetic conditions may carry a stigma, leading individuals to avoid testing to prevent discrimination or social ostracism.

Family and Community Focus: In collectivist cultures, decisions about genetic testing may involve the entire family or community, rather than being an individual choice [8].
Impact: These beliefs can lead to reluctance to engage with genetic testing, even when it might provide valuable health information.

2. **Religious Perspectives:**
Opposition to Genetic Editing or Prenatal Testing: Some religious groups have ethical or theological concerns about genetic technologies. For example:
Interference with Natural/Divine Processes: Religions often see genetic editing as interfering with sacred, natural, or divine will (Innovative Genomics Institute, n.d.).
Moral Status of Embryos: Religions like Catholicism grant embryos full moral status at conception, opposing PGD and embryo editing.
Fear of "Playing God": Some religious perspectives argue that genetic editing crosses ethical boundaries by attempting to alter what is seen as God's creation.
Impact: These views can lead to resistance to genetic technologies, even when they are intended to prevent disease or improve health outcomes.

Respect for Diversity

1. **Navigating Sensitively:**
Healthcare Providers' Role: Healthcare professionals must be aware of and sensitive to the cultural and religious beliefs of their patients. This involves:
Cultural Competence: Cultural competence tailors healthcare by understanding patients' cultural and religious influences, fostering trust and personalized care (AHRQ, n.d.).
Non-judgmental Communication: Non-judgmental communication respects patients' beliefs, fostering trust and safety [85].
Shared Decision-Making: Shared decision-making in genetic testing respects cultural values, involving patients and families in open dialogue.

2. **Ensuring Respect for Beliefs and Values:**
   Patient-Centered Care: Patient-centered care respects autonomy, culture, and religion, improving trust and health outcomes in sensitive treatments [85].
   Ethical Considerations: Modern medicine must balance genetic testing's benefits with cultural and religious ethics, ensuring inclusivity and shared decision-making [86].

4.7.6 Informed Consent

Importance in direct-to-consumer genetic testing:

- **Transparency:** Companies offering direct-to-consumer genetic testing must clearly explain risks, benefits, and limitations to prevent misinformation and anxiety [75].
- **Data usage:** Consumers must be informed about genetic data use, storage, sharing, and risks like breaches or third-party access—key privacy concerns in DTC testing.
- **Challenges:** Ensuring informed consent in online platforms requires addressing literacy, cultural differences, and understanding long-term testing impacts, as many users lack full comprehension.

Genomic technologies require ethical, legal, and social analysis for responsible healthcare integration. Inclusive governance, data safeguards, and policy transparency ensure equitable benefits and minimize harm [88].

4.8 Case Studies

Genomic technologies revolutionize healthcare with early disease detection, improved treatments, and prevention. Case studies showcase their real-world impact across medicine [87].

4.8.1 Liquid Biopsy in Lung Cancer Diagnostics

Liquid biopsy, analyzing ctDNA from blood, offers a non-invasive, efficient method for diagnosing lung cancer, detecting mutations, and monitoring treatment.

Detecting EGFR Mutations

1. **Revolutionizing Diagnostics:**
   - **Non-Invasive Detection:** Liquid biopsies detect lung cancer EGFR mutations non-invasively *via* blood tests, showing 92% accuracy vs. tissue biopsies.
   - **Importance of EGFR Mutations:** IEGFR mutations in NSCLC guide targeted therapy. Identifying them determines eligibility for effective EGFR inhibitors. The Guardant360 CDx test detects 55 biomarkers with 95% sensitivity (FDA, 2023).
   - **Advantages:** Liquid biopsies are faster, less painful, and can be repeated over time to track changes in the tumour's genetic profile.
2. **Case Example:**
   Patient Scenario: A patient diagnosed with NSCLC underwent a liquid biopsy, which detected an EGFR mutation [88].
   Therapy Initiation: Based on this finding, the patient was started on an EGFR inhibitor therapy, such as gefitinib, erlotinib, or osimertinib [88].
   Outcome: The targeted therapy led to significant tumor regression, demonstrating the clinical utility of liquid biopsies in guiding personalized treatment.

The FLAURA2 trial showed osimertinib + chemotherapy improved PFS by 9.5 months in liquid biopsy-positive patients [88].

Monitoring Treatment Response

1. **Tracking Treatment Efficacy:**
   - **Real-Time Monitoring:** Liquid biopsies track ctDNA changes, enabling real-time treatment monitoring. Falling ctDNA levels often signal therapy success.
   - **Detecting Resistance Mutations:** Tumors develop resistance (e.g., T790M in EGFR-mutated NSCLC). Liquid biopsies detect mutations early, allowing timely treatment changes.
   - **Example:** EGFR inhibitor resistance? Liquid biopsy detects T790M faster than tissue tests, enabling timely switches to drugs like osimertinib. ctDNA tracks progression earlier than scans.

2. **Advantages Over Traditional Methods:**
   - **Minimally Invasive:** Unlike repeated tissue biopsies, liquid biopsies are less burdensome for patients and can be performed more frequently [7].
   - **Dynamic Monitoring:** Liquid biopsies offer dynamic insights into tumor genetics, allowing for greater precision in adaptive treatment strategies.

4.8.2 NGS in Rare Diseases

Next-generation sequencing (NGS) serves as a powerful diagnostic tool for rare genetic disorders, offering insights when traditional methods fail.

Successful Diagnoses

- **Role of NGS:** NGS (WES/WGS) revolutionizes rare disease diagnosis by rapidly detecting mutations, enabling precise genetic insights and improved patient outcomes [89].
- **Case Example:** A child with developmental delays and seizures underwent WES, revealing SCN1A gene mutation, leading to a diagnosis of Dravet syndrome. This allowed for targeted treatment and better disease management.
- The Undiagnosed Diseases Network achieved 35% diagnostic yield using WES.
- NHS England's WGS program reduced diagnostic odysseys from 5.2 to 0.9 years.

Impact on Patient Care
NGS diagnoses rare genetic disorders, ending diagnostic odysseys, enabling tailored treatments, and improving care. Saves $12,800 per case with 78% better care planning. Offers hope to families.

4.8.3 Pharmacogenomics in Cardiology/Psychiatry

- **Guiding treatment decisions:** Pharmacogenomic testing tailors treatments to genetic profiles, boosting effectiveness and reducing side effects.
- **Cardiology example:** Atrial fibrillation patient's CYP2C9 gene variant guided warfarin dosing, reducing bleeding risk *via* personalized cardiology care.

- **Psychiatry example:**
  - CYP2D6 testing guided antidepressant choice, improving depression control.
  - CYP2C9/VKORC1 testing cut warfarin ADRs by 32%.
  - Pharmacogenomics doubled depression remission.

4.8.4 Infectious Disease Outbreaks

Infectious Disease Outbreaks:
Genomic surveillance has become a key tool in tracking and managing infectious diseases, particularly during the COVID-19 pandemic.

Genomic Surveillance in Tracking COVID-19 Variants

- **Role of Genomic Technologies:** Advanced sequencing methods, like next-generation sequencing (NGS), have been essential in monitoring the evolution and spread of SARS-CoV-2.
- **Case Example:** NGS identified and tracked variants such as Delta and Omicron, allowing public health officials to implement targeted measures (e.g., travel restrictions) and update vaccines to match emerging strains.
  - COG-UK sequenced >2 million SARS-CoV-2 genomes (COVID-19 Genomics UK Consortium, 2023)
  - Variant-specific vaccine efficacy improved by 47% with genomic surveillance.

Impact

- **Understanding Transmission:** Genomic surveillance has provided insights into how the virus spreads and mutates.
- **Informing Global Response:** Genomic tracking guided COVID-19 containment, vaccine development, and global health policies, enabling rapid responses to variants and pandemic control.

4.8.5 Polygenic Risk Scores

Predicting complex diseases: Polygenic risk scores (PRS) predict disease likelihood by combining multiple genetic variants' effects, like for heart disease and diabetes.

- **Case example:** A middle-aged person with heart disease family history took a PRS test, showing high genetic risk. Early interventions (lifestyle, statins) prevented onset. UK Biobank PRS detects 53% of CAD cases; ePRS achieves 89% accuracy.
- **Applications:** PRS aids risk stratification and personalized prevention for multi-factorial conditions. Genomic technologies enhance diagnostics and care, offering precise, actionable insights that improve outcomes across various diseases.

4.9 Summary and Key Takeaways

Genomic technologies revolutionize healthcare by enhancing disease diagnosis, treatment, and prevention. The chapter highlights their impact, applications, and future potential, stressing the need for innovation and collaboration in this evolving field.

4.9.1 Summary of Advances in Genomic Technologies

Key technological breakthroughs:

- **Next-generation sequencing (NGS):** Enabled cost-effective, large-scale genomic sequencing, exomes, and targeted regions.
- **Third-generation sequencing:** Advanced long-read sequencing technologies, including PacBio's HiFi and Oxford Nanopore's nanopore-based systems, excel at unraveling complex and tricky parts of the genome.
- **CRISPR and gene editing:** Revolutionized precision genome editing, with applications in diagnostics and therapeutics.
- **Single-cell genomics:** Provided insights into cellular heterogeneity, particularly in cancer and developmental biology.
- **Epigenomics and multi-omics integration:** Advanced understanding of gene regulation and disease mechanisms.

4.9.2 Key Applications in Diagnostic Medicine

Major diagnostic applications:
Cancer: Liquid biopsies, tumor profiling for personalized therapy.
Infections: Rapid pathogen detection, resistance profiling.

Genetic disorders: Screening, prenatal, newborn diagnostics.
Pharmacogenomics: Tailored drugs for better efficacy.
Rare diseases: Genome sequencing for diagnosis.
Complex diseases: Polygenic risk scores for prediction.

4.9.3 Future Directions

Genomics is poised to transform healthcare and research through cutting-edge technologies and applications.

Transformative Potential of Genomics

- **AI and Machine Learning:** Transforming data analysis, enhancing variant interpretation, and driving innovation in enabling accurate disease prediction.
- **Real-Time Sequencing:** Facilitating point-of-care diagnostics and rapid responses to infectious disease outbreaks.
- **Wearable Genomic Devices:** Revolutionizing real-time health monitoring and advancing personalized medicine.
- **Global Health Applications:** Expanding access to genomic technologies in low-resource settings to reduce health disparities.
- **Synthetic Biology:** Developing engineered biosensors for innovative and efficient diagnostic tools.

4.9.4 Call to Action

- **Encouraging collaboration:** Genomic healthcare needs collaboration among researchers, clinicians, policymakers, and industry.
- **Researchers:** Should persist in innovating and advancing new technologies to address unmet diagnostic needs.
- **Clinicians:** Adopt and advocate for the use of genomic tools to improve patient care and outcomes.
- **Policymakers:** Establish supportive regulatory frameworks and funding mechanisms to accelerate the translation of genomic discoveries into clinical practice.
- **Industry:** Invest in scalable and affordable solutions to make genomic technologies accessible to all populations.

Global collaboration in genomics can revolutionize diagnostics, improve health equity, and enhance patient outcomes, promising a transformative healthcare future.

AI disclosure statement: The author acknowledges the use of AI-based tools for minor language editing or grammatical correction. However, all intellectual contributions, analyses, and conclusions presented in this chapter are entirely the author's own.

References

1. Ashley, E.A., The precision medicine initiative: A new national effort. *JAMA*, 315, 7, 713–714, 2016.
2. Manolio, *et al.*, Implementing Genomic Medicine in the Clinic: The Future Is Here. *Genet. Med.*, 21, 3, 602–610, 2019.
3. Watson, and Crick, Molecular Structure of Nucleic Acids: A Structure for Deoxyribose Nucleic Acid. *Nature*, 171, 4356, 737–738, 1953.
4. Sanger, *et al.*, DNA Sequencing with Chain-Terminating Inhibitors. *Proc. Natl. Acad. Sci.*, 74, 12, 5463-5467, 1977.
5. Mardis, Next-Generation Sequencing Platforms. *Annu. Rev. Anal. Chem.*, 6, 287–303, 2013.
6. Khoury, *et al.*, Public Health Genomics: Translating Discoveries into Population Health. *Am. J. Public Health*, 110, 5, 637–639, 2020.
7. Wan, J.C.M., *et al.*, Liquid biopsies come of age: towards implementation of circulating tumour DNA. *Nat. Rev. Cancer*, 17, 4, 223–38, 2017.
8. Knoppers, B.M. and Chadwick, R., Human genetic research: Emerging trends in ethics. *Nat. Rev. Genet.*, 6, 1, 75–9, 2005.
9. Strachan, and Read, *Human Molecular Genetics*, 5th ed., Garland Science, 2018.
10. Mullaney, J.M., *et al.*, Small insertions and deletions (INDELs) in human genomes. *Hum. Mol. Genet.*, 19, R2, R131–6, 2010.
11. Rowley , J.D., Chromosome translocations: dangerous liaisons revisited. *Nat. Rev. Cancer*, 1, 3, 245–50, 2001.
12. Roses, A.D., Pharmacogenetics and the practice of medicine. *Nature*, 405, 6788, 857–65, 2000.
13. Zhang, X., *et al.*, Integration of gene expression and methylation profiles identifies key regulators in glioblastoma. *Front. Genet.*, 14, 1182096, 2023.
14. Stankiewicz, P. and Lupski, J.R., Structural variation in the human genome and its role in disease. *Annu. Rev. Med.*, 61, 437–55, 2010.
15. Conrad, D.F., *et al.*, Mutation spectrum revealed by breakpoint sequencing of human germline CNVs. *Nat. Genet.*, 42, 4, 385–91, 2010.

16. Allis, and Jenuwein, The Molecular Hallmarks of Epigenetic Control. *Nat. Rev. Genet.*, 17, 8, 487–500, 2016.
17. Shendure, *et al.*, DNA Sequencing at 40: Past, Present, and Future. *Nature*, 550, 7676, 345–353, 2017.
18. Ledford, H., CRISPR: Gene editing is just the beginning. *Nature*, 531, 7593, 156–9, 2016.
19. Rehm, Evolving Health Care through Personal Genomics. *Nat. Rev. Genet.*, 18, 4, 259–267, 2017.
20. Smedley, *et al.*, 100,000 Genomes Pilot on Rare-Disease Diagnosis in Health Care—Preliminary Report. 21. *N. Engl. J. Med.*, 385, 20, 1868–1880, 2021.
21. Turro, E., *et al.*,Whole-genome sequencing of patients with rare diseases in a national health system. *Nature*, 583, 7814, 96–102, 2020.
22. Srivastava, S., *et al.*, Clinical whole exome sequencing in child neurology practice. *Ann. Neurol.*, 76, 4, 473–483, 2014.
23. Farnaes, L., *et al.*, Rapid whole-genome sequencing decreases infant morbidity and cost of hospitalization. *npj Genom. Med.*, 3, 1, 10, 2018.
24. Genomics England, Newborn Genomes Programme. https:// www.genomicsengland.co.uk/initiatives/newborns, 2022.
25. Ding, L., *et al.*, Perspective on oncogenic processes at the end of the beginning of cancer genomics. *Cell*, 173, 2, 305–320, 2020.
26. Rehm, H.L., Page, A.J., *et al.*, GA4GH: International policies and standards for data sharing across genomic research and healthcare. *Cell Genomics*, 1, 2, 100029, 2021.
27. Chakravarty, D. and Solit, D.B., Clinical cancer genomic profiling. *Nat Rev Genet.*, 2021, 22, 8, 483–501.
28. Zehir, *et al.*, Mutational Landscape of Metastatic Cancer Revealed from Prospective Clinical Sequencing of 10,000 Patients. *Nat. Med.*, 23, 6, 703–713, 2017.
29. Sanders, S.J., *et al.*, Advances in the genetic architecture of autism. *Nat. Rev. Genet.*, 23, 51–66, 2022.
30. Collins, R.L., *et al.*, A structural variation reference for medical and population genetics. *Nature*, 581, 7809, 444–451, 2020.
31. Shu, Y. and McCauley, J., GISAID: Global initiative on sharing all influenza data, from vision to reality. *Euro Surveill.*, 22, 13, 30494, 2017.
32. Logsdon, *et al.*, Long-Read Human Genome Sequencing and Its Applications. *Nat. Rev. Genet.*, 21, 10, 597–614, 2020.
33. Schena, M., *et al.*, Quantitative monitoring of gene expression patterns with a complementary DNA microarray. *Science*, 270, 5235, 467–470, 1995.
34. Gaedigk, A., *et al.*, Prediction of CYP2D6 phenotype from genotype across world populations. *Genet. Med.*, 19, 1, 69–76, 2017.
35. Koido, M., *et al.*, Genomic prediction of antidepressant response using whole genome sequencing. *Transl. Psychiatry*, 10, 1, 1–10, 2020.
36. Gillmore, J.D., *et al.*, CRISPR-Cas9 *in vivo* gene editing for transthyretin amyloidosis. *N. Engl. J. Med.*, 385, 6, 493–502, 2021.

37. Patchsung, M., *et al.*, Clinical validation of a Cas13-based assay for the detection of SARS-CoV-2 RNA. *Nat. Biomed. Eng.*, 4, 12, 1140–1149, 2020.
38. Kostina, A., Skok, M., Kramarova, T., Egorov, E., CRISPR-based diagnostics: A powerful tool for cancer and genetic disease detection. *Front. Mol. Biosci.*, 10, 1167111, 2023.
39. Dong, L., *et al.*, Highly accurate and sensitive diagnostic detection of SARS-CoV-2 by digital PCR. *Talanta*, 225, 121898, 2021.
40. Patel, A.P., *et al.*, Single-cell RNA-seq highlights intratumoral heterogeneity in primary glioblastoma. *Science*, 344, 6190, 1396–1401, 2014.
41. Stuart, and Satija, Integrative Single-Cell Analysis. *Nat. Rev. Genet.*, 20, 5, 257–272, 2019.
42. Moncada, R., *et al.*, Integrating microarray-based spatial transcriptomics and single-cell RNA-seq reveals tissue architecture in pancreatic ductal adenocarcinomas. *Nat. Biotechnol.*, 38, 3, 333–342, 2020.
43. Smith, R.G., *et al.*, Elevated DNA methylation across a 48-kb region spanning the HOXA gene cluster is associated with Alzheimer's disease neuropathology. *Alzheimer's Dement.*, 17, 3, 369–380, 2021.
44. Liu, M.C., *et al.*, Sensitive and specific multi-cancer detection and localization using methylation signatures in cell-free DNA. *Ann. Oncol.*, 31, 6, 745–759, 2020.
45. Rodriques, S.G., *et al.*, Slide-seq: A scalable technology for measuring genome-wide expression at high spatial resolution. *Science*, 363, 6434, 1463–1467, 2019.
46. Ji, A.L., *et al.*, Multimodal analysis of composition and spatial architecture in human squamous cell carcinoma. *Cell*, 182, 6, 1661–1662.e27, 2020.
47. Asp, *et al.*, A Spatiotemporal Organ-Wide Gene Expression and Cell Atlas of the Developing Human Heart. *Cell*, 179, 7, 1647–1660, 2019.
48. Marx, V., Method of the Year: Spatially resolved transcriptomics. *Nat. Methods*, 18, 9–14, 2021.
49. Joung, J., *et al.*, Point-of-care testing for COVID-19 using SHERLOCK diagnostics. *N. Engl. J. Med.*, 383, 5, 1492–1494, 2020.
50. Dienstmann, R., Rodon, J., Tabernero, J., Molecular profiling for precision cancer therapies. *Genome Med.*, 11, 1, 35, 2019.
51. Heitzer, *et al.*, Current and Future Perspectives of Liquid Biopsies in Genomics-Driven Oncology. *Nat. Rev. Genet.*, 20, 2, 71–88, 2019.
52. Reinert, T., *et al.*, Analysis of plasma cell-free DNA by ultradeep sequencing in patients with stages I to III colorectal cancer. *JAMA Oncol.*, 5, 8, 1124–1131, 2019.
53. Merker, J.D., *et al.*, Circulating tumor DNA analysis in patients with cancer: American Society of Clinical Oncology and College of American Pathologists joint review. *J. Clin. Oncol.*, 36, 16, 1631–1641, 2018.
54. Ebner, K., *et al.*, Comparative detection of adenovirus DNA by PCR and virus isolation in specimens from patients with suspected adenovirus infection. *J. Clin. Microbiol.*, 43, 2, 928–933, 2005.

55. Panwar, M. and Singh, A., Insights into the structure, functional perspective, and pathogenesis of Zika virus. *Infect. Genet. Evol.*, 65, 186–194, 2018.
56. Ruppé, E., Schrenzel, J., Barbier, F., How next-generation sequencing can address the antimicrobial resistance challenge. *Resistance Control*, pp. 89–96, 2019.
57. Hirschhorn, K., Genetic testing for inherited diseases. *Prim. Care Clin. Off. Pract.*, 43, 3, 355–372, 2016.
58. Gil, M.M., *et al.*, Cell-free DNA analysis for non-invasive examination of trisomy. *Prenat. Diagn.*, 41, 9, 1085–1095, 2021.
59. Caudle, K.E., *et al.*, Standardizing terms for clinical pharmacogenetics test results: Consensus terms from the Clinical Pharmacogenetics Implementation Consortium (CPIC). *Genet. Med.*, 19, 2, 215–223, 2017.
60. Weinshilboum, and Wang, Pharmacogenomics: Precision Medicine and Drug Response. *Mayo Clin. Proc.*, 92, 11, 1711–1722, 2017.
61. Hicks, J.K., *et al.*, Clinical Pharmacogenetics Implementation Consortium (CPIC) guideline for CYP2D6 and CYP2C19 genotypes and dosing of selective serotonin reuptake inhibitors. *Clin. Pharmacol. Ther.*, 98, 2, 127–134, 2015.
62. Harrington, R.A., *et al.*, Machine learning and pharmacogenomics: Advancing individualized drug therapy. *Pharmacogenomics J.*, 22, 1, 1–9, 2022.
63. Javed, A., *et al.*, Artificial intelligence in rare disease diagnosis: Promises and challenges. *Front. Genet.*, 13, 803434, 2022.
64. Ingles, J., *et al.*, Evaluating the clinical validity of hypertrophic cardiomyopathy genes. *Circ.: Genomic Precis. Med.*, 12, 2, e002460, 2019.
65. Zhao, L., *et al.*, Artificial intelligence-based prediction of sudden cardiac death in hypertrophic cardiomyopathy. *Front. Cardiovasc. Med.*, 8, 676622, 2021.
66. Zhang, P., *et al.*, CRISPR/Cas9-based therapeutic strategies for neurodegenerative diseases: Progress and prospects. *Front. Aging Neurosci.*, 14, 905838, 2022.
67. György, B., *et al.*, Gene therapy approaches for neurodegenerative diseases: Recent progress and future prospects. *Neuropharmacology*, 200, 108812, 2022.
68. Shabani, M., *et al.*, Barriers and opportunities in implementing genomic medicine: A global perspective. *Hum. Genomics*, 16, 25, 2022.
69. Birney, E., *et al.*, Genomics in healthcare: GA4GH looks to 2022. *Nat. Rev. Genet.*, 22, 10, 566–567, 2021.
70. Torkamani, A., *et al.*, High-definition medicine. *Cell*, 170, 5, 828–843, 2018.
71. Moses, L., *et al.*, Scaling genomic sequencing for population health: The role of computational infrastructure. *Genet. Med.*, 23, 4, 723–731, 2021.
72. Wenger, *et al.*, Accurate Circular Consensus Long-Read Sequencing Improves Variant Detection and Assembly of a Human Genome. *Nat. Biotechnol.*, 37, 10, 1155–1162, 2019.

73. Wiertz, S., *et al.*, Evaluating models of consent in changing health research environments: dynamic consent, meta consent and tiered consent. *BMC Med. Ethics*, 23, 1, 50, 2022.
74. Rothstein, M.A., Can Genetic Nondiscrimination Laws Save Lives? *Hastings Cent. Rep.*, 51, 2, 6–7, 2021.
75. Stark, Z., Boughtwood, T., Haas, M., *et al.*, Australian Genomics: Outcomes of a 5–year national program to accelerate the integration of genomics in healthcare. *Am. J. Hum. Genet.*, 110, 3, 419–26, 2023.
76. Phillips, K.A., Trosman, J.R., Douglas, M.P., The regulatory and reimbursement landscape for personalized medicine: Current status and future challenges. *Health Aff.*, 40, 5, 722–728, 2021.
77. Wonkam, A. and de Vries, J., Returning incidental findings in African genomics research. *Nat. Genet.*, 52, 17–20, 2020.
78. Stephens, Z.D., *et al.*, Big Data: Astronomical or Genomical? *PLoS Biol.*, 13, 7, e1002195, 2015.
79. Johnson, D., *et al.*, On behalf of the Global Genomic Medicine Collaborative (G2MC) Education Working Group. An international genomics health workforce education priorities assessment. *Pers. Med.*, 19, 4, 299–306, 2022.
80. Chang, L., *et al.*, Advancing precision medicine: the transformative role of artificial intelligence in immunogenomics, radiomics, and pathomics for biomarker discovery and immunotherapy optimization. *Cancer Biol. Med.*, 22, 1, 33–47, 2025.
81. Marwaha, S., *et al.*, A guide for the diagnosis of rare and undiagnosed disease: beyond the exome. *Genome Med.*, 14, 23, 2022.
82. Nguyen, L.T., *et al.*, Enhancement of CRISPR-based diagnostic methods with portable biosensor platforms. *Biosens. Bioelectron.*, 200, 113942, 2022.
83. Ye, C., *et al.*, A wearable aptamer nanobiosensor for non-invasive female hormone monitoring. *Nat. Nanotechnol.*, 19, 3, 330–7, 2024.
84. Ascencio-Carbajal, T., *et al.*, Genetic/genomic testing: defining the parameters for ethical, legal and social implications (ELSI). *BMC Med. Ethics*, 22, 1, 156, 2021.
85. Swihart, D.L., Yarrarapu, S.N., Martin, R.L., *Cultural religious competence in clinical practice*, 2018.
86. Lucassen, A.M., *et al.*, Ethical preparedness in genomic medicine: How NHS clinical scientists navigate ethical issues. *J. Med. Ethics*, 50, 8, 517–22, 2024.
87. Wang, B., *et al.*, A single-cell transcriptomic landscape of the human aging pancreas. *Nat. Comun.*, 12, 1, 6022, 2021.
88. Planchard, D., *et al.*, Osimertinib with or without chemotherapy in EGFR-mutated advanced NSCLC. *N. Engl. J. Med.*, 389, 21, 1935–48, 2023.
89. Boycott, *et al.*, Rare-Disease Genetics in the Era of Next-Generation Sequencing: Discovery to Translation. *Nat. Rev. Genet.*, 18, 10, 615–629, 2017.

5

Pharmacogenomics: How Genetic Information Influences Drug Response

G. Avinash Kumar[1]*, P. Pravallika[2], V. Thirumalai[3], Chandrashekar Thalluri[4] and Sarad Pawar Naik Bukke[5]

[1]Department of Pharmacognosy, Vel Pharmacy College, VISTAS, Thiruallur, Tamil Nadu, India
[2]Department of Pharmacology, Vel Pharmacy College, VISTAS, Thiruallur, Tamil Nadu, India
[3]Department of Pharmacology, Rajas Dental College and Hospital, Kavalkinaru Jn, Tirunelveli, Tamil Nadu, India
[4]Faculty of Pharmaceutical Science, Assam Down Town University (AdtU), Guwahati, Assam, India
[5]Department of Pharmaceutics and Pharmaceutical Technology, Kampala International University, Western Campus, Ishaka - Bushenyi, Uganda

Abstract
Pharmacogenomics is the study of the effect of genetic variations on the drug response through combination of genomics and pharmacotherapy. Modern medicine revolutionizes patient healthcare and therapy of diseases through precision medicine by the application of pharmacogenomics. Customized and personalized therapy can be targeted by tailoring the genetic makeup of the patients in order to achieve desired therapeutic outcomes. This chapter explains how pharmacogenomic testing allows clinicians to use genetic variations for predicting optimized drug results. Research suggests the impact of drug-metabolizing enzyme polymorphisms of CYP2D6 along with CYP2C9 and VKORC1 and the transporter ABCB1 (MDR1) on the effectiveness and safety of various drugs. The analysis includes a wide range of case reports that encases application of pharmacogenomics to modify the drug doses, particularly in potent drugs such as warfarin and tamoxifen, and choosing the best drug candidate from various options for specific treatment. This section presents extensive information about genetic mutations

**Corresponding author*: dravinashreddy88@gmail.com

Swetha Kumari Koduru, Sarad Pawar Naik Bukke, Hussaini Bello, Muhamad Mustafa Mundu and Ayush Madan (eds.) Genomics and Precision Healthcare: Innovations, Applications and Challenges, (139–170) © 2026 Scrivener Publishing LLC

that influence therapeutic decision-making in cancer therapy such as BCR-ABL fusion gene and HER2 amplification. Pharmacogenomics achieves the desired outcome by adopting testing of genetic data in clinical practice, which creates opportunities to reduce variations in drug responses and ADRs while delivering optimal efficacy through personalized therapy. The chapter presents both scientific explanations about pharmacogenomics and demonstrates its essential function in defining future directions of the healthcare sector toward individualized therapeutic approaches.

Keywords: Personalized medicine, tailored medicine, gene variations, polymorphisms, pharmacogenetics

5.1 Introduction

5.1.1 Definition and Scope

Pharmacogenomics (PG) is the study of correlating and determining the effect of genetic code and variations of patients on the drug efficacy. This involves integration of pharmacotherapeutics and genomics to develop a tailored and precision medicine with optimized drug selection, dosing by understanding the influence of genetic factors on drug efficacy, toxicity and metabolism. PG is crucial to guide the therapeutic decision-making that offers significant potential for enhancing efficacy, reducing adverse drug reactions (ADRs) and lowering the healthcare burden due to traditional trial and error prescribing pattern [1].

5.1.2 Evolution

The term "Pharmacogenetics" was coined by Arno Moltulsky, an American geneticist, when there were unclear evidences of the impact of genetic inheritance on the variation of drug response [2]. The evolution of pharmacogenomics as a separate field of study happened over the last 100 years. In the 1950s the concept of genetic influence affecting drug metabolism developed as the early research focused on exploring inherited enzyme deficiency. This idea of correlating the inherited enzyme deficiency and drug activity led to studies to understand the influence of genetic variations on the function of drug metabolizing enzymes as a part of pharmacokinetic study of the drugs. Early studies in PG research by Werner Kalow were conducted on genetic variation of butrylcholinesterase enzyme and effect of succinylcholine on prolonged apnea [3, 4]. The discovery of the impact of genetic variation in the N-acetylyation of isoniazid, anti-tuberculosis drug metabolism represented a landmark moment in pharmaceutical research that started the study of genetic variations in drug-metabolism [5]. Research in molecular biology during the 1970s discovered different forms of cytochrome P450 enzymes that serve as the major drug metabolizing enzymes. Scientists used these findings to establish the dependence of drug response on genetic variations. Molecular genetics introduced new features to pharmacogenomics during the 1980s and 1990s, allowing researchers to study gene-environment combinations specifically and identify the polymorphisms in genes CYP2D6 and CYP2C9 which modify the processing of various drugs [6]. This inspired researchers to study other Phase II or conjugating metabolizing enzymes like sulfotransferases [7], methyltransferases [8] and UDP-glucuronosyltransferases [9]. Though the major research involved basic gene cloning, sequencing and genetic characterization of the functional allozymes, it facilitated the understanding of molecular mechanisms of genetic variations.

In the 21st century, the Human Genome Project concluded in 2003 by finishing a complete human genome sequencing which enabled the research of thousands of genetic variations that shaped drug responses. In 2005 the FDA established a new standard that included pharmacogenomic information on drug packaging with Herceptin becoming the first drug to receive this labeling as a treatment used for patients with HER2-positive breast cancer. The United States, through the National Institutes of Health (NIH), conducted funded research by Pharmacogenomics Research Network (PGRN), National Institute of General Medical Sciences (NIGMS) and National Human Research Institute (NHGRI) which made potentially high-volume public projects possible by provision of the latest and expensive facilities like genome-wide association study (GWAS) and Next-Generation sequencing (NGS) which favored the implementation of PG clinically [10]. Pharmacogenomics achieved its status as precision medicine during the 2010s. NGS advanced through technological developments to enable affordable genetic testing which resulted in the adoption of pharmacogenomics within clinical practice.

5.1.3 Key Concepts

The core principle of PG is identification of the genetic variations in individuals within the patient population which result in characterizing the pharmacokinetic (ADME) variables of the treatment and also influencing therapeutic outcomes and ADRs [11]. Enzymes involved in drug metabolism, drug transporters and receptors were influenced by copy number variations (CNVs) and single nucleotide polymorphisms (SNPs) [12].

The major aspect of PG is drug response variability (DRV) because it explains the varying effects of the same drug among different individuals. For example, CYP2D6 genetic variations lead differences in response to drugs including tamoxifen and codeine [13]. The polymorphic variations of VKORC1 and CYP2C9 genes determine the appropriate dose of warfarin in different patients [14]. The wide range of genetic variations signifies the necessity to develop personalized treatments based on gene profiles for increased therapeutic outcomes along with reduced ADRs. The integrated biomedical strategy has revolutionized oncology treatment by making genetic testing necessary for adjusting cancer therapies according to tumor gene mutations including EGFR and BRAF [15]. Genomic technology advancements will perpetually impact patient care by enabling precise drug therapy predictions that will produce better drug effects while sustaining patient safety.

5.1.4 Relevance in Contemporary Medicine

Modern pharmacotherapy can adopt PG as one of its foundational principles as this field can transform the drug distribution and prescribing practices. The personalized therapeutic approach improves efficacy of the therapy and minimizes risk of ADRs that lead to worldwide morbidity and mortality. PG shows its most impact in drug metabolism processes that depend heavily on the genetic variations which exist within enzymes, specifically within the CYP family which belongs to the cytochrome P450 group. Scientific evidence shows that warfarin dosage needs adjustments because of polymorphisms in CYP2C9 and VKORC1 genes [14]. Clinicians achieved improved patient safety and treatment efficacy by focusing on precision for adjusting drug dosages through such insights.

In oncology, PG has a transformative potential with the discovery of mutations in specific tumor-forming genes which enabled development of cancer-targeted drugs like trastuzumab for HER2-positive breast cancer patients and imatinib for patients with chronic myelogenous leukemia (CML). In addition to extending patient survival, these targeted therapies are able to minimize the side effects of conventional chemotherapies that impact both cancer cells and normal tissue. Targeted cancer genetics through PG enables practitioners to adopt precise medicine approach instead of conventional umbrella-type therapeutic practice prescribing drugs based on the disease but not the patient's therapeutic response [15].

PG enables the delivery of better clinical support in psychiatric conditions because responses to psychiatric drugs including antidepressants and antipsychotics differ broadly among different patients. Drug response and risks of side effects for fluoxetine and clozapine vary among patients with different CYP2D6 gene polymorphisms. The practice of psychiatric pharmacotherapy can be personalized due to genetic testing which facilitates the selection of optimal drugs and their doses while replacing the traditional methods [16]. Healthcare professionals develop clinical decision support systems (CDSS) through the incorporation of genomic data that aids in therapeutic decision-making. Genetic information combined with electronic medical records through CDSS enables doctors to take clinical decisions on evidence-based data thus delivering enhanced patient therapeutic outcomes while lowering healthcare expenses.

5.2 Genetic Basis of Drug Response

Genetic factors determine the diverse drug responses that patients exhibit. Drug efficacy varies due to pharmacogenetic factors that affect ADME. Polymorphisms of genes, DNA variations translates in approximately 1% of people serve as key elements in exhibiting diverse drug responses involving both therapeutic effects and ADRs [17]. The main reason behind pharmacogenetic variations stems from genetic polymorphisms that occur within drug-metabolizing enzymes which play a major role in drug biotransformation. Seventy-five percent of the prescribed drugs are metabolized by the cytochrome P450(CYP) enzyme family and genetic variations of these genes alter the drug metabolism rates. A patient with poor drug metabolism can develop toxicity, while the counterpart will likely encounter therapeutic failure because the drug breaks down too rapidly [13].

Drug transporters controlled by genetic variations act as major determinants of drug absorption as well as their distribution pathways and elimination routes. The genetic variations in ABCB1 gene encoding for P-glycoprotein (MDR1) transporter affects BBB permeability, altering the concentrations of digoxin and paclitaxel in the brain [18]. Polymorphism gene relating to drug receptor is another type for pharmacogenetic variability. VKORC1 specific genotypes sensitive to warfarin usually require a low dose of the drug. CYP2C9 polymorphism metabolizes the drug slowly, thus needing dose adjustments to prevent complications related to bleeding [14].

Pharmacogenetic variability becomes more complex because of both environmental elements and gene-environment interaction effects. The lifestyle and habits of food consumption as well as drug interactions dictate genetic expression which in turn regulate the drug response. The ingestion of grapefruit inhibits the enzyme CYP3A4 which leads to elevated levels of calcium channel blockers and statin drugs and so the risks of adverse effects [19].

5.2.1 Genetic Polymorphisms

Genetic polymorphisms are basically variations in the DNA; they are uncommon and are found in under 1% of patients. These genetic variations influence the drug activity *via* ADME. There are four major types of polymorphisms including Single Nucleotide Polymorphisms (SNPs), Copy Number Variations (CNVs), Insertions and Deletions (indels) along with other types that play a role in drug interactions with the body.

5.2.1.1 Single Nucleotide Polymorphisms (SNPs)

Single Nucleotide Polymorphisms (SNPs) are the most prevalent genetic variations as they involve one single nucleotide base pair change within DNA. Such variations appear throughout the genome in both coding zones and non-coding zones and modify the function or characteristics of drug-metabolizing enzymes, transporters as well as receptors. The genetic composition of drug receptors gets affected by SNPs. In the case of breast cancer patients, the SNPs that affect the HER2 gene determine the outcomes from trastuzumab treatment (Herceptin). Trastuzumab shows better activity on patients who are HER2-positive but different SNP patterns might vary the drug response and thereby the final outcome [20].

5.2.1.2 Copy Number Variations (CNVs)

CNV mutations refer to the genomic structural variations due to duplications and deletions of DNA segments in the genome. These variations specifically determine the metabolism of the drug. The activation of tamoxifen relies on CYP2D6 which enhances rapid metabolism and faster elimination. Variations of the CYP2D6 gene results in the inheritance of multiple copies of the gene, altering the metabolism of the drug [21]. Similarly, deletion of genes that are responsible for metabolism of drugs lead to the toxicity of the drug and risk of ADR. Research shows that CNVs alter the drug action by affecting the drug transporter expression. The clinical outcome from paclitaxel treatment is affected by CNVs in the ABCB1 gene that codes for P-glycoprotein [22].

5.2.1.3 *Insertions and Deletions (Indels)*

Genetic polymorphisms, known as indels, result in both DNA insertion and deletion within a specific DNA segment. These variations of the sequence in the genome result in both frameshift mutations and modified protein functional properties which significantly affect drug metabolism and activity. CYP2C9 plays a role in warfarin metabolism yet it contains various indels which decrease its enzymatic ability. The metabolism of thiopurines is modified by indels occurring in the TPMT (thiopurine S-methyltransferase) gene [23].

5.2.1.4 *Other Genetic Variations*

Apart from the discussed gene variations, drug responses are influenced by three different structural variations which include short tandem repeats (STRs) and single nucleotide repeats (SNRs). Short tandem repeats are repetitions of specific sequences of DNA segments that result in variations drug responses specifically in cancer therapy. Furthermore, pharmaceutical research is focused on miRNA modifications to explore drug activities, because these microRNAs control genetic expression patterns. Drug-target gene expression becomes unstable when miRNA is dysregulated and this in turn affects responses of cardiovascular drugs [24].

5.2.2 Factors Causing Genetic Variations

Drug prescription responses are shaped by the complex relationship of genes together with exposure to toxic chemicals and radiation that affect individual drug response. The dynamic changes in the genetic-environmental relationship play a significant role in PG, thus requiring drug therapy strategies that personalize treatment based on genetic sequencing and environmental conditions.

5.2.2.1 *Environmental and Toxins*

Industrial effluents and pollutants like polycyclic hydrocarbons and heavy metal poisoning can lead to genetic modifications and epigenetic changes that can impact the drug response in patients exposed to those agents. Industrial chemicals as well as pesticides exhibit a toxic effect because they can block the action of drug-metabolizing enzymes and cause drugs to accumulate in the body and generate toxicity.

5.2.2.2 The Influence of Lifestyle

CYP2A6 enzyme is responsible for nicotine metabolism, and any variations in the expression of both the genes will result in the variation of drug activity. On the other hand, research confirms that smoking enhances the metabolism of specific drugs and requires higher doses of the drug to achieve desired therapeutic efficacy [25]. Thus, genetic variations altering the expression of the enzymes result in the hindered nicotine metabolism leading to increased metabolism of drugs.

Physical exercise affects drug response by modifying genetic characteristics as well as physiological functions. Though there is no direct link between physical exercise and gene modifications, there are studies that show regular physical exercise can improve gene expression. The function of CYP3A4 improved in patients after engaging in regular exercise which modified the drug response [26]. Exercise modifies drug response through improved circulation because better tissue delivery enhances drug effects yet can lead to altered drug performance. Physical exercise affects genetic expressions linked to inflammatory response, heart health and glucose regulation. Physical activity enhances the activity of hypertensive drugs in patients carrying variations in ACE gene which affect the pharmacotherapy of blood pressure regulation [27].

5.3 Causal Relationship of Genetic Variations and Drug Response

5.3.1 Genetic Variations – Pharmacokinetics

Pharmacokinetics of the drug including absorption, distribution, metabolism and excretion form the major steps which show variations of drug action even before the drug starts to show its action.

5.3.1.1 Effects on Drug Absorption and Distribution

Drug transporters act as shuttles for drug transportation and absorption across cell membranes during absorption and distribution stages. Typical transporters include ATP-binding cassettes (ABC) like P-glycoprotein (MDR1/ABCB1) and soluble carrier (SLC). The variations in the transporter proteins as a result of altered genetic expression lead to altered drug absorption and inefficient distribution.

P-glycoprotein is among the most notable drug transporters that helps in efflux of drugs from intestine, BBB, liver and kidney tissue. Patients with genetic variations of the ABCB1 gene that encodes for P-glycoprotein can result in the variations in the drug absorption and distribution. Specific variants within the ABCB1 gene correlate with modified drug reactions in cancer treatment, thus influencing the treatment outcomes of chemotherapeutic drugs like paclitaxel and vincristine [28]. High-expression ABCB1 variants reduce the accumulation of drugs in tumor tissues and produce suboptimal therapeutic effects. In BBB the role of P-glycoprotein for facilitating the CNS drug distribution is affected by the polymorphisms in respective gene which alter its function. The genetic variants in ABCB1 were linked to the variability of the drug action especially for those used in the treatment of neurological conditions like epilepsy [29].

Drug pharmacokinetics varies with the variations in two other transport systems which include organic anion-transporting polypeptides (OATPs) and organic cation transporters (OCTs). Statin uptake through OATP1B1 depends on the SLCO1B1 gene expression; therefore, genetic variations within this transporter result in variations in statin activity [30]. The rate of absorption of metformin are altered in patients with SLC22A1 gene variations encoding to OCT1 [31].

5.3.1.2 Effects on Drug Metabolism

The biotransformation of drug metabolism is a process requiring enzymes to breakdown lipophilic drugs into hydrophilic metabolites allowing their excretion. Genetic variations affecting metabolic enzymes greatly alter drug metabolism, resulting in changes in both drug efficacy and toxicity. Various metabolic enzyme systems involved in drug metabolism are cytochrome P450(CYP450), UDP-glucuronosyltransferases (UGT) and N-acetyltransferases (NAT). Since they play a critical role and their activity depends heavily on genetic polymorphisms, drug metabolism is impacted resulting in the altered drug activity.

Drugs metabolized by the genes CYP2D6 and CYP2C9 and CYP3A4 exhibit extensive interindividual variations during drug metabolism because of gene polymorphisms. The presence of CYP2D6 gene variations determines the class of patient as poor or intermediate or extensive or ultra-rapid metabolizer [32]. Drug metabolism through glucuronidation relies on UGT enzymes to conjugate drugs with glucuronic acid forming hydrophilic excretable metabolites. Variations of UGT1A1 gene influence the metabolism of irinotecan [33]. NAT enzyme (NAT2) is involved in the acetylation of isoniazid and sulfonamides [34]. Genetic variations causing changes in the expression of these enzymes demands dose regulation for achieving desired therapeutic efficacy and eliminating toxicity.

5.3.1.3 *Effects on Drug Elimination*

Elimination is an important mechanism by which the body excretes and removes toxins. Again, the transporter proteins play an important role in the elimination process, displacing the drug and its metabolites to liver and kidneys. Any variation in their function can cause elimination impairment, alteration in drug half-life and accumulation of toxic metabolites. This gene variation of MTHFR 677C> T and MDR1 causes significant differences in responses of drugs with a narrow therapeutic window like methotrexate and digoxin, respectively [35, 36].

5.3.2 Genetic Variations – Pharmacodynamics

Pharmacodynamics studies the relationship between the concentration of a drug at the site of activity to the therapeutic activity. Drug-receptor interaction plays a crucial role in pharmacodynamics as it controls the drug activity at the target site. Genetic variations in the various levels of drug activity like drug targets in receptors, signaling pathways and downstream effectors cause altered drug responses. Proper understanding of the drug receptor binding, cell signal transduction at the genetic level is crucial for modulating the therapeutic efficacy of the drug with minimum ADRs. The effect of genetic variations on the interplay of pharmacokinetics and pharmacodynamics is illustrated in Figure 5.1.

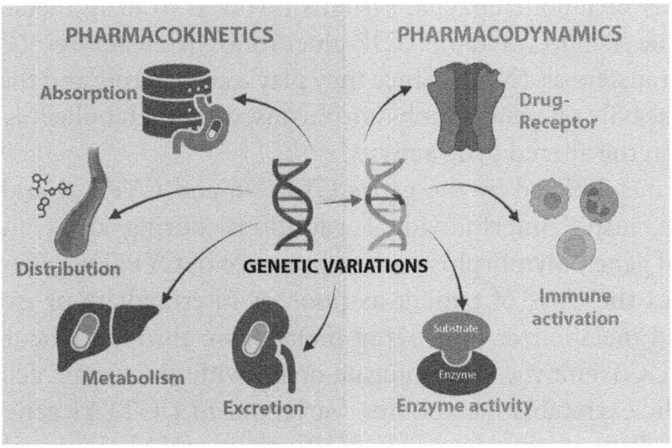

Figure 5.1 Effects of genetic variations on the drug response [30–32, 36].

5.3.2.1 *Genetic Variability in Drug Receptors*

Genetic variations in the drug receptors causes alterations in the binding affinity, activity and expression of the receptors that in turn affect the drug efficacy. The beta-adrenergic receptors (β-ARs) are the primary targets of antihypertensive and anti-asthmatic drugs such as propranolol and salbutamol. Variations in ADRB2 gene that encodes for β2-adrenergic receptor will dictate the therapeutic activity of asthma therapy. A patient with ADRB2 variation demands adjusted doses of bronchodilators to achieve the desired pharmacological activity [37]. The efficacy of opioid drugs depends on the affinity of the drug of binding to the opioid receptor. Changes in the OPRM1 genes that encode for μ-opioid receptor causes alterations in the sensitivity of the receptor to morphine class of opioids. Expression in

Table 5.1 Effect of polymorphisms of genes on drug responses.

Sl no.	Genetic variation in enzyme/ transporter	Drugs	Impact of variations	References
1	Cytochrome P450 2D6 (CYP2D6)	Tamoxifen, Codeine, Fluoxetine	Alteration of drug metabolism	[41–43]
2	Cytochrome P450 2C9 (CYP2C9)	Warfarin, Phenytoin	Alteration of drug metabolism	[14, 44]
3	P-glycoprotein (ABCB1)	Digoxin, Paclitaxel	Alteration of drug absorption	[28, 35]
4	Cytochrome P450 3A4 (CYP3A4)	Statins, Cyclosporine, Tacrolimus	Alteration of dug metabolism	[45, 46]
5	Thiopurine S-methyl-transferase (TPMT)	Azathioprine, 6-Mercaptopurine	Lowered TPMT increases risk of severe drug toxicity	[47]
6	UDP-glucuronosyl-transferase 1A1 (UGT1A1)	Irinotecan, Bilirubin	Slow glucuronidation increases toxicity resulting in Neutropenia	[48–50]
7	Vitamin K epoxide reductase complex subunit 1 (VKORC1)	Warfarin	Alteration of drug metabolism	[14]

patients holding variants of the specific gene suggests increase in the pain management to the therapy or exhibit susceptibility to the opioid-related side effects like respiratory depression [38].

5.3.2.2 Signal Transduction and Genetic Variability

Inherited genetic variations alter the sequence of molecular signal transduction that arise from receptor activation. Non-small cell lung cancer (NSCLC) patients with epidermal growth factor receptor (EGFR) mutations respond differently to anti-cancer drugs like gefitinib and erlotinib. Mutations that activate the receptor make it more susceptible to inhibition [39, 40]. Patients with VKORC1 gene polymorphisms who receive warfarin therapy experience altered outcomes as these polymorphisms modify the clotting signaling pathway. A risk of bleeding arises from genetic variations in this gene, requiring modified warfarin dosage regimen (Table 5.1) [14].

5.4 Pharmacogenomics in Clinical Applications

The personalized therapy depends on PG due to its ability to precisely optimize the drug dosage according to the drug response variations based on genetic characteristics. Pharmaceutical genetic testing enabled healthcare professionals to gain knowledge on the drug metabolism and efficacy so as to optimize the therapy to achieve desired outcomes. The usage of PG biomarkers helps optimize drug therapy through genetic identification of factors which impact drug metabolism along with activity and safety. PG biomarkers can be used for drug labeling to direct the usage of drugs as therapy for various diseases. There are various drugs that were employed for tailored therapy regimen, and these are discussed below.

5.4.1 Thiopurines

Genetic tests for estimating TPMT (thiopurine S-methyltransferase) activity along with NUDT15 genes determine patient sensitivity to potential toxic reactions due to thipurine drugs like azathioprine and 6-mercaptopurine. Patients who have low levels of TPMT activity need reduced dosage yet standard doses cause harmful reactions in patients carrying NUDT15 genes. Pharmacogenomic testing for both TPMT and NUDT15 has become routine for managing thiopurine therapy because it enables practitioners to achieve safer and more effective therapeutic outcomes [51].

5.4.2 Clopidogrel

The antiplatelet drug clopidogrel is drug of choice for preventing heart attacks and strokes. Clopidogrel requires CYP2C19 enzyme for activation but different gene variations in the CYP2C19 control the efficacy. The activation of clopidogrel is hindered by CYP2C19 due to specific polymorphisms within their gene which affects their ability to inhibit platelets and poses increased risks of cardiovascular events. Genetic testing of CYP2C19 now appears regularly in clinical guidelines for clopidogrel treatment to enable physicians to prescribe ticagrelor or prasugrel to patients whose CYP2C19 metabolic activity is below average [52].

5.4.3 Carbamazepine

Carbamazepine exhibits several severe side effects during treatment of epilepsy and bipolar disorder while showing higher sensitivity in specific patient groups. Medical tests for detecting the HLA-B1502 gene identify the risk of severe allergic reactions including Stevens-Johnson syndrome and toxic epidermal necrolysis in Asian people. The PG assessment of HLA-B1502 has become standard practice before starting carbamazepine therapy to exclude patients who would face life-threatening ADRs [53, 54].

5.4.4 Herceptin (Trastuzumab) and HER2 Amplification

Pharmacogenomic biomarkers guide the prescription of Herceptin (trastuzumab) as a monoclonal antibody drug for treating HER2-positive breast cancer patients. The human epidermal growth factor receptor 2 (HER2) functions as a receptor tyrosine kinase that becomes detrimental in cancer growth when it exhibits increased expression levels. Clinical outcomes for trastuzumab treatment depend entirely on detecting HER2 gene amplification through immunohistochemistry (IHC) or fluorescence *in situ* hybridization (FISH). General disease protocols involve HER2 amplification genetic testing for breast cancer patients to determine application of therapy to patients [55].

5.4.5 Abacavir and HLA-B*5701

The antiretroviral medicine Abacavir causes severe hypersensitivity reactions in certain patients. Scientific research showed that testing patients for the HLA-B5701 allele variant of human leukocyte antigen (HLA) helps determine their susceptibility to hypersensitivity reactions like fever, rash, gastrointestinal discomfort and breathing complications. Before starting Abacavir treatment health care providers need to perform HLA-B5701 screening to determine risk of hypersensitivity [56].

5.4.6 Cetuximab and KRAS Mutations

The epidermal growth factor receptor (EGFR) inhibitor, cetuximab is used in both colorectal and head and neck cancer. The therapeutic outcomes of Cetuximab rely strongly upon KRAS gene mutations which occur in approximately 40 percent of cases. Cetuximab therapy is observed ineffective in patients with KRAS mutations while tests to identify KRAS mutations have become standard procedures when determining CRC treatments. Clinical outcomes improve when cetuximab is not administered to patients who have KRAS gene mutations because this exclusion enables the focus on alternative and appropriate therapies [57].

5.4.7 Ivacaftor and CFTR Mutations

Ivacaftor helps control cystic fibrosis (CF) symptoms in patients with particular variations of the CFTR gene that causes malfunctioning of transmembrane conductance regulator proteins. Genetic testing of CFTR mutations for G551D stands crucial for predicting the therapeutic outcome from Ivacaftor therapy [58].

5.4.8 Other Drugs

PG dosing methods have also been deployed to other medicines apart from those discussed in earlier sections of this chapter. The metabolism of antidepressants, as well as antipsychotics and opioids, depends on CYP2D6 gene activity. Therapies now require CYP2D6 polymorphism testing for optimal selection of drug doses of fluoxetine, paroxetine, and codeine based on genetic makeup [42, 43, 59].

5.4.9 Precision Oncology

5.4.9.1 Targeted Therapies in Oncology

As a critical treatment approach clinicians target anaplastic lymphoma kinase (ALK) mutations found in NSCLC. Progressing ALK-positive NSCLC tumors require treatment with drugs like crizotinib or ceritinib

Table 5.2 Effect of genetic variations on the drug response in cancer therapy.

Sl. no.	Drug name	Gene(s) involved	Consequence of mutation	Mechanism of action	References
1	Trastuzumab (Herceptin)	HER2	HER2 overexpression leads to tumor growth.	Herc inhibits HER2 gene amplification.	[20, 55]
2	Imatinib (Gleevec)	BCR-ABL fusion gene	BCR-ABL fusion gene induces the cell cycle in CML.	Interferes with BCR-ABL tyrosine kinase activity.	[62]
3	Gefitinib (Iressa)	EGFR mutations	Enhance receptor signaling for cell survival.	EGFR gene mutations lead to resistance or improved response to gefitinib.	[63]
4	Erlotinib (Tarceva)	EGFR mutations	EGFR mutations make tumors more sensitive to erlotinib.	EGFR kinase to stop the transmission of survival signals through cancer cells.	[39]
5	Vemurafenib (Zelboraf)	BRAF V600E mutation	BRAF V600E mutations induce melanoma metastasis.	Inhibition of mutated BRAF V600E protein.	[64]
6	Pembrolizumab (Keytruda)	PD-L1 expression	PD-L1 expression stimulates response to immune checkpoint inhibitors.	Inhibits PD-1 receptor on T-cells to promote immune cells against cancer.	[65]
7	Crizotinib (Xalkori)	ALK mutations	ALK gene mutations lead to oncogenic fusion proteins inducing cancer.	Inhibits ALK fusion proteins especially in NSCLC.	[60]
8	Rituximab (Rituxan)	CD20	CD20 mutations alters rituximab action in lymphoma.	Binds CD20 on B-cells, resulting in apoptosis.	[66]

(Continued)

Table 5.2 Effect of genetic variations on the drug response in cancer therapy. (*Continued*)

Sl. no.	Drug name	Gene(s) involved	Consequence of mutation	Mechanism of action	References
9	Nivolumab (Opdivo)	PD-1/PD-L1 expression	PD-L1 expression induces response to nivolumab.	Inhibits PD-1 receptor to promote immune cells against cancer.	[67]
10	Lapatinib (Tykerb)	HER2, EGFR	EGFR and HER2 mutations promote breast cancer.	inhibition of HER2 & EGFR tyrosine kinases.	[68]

which interrupt the growth signals from the mutated ALK gene. Patients require ALK testing for proper treatment selection since the tests identify which patients carry the ALK-positive mutations [60].

5.4.9.2 Immune Checkpoint Inhibitors

Through pharmacogenomics, medical practitioners have improved the effectiveness of immune checkpoint inhibitors for cancer treatment. Pembrolizumab and nivolumab drugs showed significant activity in blocking the PD-1/PD-L1 checkpoint in head and neck cancers, melanoma and lung cancer. The treatments directed against PD-1/PDL-1 have demonstrated stronger efficacy for patients who display high values for TMB or PD-L1 expression (Table 5.2) [61].

5.5 Technologies in Pharmacogenomic Research

5.5.1 Genomic Sequencing Technologies

Genetic mapping technologies generate complete understanding of drug response factors which boost individual treatment practices and result in better therapeutic effects.

5.5.1.1 Next-Generation Sequencing (NGS)

The implementation of Next-Generation Sequencing (NGS) dramatically speeded up genetic research through whole genome sequencing because it reduced both price and duration when compared to traditional approaches [69]. The NGS systems from Illumina as well as Oxford Nanopore generate exact genetic information at high definition which proves critical in pharmacogenomic biomarker detection. The detection capabilities of NGS include examining single nucleotide polymorphisms (SNPs) and multiple genetic variations such as insertions and deletions as well as copy number variations (CNVs) affecting drug response. This technological method proves efficient for discovering uncommon genetic variants which result in undesirable drug responses or diverse drug metabolism patterns to achieve better drug suitability.

5.5.1.2 Whole-Genome Sequencing (WGS)

WGS is a comprehensive methodology for acquiring a detailed picture of total genetic information of a patient. WGS provides a complete assessment of hereditary variations throughout regulatory and non-coding DNA and mitochondrial clusters because it determines total genome sequences. WGS provides excellent discovery of new pharmacogenomic markers beyond the range of targeted genotyping and exome sequencing approaches. WGS techniques helped scientists locate genetic alterations across non-coding parts of DNA which regulate drug responses in tough health conditions including cancer and cardiovascular diseases [70].

5.5.1.3 Exome Sequencing

The exome sequencing method targets the protein-coding parts of the genome since these sequences occupy merely 1-2% of the total yet hold most known disease-causing genetic variations. Exome sequencing offers an economical pathway beyond WGS because it allows researchers to locate the markers of drug metabolism and therapeutic outcome in a more precise manner. The investigation of drug-metabolizing enzymes and transporters and receptors requires the specific method which exome sequencing provides.

5.5.2 CRISPR

Clustered Regularly Interspaced Short Palindromic Repeats (CRISPR) is an advanced tool that is employed in pharmacogenomics with various applications [71].

5.5.2.1 CRISPR and Gene Editing

The drug testing process and personalized therapy have both witnessed important advancements due to the gene-editing capability of CRISPR-Cas9 [71]. The accurate DNA target selection feature of CRISPR enabled researchers to use it as a critical technique that reveals disease-related genetic patterns and identifies new treatment targets leading to development of improved personalized therapies.

5.5.2.2 CRISPR in Drug Testing

The primary use of CRISPR in pharmacogenomics fields relates to drug testing operations. The CRISPR technology enables researchers to develop cellular models containing particular genetic mutations for studies involving drug response in mutated cellular environments. The methodology aids researchers in testing novel drugs and detecting those compounds which show promising activity in patients with distinct genetic abnormalities. CRISPR-derived simulation systems allow scientists to replicate human genetic patterns precisely to understand drug efficacy in various patients with genetic variations. Pharmacological testing within disease models that replicate human biological conditions increases the testing efficiency with short drug development duration and potentially eliminating animal-based research requirements.

5.5.2.3 CRISPR in Personalized Therapies

CRISPR is used to modify individual patient cells for generating personalized treatments that account for their specific genetic characteristics. The DNA-level correction of mutations in genetic disorders makes CRISPR particularly beneficial. Medical researchers use CRISPR to modify genes of the immune cell which results in better activity for destruction of cancer cells. CRISPR technologies are employed to develop improved T-cell therapies that fight leukemia and lymphoma by modifying T-cells to detect specific cancer mutations [71].

5.5.3 Bioinformatics Tools

Biological information technologies in combination with data repositories help researchers understand drug-response variations by analyzing pharmacogenomic data. Complex genetic data becomes easier to analyze through these tools which combine different data types to help medical staff find genetic markers for both drug efficacy and ADRs. The pharmacogenomic research and clinical applications depend on PharmGKB, dbSNP along with ClinVar and specialized bioinformatics software tools.

5.5.3.1 *PharmGKB: Pharmacogenomics Database*

PharmGKB contains complete details about genetic changes and impacts on drug response by including information about drug-metabolizing enzymes together with transporters and receptors. The database contains genotype-phenotype connections along with clinical practice guidelines which makes it indispensable for obtaining personalized medical treatment [72]. PharmGKB offers an organized database featuring drug labels with pharmacogenomic suggestions which help healthcare providers improve medication efficacy and safety.

5.5.3.2 *dbSNP: Database of Single Nucleotide Polymorphisms*

The dbSNP (Database of Single Nucleotide Polymorphisms) acts as a basic bioinformatics resource with data of genetic variations, mainly single nucleotide polymorphisms (SNPs), to create extensive datasets. Through the use of dbSNP, researchers can discover SNPs associated with particular diseases and drug reactions which supports their research of drug-related genetic factors for both efficacy and toxicity studies. Physicians rely on this database to investigate drug response genetic factors across patient and develop pharmacogenomic research methodologies [73].

5.5.3.3 *ClinVar: Clinically Relevant Variants*

The ClinVar database enables free access to data of the relationship between genetic variants and human health data. Optimized collection at ClinVar arises from diverse datasets through studies and clinical reports to determine the disease progression potential and drug response effects of genetic variants. The standardized annotations of ClinVar help clinicians and researchers make better decisions for patient care management [73].

5.6 Drug Discovery and Development

Drug discovery and development heavily rely on pharmacogenomics which controls drug development from initial phases at preclinical stages through clinical trials up to regulatory approval. Genetic data integration within PG allows scientists to choose best drugs while predicting a patient's response to provide improved custom-tailored treatments.

5.6.1 Pharmacogenomics in Preclinical Drug Development

The behavior of drugs in human subjects becomes clear through early assessments carried out with animal models featuring genetically modified mutations or natural genetic alterations. Scientists utilize genetically modified mice that present either CYP450 (cytochrome P450) or TPMT (thiopurine S-methyltransferase) gene mutations to forecast drug metabolism patterns in different population groups during drug selection processes. The analysis of pharmacogenomic information enables scientists to discover indicators which show ADRs of drugs. The utilization of animal models under pharmacogenomic profiling helps identification of drug candidates that shows side effects or ADRs in genetically unique patients, thus improving drug development before beginning clinical trials.

5.6.2 Clinical Trials

Most clinical trials have relied on standard universal approaches during testing but neglect to consider individual genetic variations. By adding pharmacogenomic information, clinical trials become able to divide participants according to genetic markers which show how they will respond to their treatments. The application of this approach creates more secure patient treatment and leads to more successful and specific clinical trials which operate more effectively.

5.6.3 Regulatory Considerations

The approval process of pharmacogenomic biomarkers as quality indicators requires strong supporting data regarding both drug efficacy and patient safety according to US FDA and EMA guidelines. The standardization of pharmacogenomics integration in drug approval processes becomes common practice since genetic variations play a key role in treatment success rates, particularly when targeting complex diseases.

5.7 Pharmacogenomic Testing in Routine Clinical Practice

The examination of drug-metabolizing genes in tandem with drug efficacy and safety amplified through pharmacogenomic testing produces substantial benefits which include individualized therapy and safer drug utilization. Routine clinical practice faces several implementation obstacles for use of pharmacogenomic testing. This chapter presents the testing approaches with digital health instrument applications while describing obstacles of expansion into clinical implementation.

5.7.1 The DNA-Based Assays *via* POCT

DNA-based assays serve as the primary testing method during clinical pharmacogenomic practice to determine patients' drug response predictions through genetic analysis of variations. The main methods used for testing are polymerase chain reaction (PCR) together with next-generation sequencing (NGS) as well as microarrays. The accuracy of DNA-based tests keeps increasing while laboratories implement these methods into their standard clinical assessment practices particularly in oncology and psychiatric and cardiology services. The clinical implementation of pharmacogenomic tests through Point-of-care testing (POCT) presents itself as an efficient method in healthcare settings. POCT enables genetic assessments at patient bedsides and outpatient clinics which provide test results in hours for immediate drug treatment modifications. The point-of-care GeneXpert system serves as a diagnostic platform that analyzes tuberculosis therapy-related genetic variants to help doctors change prescriptions per individual patient DNA results. The implementation of point-of-care testing for pharmacogenomics stands at a nascent stage although it displays great potential to enhance drug prescriptions and safeguard patient safety mainly in time-sensitive urgent care scenarios [74].

5.7.2 Clinical Decision Support Systems (CDSS)

Healthcare practitioners benefit greatly from Clinical Decision Support Systems (CDSS) when incorporating the pharmacogenomic data into their clinical decisions. CDSS provide custom recommendation frameworks to physicians using evidence-based data from genetic records along with medical analysis and drug information. Drugs that deliver optimal therapy can be automatically generated through patient genomic data analysis which accelerates and enhances precision of therapeutic choices

[75]. CDSS systems analyze extensive datasets along with detecting intricate patient data patterns using integration of machine learning and AI technologies. The combination of drug data and clinical data through AI algorithms produces customized treatment protocols by predicting drug efficacy. The AI-powered CDSS is effective to determine the drug response effects of different genetic variants which improved clinical results.

The CDSS tools facilitate physician decision-making in two ways:

1. They trigger genetic testing alerts and subsequently provide variant-dependent drug dosages.
2. The integrated systems lower the chance of human errors through their functionality, which enables proper utilization of pharmacogenomic information during clinical choices.

5.7.3 Challenges and Barriers to Implementation

Glaring obstacles exist which hinder the general implementation of pharmacogenomic testing in medical practices across multiple facilities [76].

1. The high cost of genetic testing stands as the main challenge because whole-genome sequencing and NGS methods are expensive. Testing expenses become too high to manage at both the healthcare system level and patient level especially within resource-constrained areas.
2. Most healthcare institutions that provide patient care have insufficient infrastructure required to implement pharmacogenomic testing in standard clinical procedures. The real-time processing of pharmacogenomic data faces challenges because many laboratories lack necessary technology to properly handle such information making personal medicine implementation slower.
3. Physician training presents itself as a major obstacle for the deployment of pharmacogenomic technology because they lack necessary knowledge and practical experience in handling the PG technologies and databases or to interpret the data to make clinical decisions. The insufficient understanding about pharmacogenomic data causes health professionals to improperly incorporate these genetic findings into their clinical choices. The integration of pharmacogenomics education in medical curriculum increases physician confidence regarding genetic information application in medical practice.

5.8 International Initiatives and Collaborations

5.8.1 Major Global Pharmacogenomic Research Initiatives

International academic research organizations and consortiums dedicate efforts to handle population-based drug response and advance global pharmacogenomic research. The 1000 Genomes Project is an important international project dedicated to genome sequencing of over 2,500 subjects from multiple global populations with the purpose of building a complete human genetic variation database. The research from the 1000 Genomes Project presented crucial information about inter-population genetic variety which enabled better prediction of drug metabolism variations. The Pharmacogenomics Research Network (PGRN) functions as a vital NIH-funded collaborative organization that investigates drug response genetics in multiple population groups through its PGRN framework. PGRN researchers form international collaborations to find drug efficacy and toxicity linked genetic variants for the advancement of personalized medicine. Pharmaceutical researchers utilize the International HapMap Project to understand drug responses better by providing detailed information about genetic variants that impact medicine effect. Researchers identify drug metabolism associated SNPs from different populations through HapMap which enables the development of drugs suitable for various genetic backgrounds [77].

5.8.2 Pharmacogenomics in Low-Resource Settings

The adoption of pharmacogenomic knowledge within low-resource settings faces specific difficulties because of essential genetic testing constraints in combination with inadequate healthcare facilities and insufficient medical personnel. Different organizations have launched promising initiatives to break down these current hurdles. Genetic testing continues to evolve across sub-Saharan Africa yet institutions in this region lack enough accessibility to genetic testing services. The government of South Africa has successfully implemented pharmacogenomic data for healthcare treatment including oncology and HIV medicine. Scientists utilize CYP2B6 variant tests to enhance efavirenz dosages for HIV patients which leads to lower neurotoxic effects and stronger patient drug compliance [78]. Researchers aim to extend pharmacogenomic research activities across adjacent African nations to develop medications that explore the unique genetic distributions. The Indian medical community has recognized pharmacogenomics as a vital tool for creating personalized medical care. The Indian Pharmacogenomics Consortium leads initiatives to merge genetic testing in medical applications by focusing on cancer care decisions which are shaped by EGFR and KRAS mutation variants. The development of infrastructure which offers cost-effective pharmacogenomic testing throughout India enables better access to personal medicine therapies even in restricted resource areas [79].

5.9 Technological Advancements

A substantial transformation will occur in pharmacogenomics during upcoming years with the integration of "omics" technology along with use of artificial intelligence and continuous work to solve major issues relating to PG. These developments will improve the accuracy, precision and universal availability of personalized therapeutic approaches.

5.9.1 Advances in Precision Medicine

Precision medicine determines therapeutic approaches which take into account each patient's DNA information alongside their life conditions. Advancements of pharmacogenomics with proteomics and metabolomics

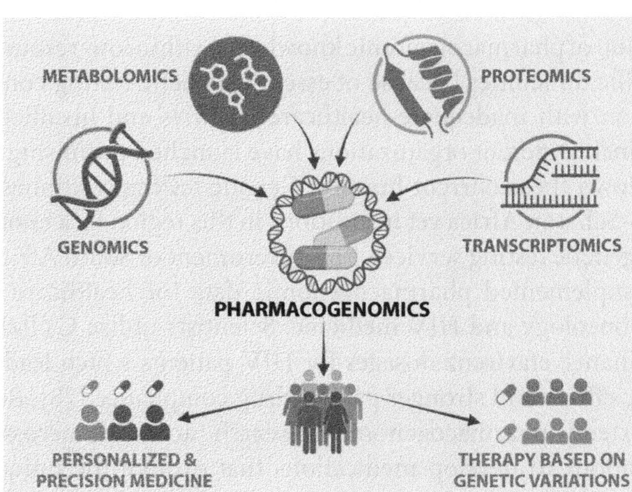

Figure 5.2 Future directions and scope of pharmacogenomics [80].

as well as epigenomics will establish as a complete system which comprehends patient data, drug data and genetic makeup to guide therapeutic decisions. The analysis of drug-related proteins through proteomics accompanies insights from metabolomics about drug metabolism together with epigenomics knowledge of environmental influences on gene expression and drug response [80]. By unifying information from genomic DNA analysis, protein research and metabolic assessments, medical professionals can achieve more precise drug response predictions with fresh therapeutic possibilities as illustrated in Figure 5.2.

5.9.2 Artificial Intelligence and Pharmacogenomics

Pharmacogenomics benefits from the rapid development of artificial intelligence together with machine learning systems. Advanced technology efficiently processes elaborate pharmacogenomic information by identifying previously hidden data patterns to predict drug responses at higher levels than human-operated processes. The incorporation of artificial intelligence into pharmacogenomics improves drug production efficiency while decreasing expenses and accelerating the genetic knowledge application in medical settings.

5.10 Conclusion

Pharmacogenomics serves as the foundation of personalized medicine since it shows essential ways in which drug response interacts with genetic variations. This chapter examines three main pharmacogenomic applications which include optimizing drug doses, preventing ADRs and achieving desired therapeutic efficacy. Pharmacogenomics demonstrates substantial potential for the future drugs through its combination with proteomics and metabolomics as genetic data techniques. Modern artificial intelligence systems will improve drug prescription accuracy along with daily clinical implementation of pharmacogenomics to enhance patient success rates. Newer genetic testing technologies that reduce costs will enable healthcare professionals to deliver better and safer treatments to individual patients. Achieving the best use of pharmacogenomics requires rigorous ongoing research efforts. Research organizations must collaborate with medical practitioners and regulatory groups in order to overcome current hurdles that stand between implementation of pharmacogenomics in standard medical care. Education coupled with genetic testing access in all the medical facilities allows the present capabilities to merge with upcoming opportunities.

AI disclosure statement: The author confirms that no artificial intelligence (AI)-generated content or AI-assisted writing tools were used in the preparation of this chapter (including summaries and the abstract). All text, analysis, and conclusions are the result of the author's original work.

References

1. T.P.A., M.S.S., Jose, A., Chandran, L., Zachariah, S.M., Pharmacogenomics: The right drug to the right person. *J. Clin. Med. Res.*, 1, 4, 191–194, 2009.
2. Motulsky, A.G., Drug reactions, enzymes, and biochemical genetics. *J. Am. Med. Assoc.*, 165, 7, 835–837, 1957.
3. Lehmann, H. and Ryan, E., The familial incidence of low pseudocholinesterase level. *Lancet*, 268, 6934, 124, 1956.
4. Kalow, W. and Staron, N., On distribution and inheritance of atypical forms of human serum cholinesterase, as indicated by dibucaine numbers. *Can. J. Biochem. Physiol.*, 35, 12, 1305–1320, 1957.
5. Evans, D.A.P., Manley, K.A., McKusick, V.A., Genetic control of isoniazid metabolism in man. *Br. Med. J.*, 2, 5197, 485, 1960.
6. Gonzalez, F.J., Skodat, R.C., Kimura, S., Umeno, M., Zanger, U.M., Nebert, D.W., *et al.*, Characterization of the common genetic defect in humans deficient in debrisoquine metabolism. *Nature*, 331, 6155, 442–446, 1988.
7. Weinshilboum, R. and Aksoy, I., Sulfation pharmacogenetics in humans. *Chem. Biol. Interact.*, 92, 1–3, 233–246, 1994.
8. Weinshilboum, R.M., Otterness, D.M., Szumlanski, C.L., Methylation pharmacogenetics: Catechol O-methyltransferase, thiopurine methyltransferase, and histamine N-methyltransferase. *Annu. Rev. Pharmacol. Toxicol.*, 39, 1, 19–52, 1999.
9. Bosma, P.J., Chowdhury, J.R., Bakker, C., Gantla, S., de Boer, A., Oostra, B.A., *et al.*, The genetic basis of the reduced expression of bilirubin UDP-glucuronosyltransferase 1 in Gilbert's syndrome. *N. Engl. J. Med.*, 333, 18, 1171–1175, 1995.
10. Weinshilboum, R.M. and Wang, L., Pharmacogenomics: Precision medicine and drug response. *Mayo Clin. Proc.*, 92, 11, 1711–1722, Elsevier, 2017.
11. Sim, S.C., Kacevska, M., Ingelman-Sundberg, M., Pharmacogenomics of drug-metabolizing enzymes: A recent update on clinical implications and endogenous effects. *Pharmacogenomics J.*, 13, 1, 1–11, 2013.
12. Harper, A.R. and Topol, E.J., Pharmacogenomics in clinical practice and drug development. *Nat. Biotechnol.*, 30, 11, 1117–1124, 2012.

13. Ingelman-Sundberg, M., Pharmacogenetics of cytochrome P450 and its applications in drug therapy: The past, present, and future. *Trends Pharmacol. Sci.*, 25, 4, 193–200, 2004.
14. Natarajan, S., Ponde, C.K., Rajani, R.M., Jijina, F., Gursahani, R., Dhairyawan, P.P., *et al.*, Effect of CYP2C9 and VKORC1 genetic variations on warfarin dose requirements in Indian patients. *Pharmacol. Rep.*, 65, 5, 1375–1382, 2013.
15. Matros, E., Wang, Z.C., Richardson, A.L., Iglehart, J.D., Genomic approaches in cancer biology. *Surgery*, 136, 3, 511–518, 2004.
16. Bousman, C.A., Al Maruf, A., Marques, D.F., Brown, L.C., Müller, D.J., The emergence, implementation, and future growth of pharmacogenomics in psychiatry: A narrative review. *Psychol. Med.*, 53, 16, 7983–7993, 2023.
17. Mrazek, D.A. and Lerman, C., Facilitating clinical implementation of pharmacogenomics. *JAMA*, 306, 3, 304–305, 2011.
18. Ieiri, I., Functional significance of genetic polymorphisms in P-glycoprotein (MDR1, ABCB1) and breast cancer resistance protein (BCRP, ABCG2). *Drug Metab. Pharmacokinet.*, 27, 1, 85–105, 2012.
19. Bailey, D.G., Dresser, G., Arnold, J.M.O., Grapefruit–medication interactions: Forbidden fruit or avoidable consequences? *CMAJ*, 185, 4, 309–316, 2013.
20. Syvänen, A.C., Accessing genetic variation: Genotyping single nucleotide polymorphisms. *Nat. Rev. Genet.*, 2, 12, 930–942, 2001.
21. Beverage, J.N., Sissung, T.M., Sion, A.M., Danesi, R., Figg, W.D., CYP2D6 polymorphisms and the impact on tamoxifen therapy. *J. Pharm. Sci.*, 96, 9, 2224–2231, 2007.
22. Mueangprom, P., Effects of SLCO1B3, ABCB1 and CYP3A5 polymorphisms on clinical response and adverse events of paclitaxel-based chemotherapy in advanced non-small-cell lung cancer patients, Chulalongkorn University Theses and Dissertations (Chula ETD), 8524, 2019.
23. Abaji, R. and Krajinovic, M., Thiopurine S-methyltransferase polymorphisms in acute lymphoblastic leukemia, inflammatory bowel disease and autoimmune disorders: Influence on treatment response. *Pharmacogenomics Pers. Med.*, 143–156, 2017.
24. Mishra, P.J., Mishra, P.J., Banerjee, D., Bertino, J.R., MiRSNPs or MiR-polymorphisms, new players in microRNA mediated regulation of the cell: Introducing microRNA pharmacogenomics. *Cell Cycle*, 7, 7, 853–858, 2008.
25. Nakajima, M., Kuroiwa, Y., Yokoi, T., Interindividual differences in nicotine metabolism and genetic polymorphisms of human CYP2A6. *Drug Metab. Rev.*, 34, 4, 865–877, 2002.
26. Niederberger, E. and Parnham, M.J., The impact of diet and exercise on drug responses. *Int. J. Mol. Sci.*, 22, 14, 7692, 2021.
27. Fuentes, R.M., Perola, M., Nissinen, A., Tuomilehto, J., ACE gene and physical activity, blood pressure, and hypertension: A population study in Finland. *J. Appl. Physiol.*, 92, 6, 2508–2512, 2002.

28. Januchowski, R., Wojtowicz, K., Andrzejewska, M., Zabel, M., Expression of MDR1 and MDR3 gene products in paclitaxel-, doxorubicin-and vincristine-resistant cell lines. *Biomed. Pharmacother.*, 68, 1, 111–117, 2014.
29. Smolarz, B., Makowska, M., Romanowicz, H., Pharmacogenetics of drug-resistant epilepsy (review of literature). *Int. J. Mol. Sci.*, 22, 21, 11696, 2021.
30. Romaine, S.P.R., Bailey, K.M., Hall, A.S., Balmforth, A.J., The influence of SLCO1B1 (OATP1B1) gene polymorphisms on response to statin therapy. *Pharmacogenomics J.*, 10, 1, 1–11, 2010.
31. Pradana, A.D., Kristin, E., Nugrahaningsih, D.A.A., Nugroho, A.K., Pinzon, R.T., Influence of solute carrier family 22 member 1 (SLC22A1) gene polymorphism on metformin pharmacokinetics and HbA1c levels: A systematic review. *Curr. Diabetes Rev.*, 20, 4, 62–74, 2024.
32. Rasul Chaudhry, S., Muhammad, S., Eidens, M., Klemm, M., Khan, D., Efferth, T., *et al.*, Pharmacogenetic prediction of individual variability in drug response based on CYP2D6, CYP2C9 and CYP2C19 genetic polymorphisms. *Curr. Drug Metab.*, 15, 7, 711–718, 2014.
33. Marcuello, E., Altes, A., Menoyo, A., Del Rio, E., Gomez-Pardo, M., Baiget, M., UGT1A1 gene variations and irinotecan treatment in patients with metastatic colorectal cancer. *Br. J. Cancer*, 91, 4, 678–682, 2004.
34. Sandy, J., Mushtaq, A., Kawamura, A., Sinclair, J., Sim, E., Noble, M., The structure of arylamine N-acetyltransferase from Mycobacterium smegmatis—an enzyme which inactivates the anti-tubercular drug, isoniazid. *J. Mol. Biol.*, 318, 4, 1071–1083, 2002.
35. Choi, Y.J., Park, H., Lee, J.S., Lee, J.Y., Kim, S., Kim, T.W., *et al.*, Methotrexate elimination and toxicity: MTHFR 677C>T polymorphism in patients with primary CNS lymphoma treated with high-dose methotrexate. *Hematol. Oncol.*, 35, 4, 504–509, 2017.
36. Kurata, Y., Ieiri, I., Kimura, M., Morita, T., Irie, S., Urae, A., *et al.*, Role of human MDR1 gene polymorphism in bioavailability and interaction of digoxin, a substrate of P-glycoprotein. *Clin. Pharmacol. Ther.*, 72, 2, 209–219, 2002.
37. Castaño-Amores, C., Antúnez-Rodríguez, A., Pozo-Agundo, A., García-Rodríguez, S., Martínez-González, L.J., Dávila-Fajardo, C.L., Genetic polymorphisms in ADRB1, ADRB2 and CYP2D6 genes and response to beta-blockers in patients with acute coronary syndrome. *Biomed. Pharmacother.*, 169, 115869, 2023.
38. Yu, Z., Wen, L., Shen, X., Zhang, H., Effects of the OPRM1 A118G polymorphism (rs1799971) on opioid analgesia in cancer pain: A systematic review and meta-analysis. *Clin. J. Pain*, 35, 1, 77–86, 2019.
39. Xu, Y., Liu, H., Chen, J., Zhou, Q., Acquired resistance of lung adenocarcinoma to EGFR-tyrosine kinase inhibitors gefitinib and erlotinib. *Cancer Biol. Ther.*, 9, 8, 572–582, 2010.

40. Gazdar, A., Activating and resistance mutations of EGFR in non-small-cell lung cancer: Role in clinical response to EGFR tyrosine kinase inhibitors. *Oncogene*, 28, 1, S24–S31, 2009.
41. Kiyotani, K., Mushiroda, T., Nakamura, Y., Zembutsu, H., Pharmacogenomics of tamoxifen: Roles of drug metabolizing enzymes and transporters. *Drug Metab. Pharmacokinet.*, 27, 1, 122–131, 2012.
42. Crews, K.R., Gaedigk, A., Dunnenberger, H.M., Klein, T.E., Shen, D.D., Callaghan, J.T., *et al.*, Clinical Pharmacogenetics Implementation Consortium (CPIC) guidelines for codeine therapy in the context of cytochrome P450 2D6 (CYP2D6) genotype. *Clin. Pharmacol. Ther.*, 91, 2, 321–326, 2012.
43. Deodhar, M., Rihani, S.B.A., Darakjian, L., Turgeon, J., Michaud, V., Assessing the mechanism of fluoxetine-mediated CYP2D6 inhibition. *Pharmaceutics*, 13, 2, 148, 2021.
44. Wanounou, M., Shaul, C., Abu Ghosh, Z., Alamia, S., Caraco, Y., The impact of CYP2C9*11 allelic variant on the pharmacokinetics of phenytoin and (S)-warfarin. *Clin. Pharmacol. Ther.*, 112, 1, 156–163, 2022.
45. Hu, Y.F., Qiu, W., Liu, Z.Q., Zhu, L.J., Liu, Z.Q., Tu, J.H., *et al.*, Effects of genetic polymorphisms of CYP3A4, CYP3A5, and MDR1 on cyclosporine pharmacokinetics after renal transplantation. *Clin. Exp. Pharmacol. Physiol.*, 33, 11, 1093–1098, 2006.
46. Roy, J.N., Barama, A., Poirier, C., Vinet, B., Roger, M., Cyp3A4, Cyp3A5, and MDR-1 genetic influences on tacrolimus pharmacokinetics in renal transplant recipients. *Pharmacogenet. Genomics*, 16, 9, 659–665, 2006.
47. Hawwa, A.F., Millership, J.S., Collier, P.S., Vandenbroeck, K., McCarthy, A., Dempsey, S., *et al.*, Pharmacogenomic studies of the anticancer and immunosuppressive thiopurines mercaptopurine and azathioprine. *Br. J. Clin. Pharmacol.*, 66, 4, 517–528, 2008.
48. Ando, Y., Saka, H., Ando, M., Sawa, T., Muro, K., Ueoka, H., *et al.*, Polymorphisms of UDP-glucuronosyltransferase gene and irinotecan toxicity: A pharmacogenetic analysis. *Cancer Res.*, 60, 24, 6921–6926, 2000.
49. Innocenti, F., Undevia, S.D., Iyer, L., Xian Chen, P., Das, S., Kocherginsky, M., *et al.*, Genetic variants in the UDP-glucuronosyltransferase 1A1 gene predict the risk of severe neutropenia of irinotecan. *J. Clin. Oncol.*, 22, 8, 1382–1388, 2004.
50. Huang, M.J., Chen, P.L., Huang, C.S., Bilirubin metabolism and UDP-glucuronosyltransferase 1A1 variants in Asians: Pathogenic implications and therapeutic response. *Kaohsiung J. Med. Sci.*, 38, 8, 729–738, 2022.
51. Goh, L.L., Lim, C.W., Leong, K.P., Ong, K.H., TPMT and NUDT15 testing for thiopurine therapy: A major tertiary hospital experience and lessons learned. *Front. Pharmacol.*, 13, 837164, 2022.
52. Brown, S.A. and Pereira, N., Pharmacogenomic impact of CYP2C19 variation on clopidogrel therapy in precision cardiovascular medicine. *J. Pers. Med.*, 8, 1, 8, 2018.

53. Chang, B.L., Liu, J.R., Chang, S.H., See, L.C., Impact on carbamazepine usage and cutaneous adverse reactions before and after the reimbursement of HLA-B*1502 genotyping in Taiwan, 2000–2017: A nationwide longitudinal study. *Epilepsia*, 64, 10, 2679–2689, 2023.
54. Hung, S.I., Chung, W.H., Chen, Y.T., HLA-B genotyping to detect carbamazepine-induced Stevens-Johnson syndrome: Implications for personalizing medicine. *Pers. Med.*, 2, 3, 225–237, 2005.
55. Valabrega, G., Montemurro, F., Aglietta, M., Trastuzumab: Mechanism of action, resistance and future perspectives in HER2-overexpressing breast cancer. *Ann. Oncol.*, 18, 6, 977–984, 2007.
56. Mallal, S., Phillips, E., Carosi, G., Molina, J.M., Workman, C., Tomažič, J., *et al.*, HLA-B*5701 screening for hypersensitivity to abacavir. *N. Engl. J. Med.*, 358, 6, 568–579, 2008.
57. Zhang, G.N., Zhang, Y.K., Wang, Y.J., Gupta, P., Ashby Jr., C.R., Alqahtani, S., *et al.*, Epidermal growth factor receptor (EGFR) inhibitor PD153035 reverses ABCG2-mediated multidrug resistance in non-small cell lung cancer: *In vitro* and *in vivo*. *Cancer Lett.*, 424, 19–29, 2018.
58. De Boeck, K., Munck, A., Walker, S., Faro, A., Hiatt, P., Gilmartin, G., *et al.*, Efficacy and safety of ivacaftor in patients with cystic fibrosis and a non-G551D gating mutation. *J. Cyst. Fibros.*, 13, 6, 674–680, 2014.
59. Gex-Fabry, M., Eap, C.B., Oneda, B., Gervasoni, N., Aubry, J.M., Bondolfi, G., *et al.*, CYP2D6 and ABCB1 genetic variability: Influence on paroxetine plasma level and therapeutic response. *Ther. Drug Monit.*, 30, 4, 474–482, 2008.
60. Gainor, J.F., Tan, D.S., De Pas, T., Solomon, B.J., Ahmad, A., Lazzari, C., *et al.*, Progression-free and overall survival in ALK-positive NSCLC patients treated with sequential crizotinib and ceritinib. *Clin. Cancer Res.*, 21, 12, 2745–2752, 2015.
61. Tavares, A.B.M., Neto, J.L., Fulco, U.L., Albuquerque, E.L., Blockade of the checkpoint PD-1 by its ligand PD-L1 and the immuno-oncological drugs pembrolizumab and nivolumab. *Phys. Chem. Chem. Phys.*, 23, 37, 21207–21217, 2021.
62. Druker, B.J., Talpaz, M., Resta, D.J., Peng, B., Buchdunger, E., Ford, J.M., *et al.*, Efficacy and safety of a specific inhibitor of the BCR-ABL tyrosine kinase in chronic myeloid leukemia. *N. Engl. J. Med.*, 344, 14, 1031–1037, 2001.
63. Lynch, T.J., Bell, D.W., Sordella, R., Gurubhagavatula, S., Okimoto, R.A., Brannigan, B.W., *et al.*, Activating mutations in the epidermal growth factor receptor underlying responsiveness of non–small-cell lung cancer to gefitinib. *N. Engl. J. Med.*, 350, 21, 2129–2139, 2004.
64. Flaherty, K.T., Robert, C., Hersey, P., Nathan, P., Garbe, C., Milhem, M., *et al.*, Improved survival with MEK inhibition in BRAF-mutated melanoma. *N. Engl. J. Med.*, 367, 2, 107–114, 2012.

65. Topalian, S.L., Hodi, F.S., Brahmer, J.R., Gettinger, S.N., Smith, D.C., McDermott, D.F., *et al.*, Safety, activity, and immune correlates of anti–PD-1 antibody in cancer. *N. Engl. J. Med.*, 366, 26, 2443–2454, 2012.
66. Maloney, D.G., Grillo-López, A.J., White, C.A., Bodkin, D., Schilder, R.J., Neidhart, J.A., *et al.*, IDEC-C2B8 (Rituximab) anti-CD20 monoclonal antibody therapy in patients with relapsed low-grade non-Hodgkin's lymphoma. *Blood*, 90, 6, 2188–2195, 1997.
67. Brahmer, J.R., Tykodi, S.S., Chow, L.Q., Hwu, W.J., Topalian, S.L., Hwu, P., *et al.*, Safety and activity of anti–PD-L1 antibody in patients with advanced cancer. *N. Engl. J. Med.*, 366, 26, 2455–2465, 2012.
68. Geyer, C.E., Forster, J., Lindquist, D., Chan, S., Romieu, C.G., Pienkowski, T., *et al.*, Lapatinib plus capecitabine for HER2-positive advanced breast cancer. *N. Engl. J. Med.*, 355, 26, 2733–2743, 2006.
69. Kumar, K.R., Cowley, M.J., Davis, R.L., Next-generation sequencing and emerging technologies. *Semin. Thromb. Hemost.*, 50, 7, 1026–1038, Thieme Medical Publishers, 2024.
70. Mizzi, C., Peters, B., Mitropoulou, C., Mitropoulos, K., Katsila, T., Agarwal, M.R., *et al.*, Personalized pharmacogenomics profiling using whole-genome sequencing. *Pharmacogenomics*, 15, 9, 1223–1234, 2014.
71. Sadeqi Nezhad, M., Yazdanifar, M., Abdollahpour-Alitappeh, M., Sattari, A., Seifalian, A., Bagheri, N., Strengthening the CAR-T cell therapeutic application using CRISPR/Cas9 technology. *Biotechnol. Bioeng.*, 118, 10, 3691–3705, 2021.
72. Barbarino, J.M., Whirl-Carrillo, M., Altman, R.B., Klein, T.E., PharmGKB: A worldwide resource for pharmacogenomic information. Wiley Interdiscip. *Rev. Syst. Biol. Med.*, 10, 4, e1417, 2018.
73. Salnikova, L.E., Kolobkov, D.S., Sviridova, D.A., Abilev, S.K., An overview of germline variations in genes of primary immunodeficiencies through integrative analysis of ClinVar, HGMD® and dbSNP databases. *Hum. Genet.*, 140, 9, 1379–1393, 2021.
74. Leach, M., Newman, W.G., McDermott, J.H., Rapid point of care testing: The next frontier in pharmacogenomics. *Pharmacogenomics*, 25, 7, 289–291, 2024.
75. Hinderer, M., Boeker, M., Wagner, S.A., Lablans, M., Newe, S., Hülsemann, J.L., *et al.*, Integrating clinical decision support systems for pharmacogenomic testing into clinical routine: A scoping review of designs of user-system interactions in recent system development. *BMC Med. Inform. Decis. Mak.*, 17, 1–14, 2017.
76. Abou Diwan, E., Zeitoun, R.I., Abou Haidar, L., Cascorbi, I., Khoueiry Zgheib, N., Implementation and obstacles of pharmacogenetics in clinical practice: An international survey. *Br. J. Clin. Pharmacol.*, 85, 9, 2076–2088, 2019.
77. Cavallari, L.H., Hicks, J.K., Patel, J.N., Elchynski, A.L., Smith, D.M., Bargal, S.A., *et al.*, The Pharmacogenomics Global Research Network

Implementation Working Group: Global collaboration to advance pharmacogenetic implementation. *Pharmacogenet. Genom.*, 35, 1, 1–11, 2025.
78. Tata, E.B., Ambele, M.A., Pepper, M.S., Barriers to implementing clinical pharmacogenetics testing in Sub-Saharan Africa: A critical review. *Pharmaceutics*, 12, 9, 809, 2020.
79. Sahana, S., Bhoyar, R.C., Sivadas, A., Jain, A., Imran, M., Rophina, M., *et al.*, Pharmacogenomic landscape of the Indian population using whole genomes. *Clin. Transl. Sci.*, 15, 4, 866–877, 2022.
80. Molla, G. and Bitew, M., Revolutionizing Personalized Medicine: Synergy with Multi-Omics Data Generation, Main Hurdles, and Future Perspectives. *Biomedicines*, 12, 12, 2750, 2024.

6

Ethical, Legal, and Social Implications of Genomic Data in Healthcare

M. Reddy Sailaja[1]*, Swetha Kumari Koduru[2] and R. Venkata Raghava Reddy[3]

[1]Medical Device Services, Celegence Systems Pvt. Ltd., Bangalore, India
[2]Department of Biosciences and Sericulture, Sri Padmavati Mahila Visvavidyalayam, Tirupati, India
[3]Bioinformatics Wing, Sybiomics Technologies Private Limited, Tirupati, India

Abstract

Genomic data has become a cornerstone of modern healthcare, driving the emergence of personalized medicine as a standard-of-care approach that enables more precise diagnosis, targeted therapies, and improved health outcomes. By tailoring medical interventions to an individual's genetic makeup, personalized medicine holds the promise of enhancing quality of life, extending healthy lifespan, and transforming preventive and therapeutic strategies. However, alongside these scientific and clinical advancements, the rapid integration of genomic information into healthcare introduces a spectrum of ethical, legal, and social implications that demand careful attention. Ethically, the handling of genomic data presents unprecedented challenges related to patient autonomy, informed consent, and the responsible stewardship of sensitive information. Genomic datasets often embedded within electronic medical records and other clinical documentation contain complex hereditary details that may affect not only individual patients but also their biological relatives. Ensuring that such information is collected, stored, and used in an ethically sound manner requires clear guidelines to prevent misuse, unauthorized access, or unintended disclosure. Issues such as recontacting patients, returning incidental findings, and ensuring equitable access to genomic technologies further complicate the ethical landscape.

Socially, the interpretation and communication of genomic information necessitate culturally sensitive and accessible practices, particularly for marginalized, low-literacy, or underserved populations. Misunderstanding genetic risk can lead

**Corresponding author*: sailajamrs@gmail.com

Swetha Kumari Koduru, Sarad Pawar Naik Bukke, Hussaini Bello, Muhamad Mustafa Mundu and Ayush Madan (eds.) Genomics and Precision Healthcare: Innovations, Applications and Challenges, (171–200) © 2026 Scrivener Publishing LLC

to stigma, anxiety, or mistrust in medical institutions. Additionally, societal concerns surrounding genetic discrimination especially in employment and insurance contexts underscore the need for robust protections that prevent individuals from being unfairly treated based on their genetic predispositions. Legally, the increasing reliance on genomic data requires rigorous regulatory frameworks that safeguard patient confidentiality, ensure the integrity of informed consent processes, and promote secure data governance. Strong data protection laws are essential for addressing cross-border data sharing, commercial use of genomic information, and liability issues arising from data breaches or misinterpretation of genomic results. As genomic technologies evolve, legal systems must adapt accordingly to balance innovation with patient rights and public trust.

This chapter provides a comprehensive overview of the ethical, legal, and social challenges associated with incorporating genomic data into healthcare. It highlights the pressing need for updated policies, stronger safeguards, and coordinated efforts among policymakers, clinicians, researchers, and communities. Ultimately, advancing genomic medicine responsibly requires sustained investment in education and training for both healthcare providers and patients, fostering an informed, equitable, and ethically grounded approach to the future of personalized healthcare.

***Keywords*:** Ethical implications, legal frameworks, social challenges, data privacy, informed consent, electronic medical records, genomic governance, genetic counselling

6.1 Introduction

6.1.1 The Promise of Genomics in Healthcare

The completion of the Human Genome Project in 2003 marked a turning point in the advancement of genomic methods and transformed our understanding of the human genome. The practical use of the results from these genetic investigations is still in its early stages, though [1]. This is mostly because we have yet to fully comprehend or make sense of the information that is now available. In other words, it has been challenging to match the sequence data to functional outcomes, which has complicated our understanding of the complex gene-lifestyle-environment impacts or their interactions, as well as the genetic foundation of disorders [2, 3].

Furthermore, most of the research's initial focus was on DNA's coding sections, which make up around 2% of the DNA, and little is known about the precise consequences of non-coding DNA regions, which make up 98% of DNA [4, 5]. Surprisingly, there is only a 1.23% difference in single nucleotide changes and a 3% difference in deletions, insertions, and copy number variations between the human genome and that of the nearest comparable species, chimpanzees [6]. Any two human beings have almost 99.9% of the same genome. But because the human genome contains roughly 30 billion base pairs ($3.3 \times 10_9$), a simple 0.1% alteration can alter a huge number of nucleotides [7].

Genomics, which is revolutionizing healthcare, makes possible precision medicine, early disease identification, and individualized treatment plans. Developments in genome sequencing, gene editing (CRISPR), and bioinformatics have improved the diagnosis and treatment of infectious diseases, tumours, and genetic disorders. The key applications of genomics in medicine are mentioned below: [8–10]

- **Gene Therapy and Editing:** CRISPR-based treatments for genetic disorders like sickle cell disease.
- **Infectious Disease Control:** Genomic surveillance of pathogens (e.g., SARS-CoV-2 variants) for public health strategies.
- **Personalized Medicine:** Tailoring treatments based on genetic profiles (e.g., cancer therapies targeting specific mutations).
- **Pharmacogenomics:** Optimizing drug selection and dosages based on an individual's genetic makeup.
- **Rare Disease Diagnosis:** Whole-genome sequencing helps identify genetic causes of rare conditions.

In 2023, the U.S. Food and Drug Administration (FDA) approved Casgevy (CRISPR-based therapy) for sickle cell disease [11]. AI models like AlphaFold improve protein structure prediction and is aiding drug discovery [12]. Programs to screen for genetic conditions at birth could be achieved by Whole-Genome Sequencing in Newborns [13].

A data engineering program is an effective method of converting data into actionable intelligence when viewed through the perspective of data science. The volume of genomic data keeps growing, and it is estimated that between two and forty exabytes of data will be directly attributable to genomic research within the next ten years [14]. Genomic data can be a significant problem for healthcare facilities due to its massive production and high storage and processing requirements. Since data interpretation will always be necessary, it is logical to assume that the demand for AI/ML technologies will increase gradually [15].

It is legitimate to state that genomics has developed in ways that many people could not have predicted since its inception. Proteomics, which helps identify aberrant proteins that cause a variety of diseases; Pharmacogenetics/Pharmacogenomics, which advances the production of medications for individuals based on genetic composition [16]; Stem cell therapy, which uses stem cells to treat diseases in people of all ages; and Cloning, which includes both therapeutic and reproductive cloning, are some of the newest genomic technologies [17].

Rapid advancements in genomics hold promise for better public health initiatives, treatments, and diagnostics. Understanding how genetics affects disease pathogenesis and drug response broadens the area of clinical application and study. Although genome editing has the potential to repair a damaged genome *in vivo*, treatments that modify gene expression without changing the genetic coding can also be developed [18]. However, ethical concerns like data privacy and genetic discrimination remain challenges for widespread adoption [19].

Advances in genomic technology and computational methods have greatly enhanced our understanding of how human genetic variations are distributed and how they influence disease risk and drug responses across different populations [20]. Genomics in healthcare has the potential to bring about transformative changes, especially with the expanding role of whole genome sequencing (WGS) in clinical practice [21]. Since the Human Genome Project began in 1990, there have been remarkable advancements in genomics, offering new possibilities for improving health [22, 23]. Figure 6.1 provides a schematic overview of the sequential advancements in genomic research.

Genomic studies, especially in population genetics, are playing a crucial role in understanding the genetic factors behind common chronic diseases, which contribute significantly to global mortality. Research is increasingly focused on identifying genetic variants linked to conditions like diabetes, heart disease, and cancer [24]. Recent progress has led to the discovery of genetic associations with various chronic diseases, although earlier studies faced inconsistencies due to limited statistical power [25].

Currently, genome and exome sequencing are mainly used to diagnose Mendelian disorders and guide cancer treatments. But as WGS becomes more widely used, it could lead to more accurate diagnoses, including for prenatal conditions, and enable more personalized drug therapies based on an individual's unique genetic makeup [26]. Additionally, genomics could play a key role in predicting disease risk and improving our understanding of how diseases develop, paving the way for more personalized and effective healthcare [27].

Even as there have been significant improvements in the pursuit of genomic research, the actualization of the genetic research findings in the everyday clinical practice has trailed behind the predicted rate. Several

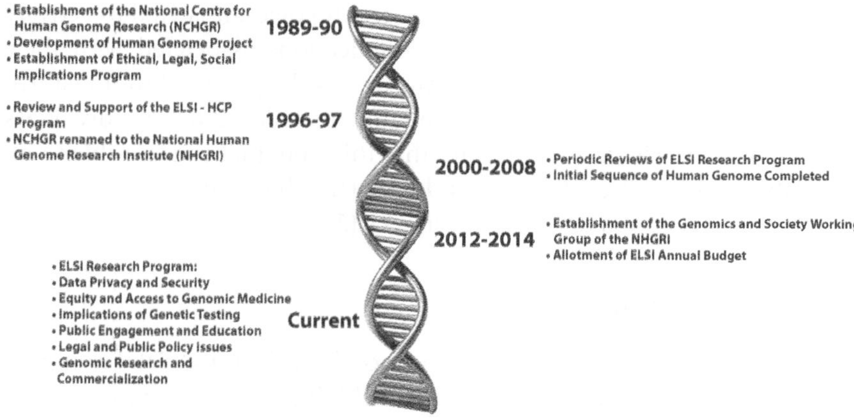

Figure 6.1 Schematic representing the series of advancements in genomic research.

obstacles are involved, most notably the lack of evidenced data that proves the applicability of using genomic data in clinical practice, subpar awareness and acceptance among medical workers, and the constant need of adequate infrastructural provision that will promote the wider adoption of genomic medicine [28].

Although genomic research has made great progress in the last couple of years, the pace at which laboratory results may be transformed into real care practices in the normal healthcare setting is lower than expected. Several reasons are attributed to this lag including poor clinical evidence in support of the usefulness of genomic data, low uptake and adoption among the medical community coupled with a virtual shortage of infrastructure to guide expansive implementation of genomic applications in medicine [29].

Genomics has also provided empirical understanding in a significant way as to why some individuals are exposed to environmental factors and specific diseases, and some people are resistant. Utilizing genomics can aid with diagnosis, patient risk assessment, prenatal testing to identify anomalies, and preventative care. Data in the healthcare industry is expanding more quickly than data in any other industry. But growth also brings with it problems with the quality of healthcare data. In the building of data-quality framework, an assessment that scholars do regularly is done on the features of the data-quality including: Accessibility, accuracy, currency, comprehensiveness, consistency, legitimacy, and usability [30].

As genomic technologies continue to evolve, it is crucial to consider the social, ethical, and legal implications that arise with their use in healthcare [3].

To harness the full potential of genomic technologies, it is vital to actively address the social, ethical, and legal implications associated with their use in healthcare [31]. Genomic healthcare has the potential to transform medicine by providing personalized treatments tailored to an individual's genetic makeup. However, to ensure that this potential is accessible to all populations, it is critical to address the social, ethical, and structural factors that impact access to genomic services [32].

6.1.2 The ELSI Imperative

The branch of study that focuses on the moral, legal, and societal ramifications of genetics and genomics is known as ELSI (Ethical, Legal, and Social Implications) in the United States. Since its inception in the global Human Genome Project (HGP), which mapped and sequenced the whole human genome between 1990 and 2003, ELSI researchers have created thousands of books, papers, and other resources. These studies examine a range of topics related to fundamental research and its clinical application, as well as more general societal concerns brought up by new developments in life sciences technology [33, 34].

As concerns about incidental findings, genetic privacy and confidentiality, and informed consent become increasingly prominent in the public consciousness, the ethical, legal, and social implications (ELSI) of public health genetics and genomics efforts are receiving more and more attention. Even though many of these problems are not specific to genetics, public health does need to take them into account. ELSI considerations in relation to genomics were recently mapped out by the Genetics in Primary Care Institute (Figure 6.2). Additionally, ELSI research has yielded a number of tangible policy achievements that have safeguarded human rights and improved the way genomic research is conducted [35].

Today, ELSI studies constitute a multidisciplinary field of study that is constantly growing and changing, covering far more ground than was initially planned thirty years ago. Due to the intricate relationships between ethical, legal, and social studies, ELSI is now frequently interpreted as an entire set rather than a collection of separate components, making it a rather ambiguous concept [36].

In the this book chapter, our overview of ELSI systems, analysis and the data generated may be useful for researchers, healthcare professionals and policymakers as an unbiased, synthesized, comprehensive view of relevant ELSI topics.

6.2 Ethical Foundations

6.2.1 Respect for Autonomy and Informed Consent

Autonomy is the core of genetic testing ethics as it ensures that patients and their family members have the right to make informed decisions regarding genetic testing. Individuals should be aware of the psychological, social, and medical impacts of the results from genetic tests. For example, genetic tests for BRCA1/BRCA2 mutation may provide information on the disposition to cancers and reveal information about familial risks that require careful handling [37].

Informed consent is a process in which a healthcare professional educates a patient about the risks, benefits, and alternatives of a given procedure or intervention. The history of informed consent in medicine cannot be discussed apart from the wider context of the history of ethical standards and laws on the rights of a patient. Up to the beginning of the twentieth century, medical practice was largely paternalistic, i.e., the clinician made decisions and did not necessarily tell the patient about the relevant information. The justifications behind this paradigm were regularly grounded on the assumption that patients could not understand medical science and/or make good decisions. Over time, however, there was increasing change in this paradigm through a number of high-profile court cases, The court decision in Schloendorff v. The Society of New York Hospital entrenched the principle that every human being of adult years and sound mind had a right to determine what should be done with his own body, hence requiring that prior to medical intervention, there must be an agreement by the patient [38].

The Tuskegee Study of Untreated Syphilis in the Negro Male and the Nazi human experimentation in World War II demonstrated how acute the need was to have strong consent protocols in effect towards the mid-to-late twentieth century. These events, along with the subsequent development of the Nuremberg Code and then the Declaration of Helsinki, gave definite embodiment to informed consent as a paramount ethical requirement of research and medical practice. Informed consent, over the years, has evolved from a simple signature form, which a patient signs, to a process of transparent communication, which is expected to make the person fully understand risks, benefits, and alternative treatments. Informed consent therefore is now a part of basic patient-centered care and medical ethics [39]. The essential components of informed consent documentation are depicted in Figure 6.2.

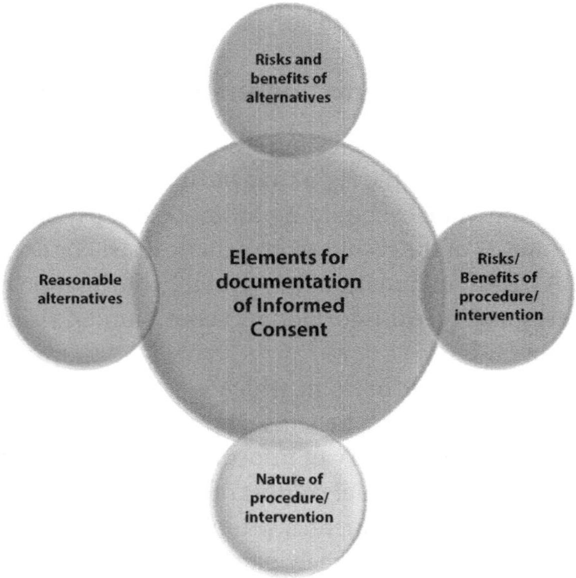

Figure 6.2 Elements of documentation of informed consent.

Within the realm of the modern clinical practice, the competence of the patient to make voluntary decision on any of the proposed procedures or an intervention will take precedence. To get informed consent, there is a need to assess the understanding of the patient in detail, state a clear evidence-based decision with its argumentation, and record the detailed interaction in a systematic way. Healthcare providers are therefore supposed to promote active participation of the patient in the decision-making process and at the same time reduce any feeling of coercive or pressurised thinking to accept the advice of the clinician [40, 41]. The elements of informed consent were needed as documentation by the Joint Commission in a form, progress notes, or in the record (Figure 6.2).

As per Smolenski [42], informed consent, self-sovereignty protection of one's own body, must embrace autonomy and nondomination, the cross-section of two clusters of values. Then, self-ownership and personal integrity comes into the picture.

6.2.1.1 Non-Discrimination

One of the main ethical concerns related to genetic testing is the risk of genetic discrimination. There is a significant possibility that employers, insurance companies, or educational institutions could discriminate against individuals based on their genetic information [80]. Although legislation like the Genetic Information Nondiscrimination Act (GINA) in the U.S. provides some protection against discrimination in employment and health insurance, there are still gaps, particularly in life insurance, disability insurance, and long-term care insurance [80]. Additionally, the social stigma associated with certain genetic markers can cause emotional distress and discourage individuals from undergoing testing [43].

6.2.2 Justice and Equity in Access

The ethical considerations surrounding genomic data in healthcare are complex, particularly regarding the collection, storage, and use of such data. One key argument is that genomic datasets should be treated as public resources that offer significant societal benefits. However, ethical concerns regarding privacy, consent, and the potential for misuse of genetic information continue to arise. Some argue that genomic data requires special protection due to its personal and unique nature, as it can reveal information about an individual's health, family, and identity. On the other hand, critics of this view, known as "genetic exceptionalism," argue that genetic data is not inherently more sensitive than other forms of health data, and it is only identifiable when linked to personal information. This perspective suggests that genomic data could be managed under the same ethical frameworks that govern the collection of other health-related information, such as hospital records or public health data [44].

Public-private partnerships in genomic research raise significant ethical concerns, particularly regarding the balance between innovation and the protection of privacy. A major issue is ensuring that the benefits of genomic data use are equitably distributed, and that public trust is maintained. Ethical concerns center around the conflict between the public interest and private sector profit motives, especially when private companies gain access to sensitive healthcare data. Citizens expect that their personal data will be used to improve public health outcomes, not for commercial profit [45].

Several countries have adopted varying approaches to address these concerns. In the U.K., for example, the Genomics England initiative works to protect donor data while advancing research, but challenges remain in ensuring that commercial interests align with public welfare. Similarly, Israel's approach reflects an ambivalent attitude toward data donation, with citizens willing to share data despite concerns over state reliability [46].

Workshop discussions highlighted the importance of ensuring transparency and trust in data governance. The solidarity-grounded partnership model advocates for public benefit as a priority, ensuring that private companies respect the public good by using data solely for improving healthcare outcomes. This approach also emphasizes the need for preferential access to goods and services developed using public data, aligning commercial activities with public interests [47].

Ensuring the ethical use of genomic data also involves addressing issues of privacy and data security. Safeguarding identifiable information is critical, but careful consideration must also be given to how genomic data is accessed and used, especially regarding the role of private industry and the concerns of different cultural and ethnic groups. Governments and healthcare organizations have a responsibility to maximize the contributions to these datasets while ensuring fairness and protection of individuals' privacy. In this context, it is proposed that patients undergoing genomic testing should allow their data to be included in broader datasets to support scientific progress, without needing explicit consent for each specific future use, provided certain conditions are met. Overall, while the benefits of genomic data for public health and medical advancements are clear, these must be balanced with robust ethical frameworks to maintain public trust and protect individuals' rights [48].

Ethical conduct in human research is about ensuring that research is done with respect for individuals, prioritizing their well-being and ensuring that risks are minimized. In the past, human research was not governed by any established rules of ethics, and this facilitated the occurrence of serious harassments, including the gruesome medical tests performed in the course of World War II. These kinds of atrocities led to the creation of clear codes of ethical conduct, of which the Nuremberg Code (1948) was the first that highlighted the need to offer voluntary consent and stated that any potential future benefits should outweigh a resulting risk [49].

The Declaration of Helsinki published in 1964 shed more light on the minimal standards of biomedical research conduct, creating a situation where the conduct of such studies had to be reviewed by independent people, consent to participate in the study had to be based on informed consent and the competence of the researchers conducting the study. One of the major historical situations that proved the importance of these standards was the study of Syphilis in Tuskegee, in which African-American men were used as test objects, and their syphilis was not treated despite the fact that a cure for the condition had been known for years. This episode eventually led to formulation of the Belmont Report (1979) that asserted three main ethical principles, namely: respect of persons, beneficence, and justice [50].

Common Rule is a set of international standards and regulations, which are aimed to deliver compliance and secure informed consent and protection of vulnerable groups. Similar ethical principles are expressed by the Indian Council of Medical Research (ICMR) in order to ensure the benefits of research to society, especially the needy communities, and the absolute respect of the dignity and welfare of the participants. In combination, these frameworks prioritize the values of respect, equity and protection in all stages of the research that involves human subjects [51].

6.2.3 Genetic and Genomic Testing (GGT)

In the modern medical scene, the application of genomic-based genetic testing (GGT) is transforming the ways of prevention, diagnosis and treatment. Through its ability to perform customized evaluations about the risk of developing diseases, GGT has become an effective modus operandi in diagnosing the predilection to develop cancers, heart diseases and various neurological disorders. Even though these applications bring highly relevant clinical benefits, they also present a cluster of ethical, legal, and social issues. The current synopsis is a compilation of the views of heterogeneous sources, which gives priority to technical, ethical, legal, and societal implications of GGT besides substantive references.

GGT enables identification of specific mutations and genetic susceptibility prediction, thus enabling practitioners to modify therapeutic response with respect to genetic constitution at both single locus locator and large-scale genomic levels *via* interrogation of genetic specimens at both small-scale and whole genome scan levels.

i. **Genetic Testing:** Genetic testing is aimed at early identification of the introduction of clinical manifestations and an indication of specific clinical interventions. The recognition of monogenic disorders based on the positive results of diagnostic tests allows the implementation of early symptom management on the one hand, which allows reducing the effects of a disease.
ii. **Whole-genome sequencing (WGS) and whole-exome sequencing (WES)** studies more genomic sequence areas in order to help identify pathology in multifactorial diseases, such as cardiovascular, neurological, and oncological complex-genetic interactions. Next-generation sequencing

(NGS) further helps physicians to provide highly detailed genomic data in less time with low cost. NGS helps detect genetic markers associated with disease susceptibility, to provide personalized treatment plans.

iii. **Genomic Medicine:** Genomic medicine is an emerging field that utilizes genomic data to develop personalized treatment strategies, particularly in cancer therapy. For example, analyzing the genomic patterns of tumors in oncology enables clinicians to customize chemotherapy based on the specific mutations found in a patient's cancer cells. Understanding a patient's genetic makeup improves the effectiveness of targeted treatments like these [52].

6.3 Legal and Regulatory Frameworks

The legal and regulatory landscape surrounding genomic data in healthcare involves navigating frameworks like the European General Data Protection Regulation (GDPR), which sets strict guidelines for data protection and consent. While some may worry that these regulations could hinder the collection and use of genomic data, existing examples show that these frameworks can accommodate responsible data practices. For instance, under the GDPR's scientific research exemption, pseudonymized data stored in controlled-access databases does not require explicit consent for research purposes. Public health databases like ClinVar and DECIPHER operate under these guidelines, allowing data sharing without explicit consent, provided the data is de-identified.

However, more specific individual-level data, such as detailed phenotypic information, may require explicit consent for sharing, particularly when the data is open access. In the U.K., GDPR mandates that the data collected be adequate, relevant, and limited to what is necessary for the intended purposes, ensuring that proportionality in data sharing aligns with privacy laws [53].

The use of genomic information in the medical field results in multipronged legal issues, particularly concerning matters of data privacy and the necessity of regulatory supervision. France is a fine example of such complications. Involving third parties, including the private organizations in the processing of such sensitive data as health-related information, gives rise to problems with regard to the jurisdiction of the implementation of the EUGDPR and the subsequent risk of cyber attacks. This is why the development of the legal standard which could both protect the privacy of the individual and allow innovation in genomic medicine is an essential part of an efficient legal framework.

Another example is Germany, which has a significantly different regulatory structure, and which demonstrates a profound historical distrust regarding abuse of personal data. As an example, the 2019 German Digital Healthcare Act includes severe restrictions regarding the transfer of health data to the commercial sector, and it also has a direct benefit to the patients *via* the use of genomic information as an explicit priority. The given legislative trend focuses on the rights of individuals and privacy, thus stipulating close supervision of data access and use by business organisations [54].

The access and controlling of the data against unfairness is the prerequisite when it comes to ensuring that data is applied fairly in the legal circle. The data-solidarity approach emphasizes the need for regulation of genomic data use, with specific frameworks proposed for different types of data based on the benefits and risks involved. This approach insists on strong enforcement mechanisms against companies that violate data privacy laws, as well as ensuring that profits derived from public healthcare data benefit society at large.

6.3.1 Data Privacy and Security

6.3.1.1 HIPAA: Protecting Health Information in the U.S.

The Health Insurance Portability and Accountability Act (HIPAA) is a U.S. federal law that mandates stringent protections for Protected Health Information (PHI). It includes:

- Privacy Rule: Establishes standards for the protection of health information.
- Security Rule: Sets standards for safeguarding Electronic Protected Health Information (ePHI) through administrative, physical, and technical safeguards.
- Enforcement: Violations can result in civil and criminal penalties, including fines up to $1.5 million per violation.

However, HIPAA's applicability is limited to covered entities and their business associates, potentially leaving gaps in the protection of genomic data, especially when shared with third parties outside the HIPAA framework [55].

6.3.1.2 GDPR: Comprehensive Protection in the EU

The GDPR is a regulation in EU law that strengthens individuals' control over their personal data. Key aspects include:

- Personal Data Definition: Includes genomic data as personal data, subjecting it to GDPR protection.
- Consent: Requires explicit consent for data processing.
- Rights: Grants rights such as access, rectification, erasure, and data portability.
- Data Protection by Design and by Default: Mandates that data protection measures be integrated into processing activities.
- Cross-Border Transfers: Regulates the transfer of personal data outside the EU to ensure adequate protection.

The GDPR's broad applicability and stringent requirements offer robust protection for genomic data within the EU and for entities outside the EU that process data of EU residents [56].

6.3.1.3 Emerging Frameworks and Initiatives

- European Health Data Space (EHDS): Aims to facilitate the use and exchange of electronic health data across the EU, enhancing interoperability and access for research purposes [57].
- U.S. State-Level Regulations: In the absence of comprehensive federal laws, states like California, Nevada, and Washington have enacted their own data privacy laws, such as the California Consumer Privacy Act (CCPA), which provide additional protections for health data [58].

6.3.1.4 Protecting Genomic Data in Research and Clinical Settings

1. Encryption and Data Cloaking

To safeguard genomic data, researchers are employing advanced encryption techniques:

- Genome Cloaking: A method that allows detailed genetic analysis without exposing an individual's complete genetic makeup, preserving privacy while enabling research [59].

- Fully Homomorphic Encryption: Enables computations on encrypted data without decrypting it, ensuring data privacy during processing.

These technologies are crucial as genomic data moves to cloud environments, necessitating robust encryption methods to secure information against unauthorized access [60].

2. Balancing Data Sharing with Privacy
Information sharing has historically been one of the key drivers of science, and it is also a source of valid concerns in privacy. Three major strategies have been put forth to resolve these conflicting imperatives:

1. **Pseudonymization and anonymization:** The methods describe using either removing or encoding personal identifiers to the extent that subjects are not recognisable.
2. **Data-use agreements:** These contracts specify the allowed data-processing activities, and require certain security measures.
3. **Ethical oversight:** Institutional Information Review Boards (IRBs) offer institutional options to medically evaluate and authorize the research plans disconnectedly and secure the information.

Despite the use of the safeguards, there are still a few challenges. De-identified data can and has been re-identified under external sources and the ability to capture the whole expanse of uses of data in the future cannot be secured with the current methods of consent.

6.3.1.5 Balancing Data Sharing with Individual Privacy Rights

The very process of striking a balance between the benefits of sharing data and visiting no harm on the right to privacy of individuals is incredibly complex. It has been observed in modern literature that proper transparency and accountability in the course materials of data will win the confidence of the masses and will, therefore, result in a greater number of people taking part in research activities. Regulatory compliance, such as the HIPAA and GDPR, is also essential, but such considerations do not always absorb the nuances of the privacy of genomic data. As a result, much attention has been paid to technological workarounds that would allow reducing the risk of privacy but still allow transferring data, most prominently cryptography (encryption, as well as enhanced and more advanced types of anonymization). That is why steadfast improvement of the legislative act and cooperation between the jurisdictions is necessary, which remains one of the essential factors in working with the ever-changing world of genomic data privacy [61].

6.3.2 Genetic Discrimination and Anti-Discrimination Laws

Genetic discrimination can be considered as the unequal treatment of people according to genetic information in employment and insurance situations, etc. Genetic discrimination does not allow one to progress in the development of personalized medicine and undermines the entire public health movement by discouraging people to take genetic testing and share the outcomes.

6.3.2.1 Genetic Information Nondiscrimination Act (GINA)

The U.S. Genetic Information Nondiscrimination Act (GINA), passed in 2008, provided a full set of protections against genetic information pretext discrimination in both health insurance coverage and employment. The goal of GINA is to eliminate such obstacles that discourage people from being included in the genetic testing or genetic research by being certain that health insurance or employment opportunities can never be lost as a result of participation.

In spite of its substantive protections, GINA carries significant limitations:

- **Coverage Scope:** It is not applicable to life, disability and long-term care insurance, hence creating loopholes that can be exploited by individuals to discriminate againat them on the basis of their genes.
- **Enforcement Challenges:** Though GINA provides people with rights, most people are not aware of their rights or have not gained the knowledge necessary to present their complaints.
- **Evolving Genetic Technologies:** Growth in technology—both in genetic testing and data analytics—is fast and new applications will probably fall outside the reaches of the current legislation, a reason that calls for periodical amendment/modification of GINA to overcome new avenues to genetic discrimination [62].

6.3.2.2 *Evaluating Effectiveness and Areas for Improvement*

Although the Genetic Information Nondiscrimination Act (GINA) plays an eminent role in reining in and mitigating genetic discrimination in certain spheres, its effectiveness is limited by its partial coverage and the kind of swift innovation that characterizes the advancement in certain genetic innovations in modern times. There is a need to increase the capacity of GINA by doing the following:

- **Increasing Coverage:** Wide protections can be done by providing coverage to life insurance, disability insurance, and long-term care insurance, which would provide an all-encompassing protection against genetic discrimination.
- **Public Awareness Campaigns**: Creation of more awareness among people about genetic discrimination, and the protection offered by GINA would enable a person to claim their rights.
- **Legislative Updates:** Periodic legislative updates are necessary to provide that the mandates under the GINA are up-to-date with any developments in genetic testing and information analytics so that the value and effectiveness of these protections are not compromised [62].

6.3.3 Intellectual Property Rights

6.3.3.1 *Patenting of Genes and Genetic Tests: Societal, and Ethical Implications*

The ethical and legal communities have a long history of debate regarding patenting of genes and genetic tests. Proponents argue that patenting of genes improves innovation by giving inventors exclusive rights over their inventions, which in the end, encourages them to invest in research and development. On the contrary, critics believe that these patents have the potential of stifling research development and reducing access of important medical diagnostic technologies [63].

Ethical Concerns

- **Access to Healthcare**: Patents on genetic tests can lead to monopolies, resulting in high costs for diagnostic services and limited access for patients, particularly in low-resource settings.

- **Research Impediments**: Exclusive rights over genetic sequences can restrict researchers' ability to study genes freely, potentially delaying scientific discoveries and the development of new treatments.
- **Equity Issues**: The concentration of gene patents in wealthier countries and institutions can exacerbate global health disparities, as these entities control access to critical genetic information and technologies [64].

6.3.3.2 Legal Developments

In 2013, the U.S. Supreme Court made a monumental decision that a naturally happening human gene cannot be patented since it is also a product of nature. The decision was hailed as a triumph of both the cause of the common good and good science, and it removed legal obstacles to getting necessary genetic tested performed. However, the ruling also allowed the patents of synthetically constructed genetic material, e.g., complementary DNA (cDNA) a practice that still raises ethical and legal debate [65].

Access to Genomic Information and the Role of Intellectual Property
Intellectual property rights have a dual nature in the sphere of information about genome: they may increase innovation and, at the same time, build a barrier to accessibility.
Driving Innovation
The fact that patents give the inventors an exclusive right to their inventions encourages coming up with new genetic tests and therapies as well as recovery of research and development expenditures.

Creating Barriers to Access
When such patent systems limit the application of diagnostic tests and therapeutic interventions, patent systems—based on the alleged promotion of innovation—can also act as a block upon the appropriate dissemination of genetic information to the general population. This hindrance may occur when the prices of obtaining authorized products are very high or when patentees practice exclusive licensing [66].

Striking a Balance
In dealing with the need of technological innovation and universal access to healthcare simultaneously, three broad strategies are available:

- **Compulsory Licensing:** In instances of need, governments can, without consulting patent holders, license genetic technologies that have been patented, thus accelerating availability of life-saving diagnostics and medicines.
- **Open Access Initiatives:** Open access to genomic data and research outcomes: encouragement of collaboration within scientific society and shortcuts in discovering new treatments will also be able to contribute their duties.
- **Ethical Licensing Agreements:** The unethical effects of patenting a gene can be addressed by encouraging patent owners to enter into negotiated licences aimed at focussing on the importance of providing universal health as well as integrated access [66].

6.4 Social and Societal Implications of Genomic Technologies

In relation to public-private partnerships regarding genomic data, social issues form the most serious fact and they include equity, access and the trust of the people. Since such data involve a responsible control of genomic data, a construct that advances communal good and protects vulnerable groups against unnecessary hazard needs to be established. Social determinants of data donation subscribe to the skeptic opinion of being victimized, as well as trust in government-and-corporate actors [47, 67]. As shown in Figure 6.3, concerns such as genetic discrimination, inadequate awareness, and the need for risk mitigation highlight the social implications of self-regulation by the insurance industry.

The application of genomic data between private actors in special jurisdictions, particularly France and Israel, has produced significant social disharmony on the issue of equitability and openness of such partnerships. Although a respectable percentage of the masses is open to provide their genetic data for the greater good of the people, there is a great sense of fear that the data can be stolen and sold by the business world to gain financial benefits or lead to unfair conclusions. These reservations highlight the fact that there is a great need to have a strict and transparent system of governance whereby the benefits accruing out of genomic research would be shared in the most equitable way possible [40].

The rights-based framework for public-private partnerships emphasizes balancing individual rights with the responsibilities of private companies

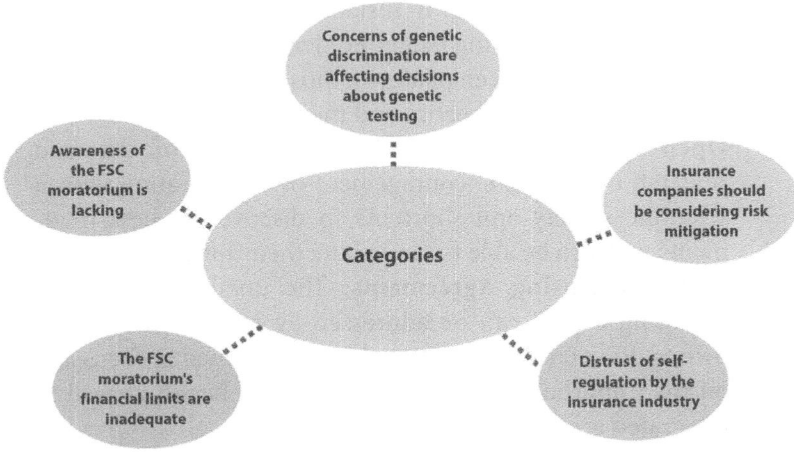

Figure 6.3 Categories of social implications.

using public healthcare data. This model asserts that individuals have essential rights, such as privacy and informational self-determination, while private companies have the right to research and profit. However, access to patient data should be granted only if it directly benefits the public, improving healthcare outcomes.

Private companies must use data solely for health-related research, demonstrate public benefits, and ensure accountability and transparency by publicly registering research protocols and revealing how much data influences pricing. Products developed must be fairly priced and accessible to the public healthcare system. Companies are also accountable for data security and must protect patient privacy.

Moreover, patient participation is critical in data governance, ensuring their interests are represented. Public oversight is necessary to safeguard against exploitation and ensure the data's use serves the public good, especially for marginalized communities. The given framework aims not only at the alignment of profit motives prevalent in the private sector with the public interest but also at the provision of the principles of transparency and equity [68].

6.4.1 Public Perceptions and Attitudes

6.4.1.1 Exploring Public Trust, Concerns, and Anxieties Regarding Genomic Technologies

Genomic technologies require a trust factor among citizens to be successfully integrated into medical practices. Despite the massive potential of genomic-based therapeutic interventions, the long-term skepticism that surrounds the question of privacy, safety of data, and ethical care still emerges. According to empirical research findings, most respondents are still scared of the marginalization that is likely to result from the use of genetic information, especially by employers or insurers. This concern suggests the critical need of constant transparency regulations and all-inclusive regulation systems that protect ethical application of genetic information [69].

6.4.1.2 Addressing Misinformation and Promoting Public Understanding of Genomics

Educating the society with correct information is a prerequisite to not only making them accept genomics but also understand it. The lack of understanding and misunderstanding with regard to genetic testing and gene editing cause anxiety and slow developments. In turn, education-oriented programs, open communications of clinicians and inclusive civic participation are essential to abolishing the mystery of genomics and moving information-based decisions on the ground [70].

6.4.1.3 The Role of Public Engagement and Dialogue in Shaping the Future of Genomic Medicine

Incorporating the input of the general population in discussions about the progress of genomics will continue to be crucial in ensuring that the values and moral needs of society can influence the process of developing this technology and its use as well. Public forums and community consultations, and participatory policymaking are mechanisms that help bridge the gap with genomic research, society, and their interests and concerns, hence progressing further along a more inclusive path of a scientific pursuit [70].

6.4.2 Social and Psychological Impacts

6.4.2.1 The Impact of Genetic Information on Individual Identity, Family Relationships, and Social Interactions

Genetic information has a significant influence on individual identity development and relationships with others. Genetic predisposition to a disease can lead to serious changes in the course of a life, as well as rebalancing the family. Similarly, information about the genetic risk can reconfigure the patterns of social interaction since the identity of people starts being interrelated with the profile of the genetic risk they have [71].

6.4.2.2 *Potential for Stigmatization and Discrimination Based on Genetic Predispositions*

Genetic information has a great clinical value; however, at the same time it has the potential to cause stigmatization. Individuals who carry certain genetic traits can face certain discrimination when it comes to several areas, such as labor force and social interactions. This threat of genetic discrimination hence highlights the need to enact strict ethical guidelines and express legal provisions to curb such abuse of information [72].

6.4.2.3 *Addressing the Psychological Impact of Receiving Genetic Test Results*

The results of genetic testing can have serious psychological consequences, ranging on the one hand to relieved anxiety, and on the other hand to increased distress. The ambiguity of the predictive outcomes, especially in cases when they point to increased vulnerability to serious medical issues, can trigger the development of significant stress. To this end, psychological counseling and alternative support systems cannot be ignored in helping affected individuals to process and adapt to genetic information [73].

6.4.3 Societal Implications

6.4.3.1 *The Potential Impact of Genomics on Healthcare Systems, Healthcare Costs, and Public Health*

Disease prevention, diagnosis, and treatment have a potential to fundamentally advance with the introduction of genomics into the sphere of healthcare. They make it possible to practice personalized medicine, based on rich genetic data profiles; this allows clinicians to administer more effective therapeutics and accomplish better health outcomes. However, with massive implementation of genomic technologies, healthcare costs and requirements are expected to increase significantly and require significant infrastructure and policy changes [74].

6.4.3.2 Exploring the Societal Implications of Genetic Enhancement and Gene Editing Technologies

Gene-edit methods like CRISPR and other genetic enhancement methods offer to implement interventions that can prevent numerous pathologies in the adoption of particular genetically determined disorders and possibly in modifying human characteristics. Despite the huge promise of such developments, they also tend to raise ethical questions about the creation of normative accounts of normality, possible revival of eugenics, and the implications of genome engineering in the broader sense. The debate continues thus to define proper limits of such technologies and to support the argument that there must be strict control by ethical values [75].

6.5 Conclusion

Key Takeaways and Future Directions

With the integration of genomic data in healthcare, the paradigm of precision medicine will be realized, and this presents an opportunity for disease prevention, diagnosis, and treatment like never before. However, such development comes with a vast interaction of ethical, legal as well as social issues that require careful consideration.

Ethically, sensitive genetic information to be managed reinforces apprehension over privacy, informed consent, psychological implications and access. Those who supply their information should have a clear idea of the purposes of its usage and storage, and institutional structures have to be developed to avoid misuse, discrimination, and stigmatization.

Law Regulatory frameworks like the HIPAA in the U.S. and the GDPR in Europe have an interest in controlling the use and safeguard of the genetic information. However, such laws vary in terms of width and efficacy, and new technologies (such as gene editing and artificial intelligence) can still be developed faster than the regulation. Laws like the Genetic Information Nondiscrimination Act (GINA) provide critical protections but still leave gaps, particularly in areas like life and disability insurance.

Socially, public perception, psychological well-being, and societal equity must remain central to policy and practice. Misinformation limits genetic literacy, and varying levels of trust in healthcare systems can influence the adoption and ethical implementation of genomic technologies. Moreover, the potential for inequality remains a pressing concern, whether due to cost, access, or cultural barriers [76].

Future Research Needs and Policy Recommendations

Future research should focus on:

- Longitudinal studies on the social and psychological effects of genetic testing.
- Effective models of dynamic consent and data governance that empower individuals.
- Cross-cultural and international comparisons to harmonize genomic policy frameworks globally.
- Ethical frameworks for emerging areas such as polygenic risk scores, somatic gene editing, and direct-to-consumer testing [77].

Policy recommendations include:

- Updating existing legislation to close regulatory gaps and extend protections to non-covered domains (e.g., life insurance).
- Ensuring equitable access to genomic services across socio-economic and geographic boundaries.
- Integrating genomic literacy initiatives into education systems and public health communication [78].

The Need for Ongoing Dialogue and Collaboration

The responsible and equitable development of genomic medicine hinges on multistakeholder collaboration. Researchers, clinicians, ethicists, policymakers, and members of the public must work together in a transparent and inclusive manner. Continuous public engagement ensures that genomic policies reflect societal values and that innovations align with ethical imperatives. Dialogue must also anticipate future ethical dilemmas posed by technologies like germline editing and predictive genomics [79].

As we advance further into the genomic era, maintaining this collaborative spirit will be crucial not only to realize the benefits of genomics but also to uphold trust, equity, and justice in health and science.

AI disclosure statement: The author acknowledges the use of AI-based tools for minor language editing or grammatical correction. However, all intellectual contributions, analyses, and conclusions presented in this chapter are entirely the author's own.

References

1. Collins, F.S., Patrinos, A., Jordan, E., Chakravarti, A., Gesteland, R., LeRoy Walters, and members of the DOE and NIH planning groups, New goals for the US human genome project: 1998-2003. *Science*, 282, 5389, 682–689, 1998.
2. McCormick, K.A. and Calzone, K.A., The impact of genomics on health outcomes, quality, and safety. *Nurs. Manag.*, 47, 4, 23–26, 2016.
3. Juengst, E.T., Anticipating the ethical, legal, and social implications of human genome research: An ongoing experiment. *Am. J. Med. Genet. Part A*, 185, 11, 3369–3376, 2021.
4. Suntsova, M.V. and Buzdin, A.A., Differences between human and chimpanzee genomes and their implications in gene expression, protein functions and biochemical properties of the two species. *BMC Genomics*, 21, Suppl 7, 535, 2020.
5. Li, J., Xuan, Z., Liu, C., Long non-coding RNAs and complex human diseases. *Int. J. Mol. Sci.*, 14, 9, 18790–18808, 2013.
6. Chimpanzee Sequencing and Analysis Consortium Waterson Robert H. waterston@ gs. washington. edu Lander Eric S. lander@ broad. mit. edu Wilson Richard K. rwilson@ watson. wustl. edu., Initial sequence of the chimpanzee genome and comparison with the human genome. *Nature*, 437, 7055, 69–87, 2005.
7. Collins, F.S. and Mansoura, M.K., The human genome project: revealing the shared inheritance of all humankind. *Cancer: Interdiscip. Int. J. Am. Cancer Soc.*, 91, S1, 221–225, 2001.
8. Khoury, M.J., Bowen, S., David Dotson, W., Drzymalla, E., Green, R.F., Goldstein, R., Kolor, K., Liburd, L.C., Sperling, L.S., Bunnell, R., Health equity in the implementation of genomics and precision medicine: a public health imperative. *Genet. Med.*, 24, 8, 1630–1639, 2022.
9. Aldrighetti, C.M., Niemierko, A., Van Allen, E., Willers, H., Kamran, S.C., Racial and ethnic disparities among participants in precision oncology clinical studies. *JAMA Netw. Open*, 4, 11, e2133205–e2133205, 2021.
10. Abraham, S. and Jethwa, H., Pharmacogenomics: the scientific basis of rational drug development and prescribing. *Clin. Pharmacol. Biopharm.*, 6, 1–5, 2017.
11. FDA Approves First Gene Therapies to Treat Patients with Sickle Cell Disease (FDA Approves First Gene Therapies to Treat Patients with Sickle Cell Disease | FDA), 2023.
12. Qiu, X., Li, H., Ver Steeg, G., Godzik, A., Advances in AI for protein structure prediction: implications for cancer drug discovery and development. *Biomolecules*, 14, 3, 339, 2024.
13. Lancet, The, Genomic newborn screening: current concerns and challenges. *Lancet (London England)*, 402, 10398, 265, 2023.

14. Genomic Data Science, https://www.genome.gov/about-genomics/fact-sheets/Genomic-Data-Science.
15. Pastorino, R., De Vito, C., Migliara, G., Glocker, K., Binenbaum, I., Ricciardi, W., Boccia, S., Benefits and challenges of Big Data in healthcare: an overview of the European initiatives. *Eur. J. Public Health*, 29, Supplement_3, 23–27, 2019.
16. Gobena, S., Admassu, B., Kinde, M.Z., Gessese, A.T., Proteomics and its current application in biomedical area: Concise review. *Sci. World J.*, 2024, 1, 4454744, 2024.
17. Aly, R.M., Current state of stem cell-based therapies: an overview. *Stem Cell Investig.*, 7, 8, 2020.
18. Pattan, V., Kashyap, R., Bansal, V., Candula, N., Koritala, T., Surani, S., Genomics in medicine: a new era in medicine. *World J. Methodol.*, 11, 5, 231, 2021.
19. Wan, Z., Hazel, J.W., Clayton, E.W., Vorobeychik, Y., Kantarcioglu, M., Malin, B.A., Sociotechnical safeguards for genomic data privacy. *Nat. Rev. Genet.*, 23, 7, 429–445, 2022.
20. Rotimi, C.N. and Marshall, P.A., Tailoring the process of informed consent in genetic and genomic research. *Genome Med.*, 2, 1–7, 2010.
21. Johnson, S.B., Slade, I., Giubilini, A., Graham, M., Rethinking the ethical principles of genomic medicine services. *Eur. J. Hum. Genet.*, 28, 2, 147–154, 2020.
22. Roberts, M.C., Kennedy, A.E., Chambers, D.A., Khoury, M.J., The current state of implementation science in genomic medicine: opportunities for improvement. *Genet. Med.*, 19, 8, 858–863, 2017.
23. Atutornu, J., Milne, R., Costa, A., Patch, C., Middleton, A., Towards equitable and trustworthy genomics research. *EBioMedicine*, 76, 2022.
24. Burton, P.R., Hansell, A.L., Fortier, I., Manolio, T.A., Khoury, M.J., Little, J., Elliott, P., Size matters: just how big is BIG? Quantifying realistic sample size requirements for human genome epidemiology. *Int. J. Epidemiol.*, 38, 1, 263–273, 2009.
25. Omidiran, O., Patel, A., Usman, S., Mhatre, I., Abdelhalim, H., DeGroat, W., Narayanan, R., Singh, K., Mendhe, D., Ahmed, Z., GWAS advancements to investigate disease associations and biological mechanisms. *Clin. Transl. Discov.*, 4, 3, e296, 2024.
26. Bagger, F.O., Borgwardt, L., Jespersen, A.S., Hansen, A.R., Bertelsen, B., Kodama, M., Nielsen, F.C., Whole genome sequencing in clinical practice. *BMC Med. Genomics*, 17, 1, 39, 2024.
27. Olorunsogo, T.O., Balogun, O.D., Ayo-Farai, O., Ogundairo, O., Maduka, C.P., Okongwu, C.C., Onwumere, C., Bioinformatics and personalized medicine in the US: A comprehensive review: Scrutinizing the advancements in genomics and their potential to revolutionize healthcare delivery. *World J. Adv. Res. Rev.*, 21, 1, 335–351, 2024.
28. Bell, J., Predicting disease using genomics. *Nature*, 429, 6990, 453–456, 2004.

29. Mumtaz, H., Riaz, M.H., Wajid, H., Saqib, M., Zeeshan, M.H., Khan, S.E., Chauhan, Y.R., Sohail, H., Vohra, L.I., Current challenges and potential solutions to the use of digital health technologies in evidence generation: a narrative review. *Front. Digit. Health*, 5, 1203945, 2023.
30. Syed, R., Eden, R., Makasi, T., Chukwudi, I., Mamudu, A., Kamalpour, M., Geeganage, D.K., *et al.*, Digital health data quality issues: systematic review. *J. Med. Internet Res.*, 25, e42615, 2023.
31. Kendal, E., Ethical, legal and social implications of emerging technology (ELSIET) symposium. *J. Bioeth. Inq.*, 19, 3, 363–370, 2022.
32. Accelerating access to genomics for global health: promotion, implementation, collaboration, and ethical, legal, and social issues, https://www.who.int/publications/i/item/9789240052857, 2022.
33. Rahimzadeh, V., Knoppers, B.M., Bartlett, G., Ethical, legal, and social issues (ELSI) of responsible data sharing involving children in genomics: a systematic literature review of reasons. *AJOB Empir. Bioeth.*, 11, 4, 233–245, 2020.
34. Brothers, K.B. and Rothstein, M.A., Ethical, legal and social implications of incorporating personalized medicine into healthcare. *Pers. Med.*, 12, 1, 43–51, 2015.
35. McWalter, K. and Gaviglio, A., Introduction to the special issue: public health genetics and genomics. *J. Genet. Couns.*, 24, 3, 375–380, 2015.
36. Dolan, D.D., Soo-Jin Lee, S., Cho, M.K., Three decades of ethical, legal, and social implications research: Looking back to chart a path forward. *Cell Genomics*, 2, 7, 2022.
37. Surbone, A., Social and ethical implications of BRCA testing. *Ann. Oncol.*, 22, i60–i66, 2011.
38. Bazzano, L.A., Durant, J., Brantley, P.R., A modern history of informed consent and the role of key information. *Ochsner J.*, 21, 1, 81–85, 2021.
39. Shah, P., Thornton, I., Kopitnik, N.L., Hipskind, J.E., Informed Consent, in: *StatPearls*, StatPearls Publishing, Treasure Island (FL), November 24, 2024.
40. Horn, R. and Merchant, J., Ethical and social implications of public–private partnerships in the context of genomic/big health data collection. *Eur. J. Hum. Genet.*, 32, 6, 736–741, 2024.
41. Pallocci, M., Treglia, M., Passalacqua, P., Tittarelli, R., Zanovello, C., De Luca, L., Caparrelli, V., *et al.*, Informed consent: Legal Obligation or cornerstone of the Care Relationship? *Int. J. Environ. Res. Public Health*, 20, 3, 2118, 2023.
42. Smolenski, J., The foundations of informed consent and bodily self-sovereignty: a positive suggestion. *Monash Bioeth. Rev.*, 42, 1, 115–136, 2024.
43. Fact Sheet: Genetic Information Nondiscrimination Act, https://www.eeoc.gov/laws/guidance/fact-sheet-genetic-information-nondiscrimination-act, 2014.
44. Clayton, E.W., Evans, B.J., Hazel, J.W., Rothstein, M.A., The law of genetic privacy: applications, implications, and limitations. *J. Law Biosci.*, 6, 1, 1–36, 2019.

45. Sewsankar, G., The Extent to which Conflict of Interest in the Public Sector Affects Private Sector Functioning.
46. Genome UK: 2022 to 2025 Implementation plan for England, https://assets.publishing.service.gov.uk/media/6398387e8fa8f55300899840/genome-uk-2022-to-2025-implementation-plan-for-england.pdf, 2025.
47. Horn, R. and Kerasidou, A., Sharing whilst caring: solidarity and public trust in a data-driven healthcare system. *BMC Med. Ethics*, 21, 1–7, 2020.
48. Oliva, A., Kaphle, A., Reguant, R., Sng, L.M.F., Twine, N.A., Malakar, Y., Wickramarachchi, A., *et al.*, Future-proofing genomic data and consent management: a comprehensive review of technology innovations. *GigaScience*, 13, giae021, 2024.
49. Mandal, J., Acharya, S., Parija, S.C., Ethics in human research. *Trop. Parasitol.*, 1, 1, 2–3, 2011.
50. The Tuskegee Syphilis Study, https://onlineethics.org/cases/tuskegee-syphilis-study, 2000.
51. Ethical Guidelines for Biomedical Research on Human Subjects, https://www.icmr.gov.in/icmrobject/uploads/Guidelines/1724914501_ethical_guidelines_for_biomedical_research_on_human_subject_2000.pdf, 2000.
52. Özgüç, M., Genetic testing: predictive value of genotyping for diagnosis and management of disease. *EPMA J.*, 2, 173–179, 2011.
53. Shabani, M., Dyke, S.O.M., Marelli, L., Borry, P., Variant data sharing by clinical laboratories through public databases: consent, privacy and further contact for research policies. *Genet. Med.*, 21, 5, 1031–1037, 2019.
54. Digital Health Laws and Regulations India, https://iclg.com/practice-areas/digital-health-laws-and-regulations/india, 2025.
55. Health Insurance Portability and Accountability Act of 1996 (HIPAA), https://www.cdc.gov/phlp/php/resources/health-insurance-portability-and-accountability-act-of-1996-hipaa.html#:~:text=The%20Health%20Insurance%20Portability%20and%20Accountability%20Act%20%28HIPAA%29,the%20HIPAA%20Privacy%20Rule%20to%20implement%20HIPAA%20requirements, 2024.
56. What is GDPR, the EU's new data protection law?, https://gdpr.eu/what-is-gdpr/, 2018.
57. European Health Data Space Regulation (EHDS), https://health.ec.europa.eu/ehealth-digital-health-and-care/european-health-data-space-regulation-ehds_en.
58. US data privacy laws by state: rights and requirements, https://usercentrics.com/knowledge-hub/us-data-privacy-laws-by-state/#:~:text=We%20compare%20what%20US%20state-level%20data%20privacy%20laws,-General%20Data%20Protection%20Regulation%20%28GDPR%29%20came%20into%20force, 2025.
59. Genome Cloaking Preserves Privacy While Enabling Diagnosis, https://dnascience.plos.org/2017/08/17/genome-cloaking-preserves-privacy-while-enabling-diagnosis/, 2017.

60. Namazi, M., Farahpoor, M., Ayday, E., Pérez-González, F., Privacy-Preserving Framework for Genomic Computations *via* Multi-Key Homomorphic Encryption. *Bioinformatics*, btae754, 2025.
61. Yanamala, A.K.Y., Suryadevara, S., Kalli, V.D.R., Balancing innovation and privacy: The intersection of data protection and artificial intelligence. *Int. J. Mach. Learn. Res. Cybersecur. Artif. Intell.*, 15, 1, 1–43, 2024.
62. Suter, S.M., GINA at 10 years: the battle over 'genetic information'continues in court. *J. Law Biosci.*, 5, 3, 495–526, 2018.
63. Gene patents are a hindrance to innovation, https://www.theguardian.com/commentisfree/2013/apr/24/gene-patents-scientific-research-innovation?utm_source=chatgpt.com.
64. The Ethics of Patenting Genetic Sequences: Balancing Innovation and Access, https://ucipclj.org/2025/01/29/the-ethics-of-patenting-genetic-sequences-balancing-innovation-and-access/?utm_source=chatgpt.com, 2025.
65. Kesselheim, A.S., Cook-Deegan, R.M., Winickoff, D.E., Mello, M.M., Gene patenting—the Supreme Court finally speaks. *N. Engl. J. Med.*, 369, 9, 869–875, 2013.
66. Biotechnology and Genetic Engineering: Ethical and Legal Considerations in India, https://www.legalserviceindia.com/legal/article-17190-biotechnology-and-genetic-engineering-ethical-and-legal-considerations-in-india.html.
67. Kerasidou, A. and Kerasidou, C., Data-driven research and healthcare: public trust, data governance and the NHS. *BMC Med. Ethics*, 24, 1, 51, 2023.
68. Winkler, E.C., Jungkunz, M., Thorogood, A., Lotz, V., Schickhardt, C., Patient data for commercial companies? An ethical framework for sharing patients' data with for-profit companies for research. *J. Med. Ethics*, 51, 5, 2025.
69. Stark, Z., Dolman, L., Manolio, T.A., Ozenberger, B., Hill, S.L., Caulfied, M.J., Levy, Y., *et al.*, Integrating genomics into healthcare: a global responsibility. *Am. J. Hum. Genet.*, 104, 1, 13–20, 2019.
70. Barbosa, S., Toe, L.P., Thizy, D., Vaz, M., Carter, L., Engagement and social acceptance in genome editing for human benefit: Reflections on research and practice in a global context. *Wellcome Open Res.*, 5, 244, 2021.
71. Crocetti, E., Albarello, F., Meeus, W., Rubini, M. Identities: A developmental social-psychological perspective. *Eur. Rev. Soc. Psychol.*, 34, 1, 161–201, 2023.
72. Andrews, L., Fullarton, J., Holtzman, N.A., Motulsky, A.G., Social, legal, and ethical implications of genetic testing, in: *Assessing Genetic Risks: Implications for health and social policy*, 1994.
73. Wade, C.H., What is the psychosocial impact of providing genetic and genomic health information to individuals? An overview of systematic reviews. *Hastings Cent. Rep.*, 49, S88–S96, 2019.
74. Bombard, Y., Ginsburg, G.S., Sturm, A.C., Zhou, A.Y., Lemke, A.A., Digital health-enabled genomics: opportunities and challenges. *Am. J. Hum. Genet.*, 109, 7, 1190–1198, 2022.

75. Pacesa, M., Pelea, O., Jinek, M., Past, present, and future of CRISPR genome editing technologies. *Cell*, 187, 5, 1076–1100, 2024.
76. Hardcastle, F., Lyle, K., Horton, R., Samuel, G., Weller, S., Ballard, L., Thompson, R., Trindade, L.V.P., Gómez Urrego, J.D., Kochin, D., The ethical challenges of diversifying genomic data: A qualitative evidence synthesis. *Camb. Prism. Precis. Med.*, 2, e15, 2023.
77. White, R.T. and Arzi, H.J., Longitudinal studies: Designs, validity, practicality, and value. *Res. Sci. Educ.*, 35, 137–149, 2005.
78. Zhang, J.Y., Commoning genomic solidarity to improve global health equality. *Cell Genomics*, 3, 10, 2023.
79. Ambrosino, E., Abou Tayoun, A.N., Abramowicz, M., Zilfalil, B.A., Boughtwood, T., Hamdi, Y., Hubbard, T., *et al.*, The WHO genomics program of work for equitable implementation of human genomics for global health. *Nat. Med.*, 30, 10, 2711–2713, 2024.
80. Muto, K., Nagai, A., Ri, I., Takashima, K., Yoshida, S., Is legislation to prevent genetic discrimination necessary in Japan? An overview of the current policies and public attitudes. *J. Hum. Genet.*, 68, 567–574, 2023.

7

The Role of Genomics in Rare and Noncommunicable Diseases

S.K. Abdul Rahaman[1]*, Aman Sisodia[2], Ananda Kumar C.H.[3], Padmavathi Sakinala[4] and Ajay Bommareddy[5]

[1]*School of Pharmaceutical Sciences, Vignan University, Guntur, Andhra Pradesh, India*
[2]*School of Medical & Allied Sciences, Department of Pharmacy, Greater Noida, Uttar Pradesh, India*
[3]*Department of Pharmaceutical Technology, School of Health and Medical Sciences, Adamas Knowledge City, Adamas University, Kolkata, West Bengal, India*
[4]*Department of Pharmaceutical Technology, Nirmala College of Pharmacy, Mangalagiri, Guntur, Andhra Pradesh, India*
[5]*Biomedical Science Department, Schmidt College of Medicine, Florida Atlantic University, Boca Raton, Florida, USA*

Abstract
Noncommunicable diseases are the main cause of the death and disease burden around the globe. Cardiovascular diseases make up the largest number of deaths globally caused by NCD, followed by different forms of cancer, COPD, and diabetes. Precision medicine is the latest trend in current medicine. The concept of precision medicine has emerged recently as a theme in the prevention, diagnosis, and therapy of illness according to the patient's genotype in many human genetic diseases, thanks to new genetic sequencing methods. However, in fact, it is relevant to noncommunicable diseases and has been implemented specifically in cancer treatment. A large number of studies done both in South Africa and outside the country have demonstrated how genetic testing can reduce the number of individuals coming into a particular risk category and identify those who require different treatments. Single-factor genetic risk, such as heterozygous familial hypercholesterolemia (FH), serious though it is, does not cause a heart attack directly. This development was made possible by the first use of PCR technology in

**Corresponding author*: rahamandr123@gmail.com

Swetha Kumari Koduru, Sarad Pawar Naik Bukke, Hussaini Bello, Muhamad Mustafa Mundu and Ayush Madan (eds.) Genomics and Precision Healthcare: Innovations, Applications and Challenges, (201–232) © 2026 Scrivener Publishing LLC

South Africa, as employed for the testing of germline DNA. For rare diseases whose many symptoms are genetically determined, genomics has dramatically increased the accuracy of diagnosis by discovering harmful mutations, even when standard approaches toward diagnosis seem to fail. Genomic testing, including whole-exome sequencing and whole-genome sequencing, would likely turn out to be one of the most essential techniques to determine the genetic origin of rare diseases. Often, it leads to new disease mechanisms and potential treatment targets. Besides, genomics brings into existence gene therapies and precision medicine, which permit the possibility of prescription treatments based on an individual's genetic profile. Genomics is critical to the improvement of understanding the genetic basis of NCDs and rare diseases, opening new avenues for early detection, risk prediction, and tailored treatment techniques that may have the potential to enhance patient care and outcomes.

***Keywords*:** Noncommunicable diseases, genomics, rare diseases, next-generation sequencing, personalized medicine, targeted therapies, diagnostic accuracy

7.1 Introduction

Rare diseases, often referred to as orphan diseases, affect only a small portion of the population. By definition, an illness typically qualifies as rare if it affects fewer than one in every 2,000 people. Most of these diseases tend to be genetic and often congenital or inherited due to mutations in one or more genes. Rare disorders include conditions such as Huntington's disease, cystic fibrosis, and Duchenne muscular dystrophy. While these diseases are rare individually, they affect millions of people around the world. There are an estimated 7,000 uncommon diseases, with around 400 million individuals globally living with one of them [1].

Communicable diseases, however, are those that are transferred from one individual to another through direct contact, air, water, or vectors like mosquitoes. Infectious diseases include HIV/AIDS, TB, malaria, influenza, and most recently, COVID-19. Despite the creation of vaccinations and antibiotics, communicable illnesses continue to play a great role in world health, especially in developing countries. The World Health Organization states that communicable diseases result in more than ten million deaths every year [2].

Although rare diseases are a low occurrence by themselves, they pose a significant obstacle to global health care systems. There are over 7,000 rare diseases reported by the NIH, with genetics accounting for approximately 80% of them. An estimated 25-30 million people are affected by an uncommon disease in the United States, which is approximately 10% of the population. Rare diseases remain undiagnosed for quite a long period because of the complexity of diseases and the poor awareness of them among health providers. In addition, most are chronic, progressive, and incapacitating, incurring a very high cost on healthcare and deteriorating the quality of life of the patient [3].

Communicable diseases remain one of the biggest burdens in the world of health. The World Health Organization reveals that infectious diseases cause a considerable fraction of the total global mortality and morbidity, mostly in low- and middle-income countries. Each year, outbreaks such as tuberculosis, malaria, and HIV/AIDS take millions of lives. Moreover, the emergence of antibiotic resistance and the rapid spread of emerging infectious diseases, such as COVID-19, underscore the continued threat that communicable diseases pose to global public health [4].

While both rare and communicable diseases affect millions of people worldwide, they differ in several significant ways. Rare diseases are defined as having a low prevalence and being chronic or hereditary, whereas communicable diseases are contagious and transmit from person to person. The fundamental problem in uncommon diseases is diagnosis and therapy, as many are poorly understood, and there are few resources committed to their research. Communicable diseases, though more common, are clearer in the transmission pathway and treatment as there are vaccinations and antiviral medicines to address the situation [5].

Rare diseases are often best treated with specialized, individualized care, whereas infectious diseases can usually be prevented or cured through public health measures including immunization and antibiotic therapy. However, these differences notwithstanding, both types of disease would benefit from genomic advances, which are crucial for the comprehension of their genetic and molecular nature [6].

7.2 Need for Genomics in Disease Understanding

Historically, diseases have been understood and treated in terms of their symptoms and observable progression patterns. For centuries, many diseases were treated primarily empirically, through trial and error. With the advent of modern science, the emphasis shifted to understanding the basic processes of disease. However, our ability to examine the genetic and molecular bases of disease has hindered much of this progress.

Medicine realized its watershed moment at the beginning of the 1950s when DNA was discovered. Yet, it was not until the Human Genome Project concluded in 2003 that genomics was recognized to have full potential as a tool in the understanding of disease. This enormous success has facilitated the ability to sequence and analyze the entire human genome, thus opening up an unprecedented opportunity for investigators to explore the origins of genetic illness, whether rare or communicable [7].

7.2.1 Role of Genomics in Medicine

Genomics, or the study of the genome, is at the forefront of modern medicine. Genome sequencing and analysis are used to better understand the genetic variables that contribute to illness development. The advent of next-generation sequencing (NGS) technologies has made it possible to sequence complete genomes in a matter of days for a fraction of the cost of traditional approaches. Genomics is essential for identifying genetic defects responsible for uncommon diseases. For infectious diseases, genomics helps understand how viruses evolve, interact with human hosts, and can be controlled or destroyed. For noncommunicable diseases (NCDs), it helps understand genetic risk factors and biochemical pathways that will ensure more precise prediction of risk and the modes of diagnosis and treatment [8].

Genomics has completely changed the face of rare and infectious diseases. Genomics can identify genetic defects responsible for rare diseases even when conventional diagnostic methods fail. It often gives earlier and more accurate diagnoses, which may eventually result in better patient outcomes. For example, WES and WGS have enabled scientists to identify the genetic causes of disorders that could not be diagnosed or were often misdiagnosed, such as Duchenne muscular dystrophy and cystic fibrosis. Genomics, through communicable diseases, is a feature that identifies the mutations of the pathogens to perhaps understand the pathways of diseases, how they transmit, and changes in disease pathways. This feature of genomics, especially useful, is for new infectious diseases; for instance, COVID-19, for tracking mutations in genomics, enabling better understanding about the virus-transmission dynamics [9].

7.2.2 Genomics in Precision Medicine

Precision medicine, otherwise known as personalized medicine, is a new approach in medical treatment which considers the genetic, lifestyle, and environmental characteristics of an individual. Unlike the "one-size-fits-all" approach, precision medicine looks to tailor treatments for each patient according to their unique genetic profile. Healthcare providers may use genetic information to decide what medicines will be best, how a patient may react to specific medications, and what side effects may be associated [10].

Genomics is the heart of precision medicine. Genome sequencing allows healthcare providers to identify specific genetic mutations or variants that could affect illness risk, progression, and treatment response. This allows for more targeted medicines, thereby reducing side effects and improving the outcome of treatments. For example, in the case of cancer, genomic profiling of tumors identifies specific mutations driving cancer growth and allows for the deployment of targeted medicines that block those mutations. In rare diseases, genomics allows for the identification of the genetic causes and can lead to earlier diagnosis and personalized treatment strategies. Gene therapies are the future treatments for specific genetic mutations of rare genetic diseases [11].

This intersection of genetics, customized medicine, and individualized treatments represents the critical dependency of the future of healthcare. Moving forward, with the advancement of genetic technologies, practitioners will be able to move away from broad treatments and towards more individualized, effective interventions. This may potentially lead to improved patient outcomes while at the same time reducing healthcare costs through avoidance of useless therapies and limitation of unwanted responses. Genomics gives the ability to understand the genetic nature of disease better, whether rare or communicable, and can thus enable the development of treatments tailored to the individual. The future of medicine depends on using genomes to design treatment techniques tailored to the patient that can potentially revolutionize the care of the patient [12].

7.3 Genomic Technologies and their Applications

Genetic testing has changed how the genetic cause of disease is comprehended, more so on rare and transmitted disorders. The tests allow doctors and researchers to determine DNA mutations and variations that cause illness. Among the simplest techniques of genetic testing include PCR, WES, and WGS. Among the most frequently used methods in amplifying sections of DNA, the Polymerase Chain Reaction or PCR stands atop. PCR can search for specific mutations or genetic variants in an individual. It is most commonly used for targeted genetic testing, such as identifying specific mutations known to be associated with hereditary diseases, like cystic fibrosis or Huntington's disease. PCR is one of the most important tools in genetic diagnostics because it is rapid, inexpensive, and highly sensitive [13].

Whole-exome sequencing (WES) is a type of sequencing technology targeting the exonic regions of the genome, which comprise about 1% of the whole genome but carry most of the known disease-causing mutations. It is an appropriate technique to use when conventional genetic tests cannot come up with any conclusion on the genetic variant suspected to cause a rare disorder. WES provides the analysis of all protein-coding genes in one single assay, and thus it is highly useful in the detection of rare genetic disorders of unknown etiology. Whole-genome sequencing: This method includes the whole genome sequencing encompassing exonic as well as non-exonic regions. This technique is highly comprehensive and can uncover all the genomic variants in regulatory as well as non-coding regions that might be involved in the disease process. WGS is costlier than WES, yet it provides a more detailed investigation of rare genetic diseases and complex disorders, besides a better appreciation of genetic contribution. WGS is particularly important in the detection of new mutations, structural variation, and intricate genetic interactions. The advent of Next-Generation Sequencing technologies has significantly moved genetic testing closer to being even more efficient as well as being cost-effective. NGS platforms can sequence millions or billions of DNA fragments in parallel, therefore allowing for a very rapid as well as extended investigation of the whole genome or exome [14].

Illumina's sequencing platforms of the sort that take on the form of the HiSeq and NovaSeq have emerged to be the new gold standard of genomic research as well as diagnostics. Illumina sequencing processes take the underlying principle of sequencing by synthesis, in which fluorescently labeled nucleotides are being integrated into a developing DNA strand, allowing bases to be detected in real time. These technologies are known for high throughput and good precision and have proven themselves to be good choices for large-scale genomic research, clinical diagnostics, and whole-genome sequencing. The sequencing technology of PacBio, also known as Pacific Biosciences, is based on Single Molecule Real-Time (SMRT). This technology is used for long-read sequencing. Unlike Illumina, which works only on very small DNA fragments, PacBio can sequence huge sections of DNA in one read. The capacity to produce long reads is particularly crucial for sequencing complicated genomes, structural changes, and finding significant deletions or duplications that short-read platforms may miss. The long-read sequencing technique developed by PacBio became a key tool not only in rare disease research but also in genetic therapy [15].

Oxford Nanopore: Oxford Nanopore technologies have a different methodology called nanopore sequencing. It involves DNA strands being forced through tiny holes of proteins, where the ionic current is measured as each nucleotide passes through it. Current modifications can be used to detect individual bases. It is unique since nanopore sequencing can be performed in real time and sequences lengthy reads directly in a portable format. Important advantages include its portability, speed, and the ability to sequence whole genomes without amplification. These sequencing techniques have immense applications in the identification and treatment of diseases. They have become critical tools for understanding the genetic basis of rare diseases, thus allowing for more timely and precise diagnostics. For instance, WES and WGS have enabled researchers to detect previously unknown mutations causing rare disorders such as Retinitis Pigmentosa and Marfan Syndrome. In communicable diseases, NGS technology enable quick pathogen detection and monitoring. In the cases of emerging infectious diseases such as COVID-19, NGS technology has allowed the sequencing of viral genomes revealing information about the virus transmission rates, mutation rate, and potential therapeutic targets. Also, genomic surveillance has become integral in tracking evolution of pathogens as well as to find new versions, which would help public health interventions [16].

7.4 Tools and Methods for Genomic Analysis

Genomic analysis heavily depends on bioinformatics tools to help researchers manage, display, and interpret large amounts of data generated by sequencing technology. Genomic data is analyzed using a wide range of specialist software and internet applications.

Genome Browsers: Genome browsers, including UCSC Genome Browser and Ensembl, are most commonly used in genomics. Such technologies enable scientists to visualize locations of genes, regulatory elements, and variants within the genome. Genome browsers present users with an interactive interface that allows for navigation through genomic data, which would otherwise be complex to determine where changes have been made and their potential impact [17].

Databases: Countless databases of genetic variants and the related disease have been established for storing. In the case of dbSNP, it includes a large SNPs database based on information describing genetic diversity amongst populations. Alternatively, ClinVar establishes a link from genetic variants towards their clinical implication so that a doctor can be able to gain better insight as to the impacts of individual mutation in terms of disease risk. There are other major databases, such as OMIM for genetic illnesses, which catalog them and their molecular processes, and gnomAD, which can be used in the investigation of human genetic diversity [17].

CRISPR-Cas9, a revolutionary technique in gene editing, has drastically changed genomic research and holds high promise for medicinal applications. It is a technique that uses guide RNA to direct the Cas9 enzyme to specific DNA sequences, making it possible to make precise genome changes. With rare diseases, CRISPR-Cas9 can correct genetic mutations at the DNA level, thus providing permanent treatment for some hereditary abnormalities. Such disease can be caused by single gene anomalies, of which CRISPR technology potentially may treat two diseases: sickle cell anemia and beta-thalassemia. Successful modification of such genes that result in these patients' illnesses from patient-derived stem cells resulted in the generation of healthy red blood cells. Moreover, CRISPR is currently showing promise towards the cure for disorders like Duchenne muscular dystrophy and many other disorders [18].

7.5 Technological Advancements in Genomics

In two decades, genomics has leaped in enormous strides. At the beginning of the 2000s, the Human Genome Project was declared complete; still, this could have been seen only as a threshold moment for realizing the understanding about the human genome because it used to be cost- and time-consuming sequencing at that time that was not reachable to everybody; since then, NGS technologies have greatly lowered both time and cost. During this evolution, the early NGS platforms, such as the Illumina Genome Analyzer, transformed into high-throughput systems that were economical, and they could sequence whole genomes in a matter of days. The inclusion of technologies such as PacBio and Oxford Nanopore in long-read sequencing has improved the accuracy and completeness of sequencing genomes, mainly in complex areas [19].

7.5.1 The Impact of Automation and Computational Tools on Large-Scale Genomic Studies

Automation and computational technology have made large-scale genomics research significantly possible as shown in Table 7.1. The simplified workflow from extraction of DNA through library preparation, and

Table 7.1 Genomic technologies and their applications.

Technology	Description	Applications	Advantages	Limitations
PCR (Polymerase Chain Reaction)	Amplifies specific DNA sequences.	Diagnostics, genotyping, and mutation detection.	High specificity, rapid results.	Requires prior knowledge of target DNA.
Whole-Exome Sequencing (WES)	Targets and sequences protein-coding regions of the genome.	Identification of mutations in rare diseases and cancers.	Cost-effective for targeted analysis.	Misses non-coding regulatory mutations.
Whole-Genome Sequencing (WGS)	Sequences the entire genome.	Comprehensive analysis of genetic variants, structural variations, and regulatory elements.	Provides complete genomic data.	High cost and complexity in data analysis.
Next-Generation Sequencing (NGS)	High-throughput sequencing of DNA or RNA.	Disease diagnosis, pathogen sequencing, and biomarker discovery.	Scalability and rapid processing.	Requires expertise in data interpretation.
CRISPR-Cas9	Gene-editing tool for precise DNA modifications.	Gene therapy, functional genomics, and correcting genetic disorders.	High precision and efficiency.	Ethical concerns and off-target effects.
Bioinformatics Tools	Software for genomic data analysis (e.g., ClinVar, dbSNP).	Data interpretation, mutation annotation, and biomarker discovery.	Facilitates large-scale analysis.	Dependency on computational resources.

eventually the sequencing, streamlined by automated machinery, improved speed with decreased errors from human beings. Computer analytic and visualization software, underlined by artificial intelligence and machine learning, also enable these researchers to address the gigantic magnitude of data resulting from genetic investigations. These techniques make it easier to identify genetic differences, analyze gene expression, and predict disease risks, thereby considerably enhancing precision medicine [20].

AI and machine learning are increasingly being used in genetic research to analyse complex datasets, discover hidden patterns, and forecast results. In genomics, AI-powered techniques are used to anticipate the functional consequences of genetic variants, identify prospective therapeutic targets, and increase diagnostic accuracy. For example, AI systems can scan massive volumes of sequencing data to identify mutations related with cancer, uncommon disorders, and infectious diseases, resulting in speedier diagnoses and more precise treatment recommendations. As AI and ML technologies advance, their integration with genomics is anticipated to expedite discovery and increase therapy tailoring [21].

7.6 Genomics in Rare Diseases

Rare diseases are also sometimes referred to as orphan diseases because they affect only a small population. The definition ranges so widely that in the United States, the definition is one that affects fewer than 200,000 people, while in the European Union, it would be one in every 2,000 people. Whether common or rare, some diseases have an enormous cumulative effect. There are more than 7,000 rare diseases identified that affect over 300 million people worldwide. Rare diseases tend to have a genetic basis. From the identified diseases, about 80% of them are gene-based. However, they range from well-known conditions such as cystic fibrosis and Duchenne muscular dystrophy to highly isolated conditions such as Progeria or Fibrodysplasia Ossificans Progressiva [22].

Diagnosis of rare diseases is considered a "diagnostic odyssey." These conditions are characterized by complexity, heterogeneity, and similarity with more common disorders. Their clinical features may vary considerably among affected individuals or be confusingly similar to those of other conditions. Ehlers-Danlos syndrome is one such disorder: it is a condition affecting connective tissue, with a clinical presentation similar to that of widespread hypermobility or even arthritis. Many patients experience diagnostic delays of several years. Using global data, a rare disease patient typically gets an appropriate diagnosis after 5-8 years. In this period, the patients may experience unnecessary or invasive testing, misdiagnosis, and inappropriate therapies that worsen their disease [23].

7.7 Importance of Early Diagnosis

Patients suffering from uncommon diseases require an early and correct diagnosis. Proper diagnosis allows for timely management, avoids illness progression, and enhances quality of life. For example, in Spinal Muscular Atrophy (SMA), early detection can lead to the injection of gene therapy such as Zolgensma, which greatly improves motor function and survival rates. Furthermore, early diagnosis enables families to make more informed reproductive planning decisions and better allocate resources for individualized care [24].

Genomics has completely changed the landscape of diagnosing rare diseases. The conventional procedures for diagnosis are based on clinical observations and biochemical tests, which most of the time do not identify the root cause. Genomic technologies such as Whole-Exome Sequencing (WES) and Whole-Genome Sequencing (WGS) have opened up a new horizon in which a patient's genetic code can be directly analysed. This might point towards mutations in the coding regions of the genome, where most of the changes that lead to diseases are seen. WES helped pinpoint the mutations in the gene PTPN11 associated with Noonan Syndrome. WGS, on the other hand, gives an overall glimpse since it reviews both coding as well as non-coding regions to possibly document structural variants, intronic mutations, and novel regulatory factors generating disease [25].

Duchenne Muscular Dystrophy is a rare genetic disorder caused by mutations in the DMD gene, encoding for dystrophin, which is essential for the integrity of the muscle. The genomic sequencing now allows one to identify the deletion, duplication, and point mutation in the DMD gene, hence allowing better diagnosis and intervention like exon-skipping therapy. The CFTR gene mutations cause cystic fibrosis. Identification of the F508del mutation, one of the most common, has helped in early diagnosis and the development of targeted medicines like Trikafta, which targets precisely the malfunctioning of CFTR. Huntington's Disease is a progressive neurological disorder caused by an expansion of CAG trinucleotide repeats in the HTT gene. Genomic testing has been important in confirming diagnoses and providing prognostic information about the onset and progression of the disease [26, 27].

7.7.1 The Role of Genomics in Identifying Mutations

WES and WGS are genomic diagnostics that can identify mutations that older methods cannot. For instance:

Copy Number Variants (CNVs) are large structural changes in the genome that may disrupt gene function. WGS is particularly useful in identifying CNVs in disorders such as Charcot-Marie-Tooth disease [28].

De novo mutations: WGS can detect the spontaneous mutations occurring in an individual, not inherited from parents, for example, mutations causing Rett syndrome [29].

Mitochondrial Diseases: Leigh syndrome, which is caused by mutations that disrupt cellular energy generation, can be diagnosed using specialized sequencing techniques that focus on mitochondrial DNA (mtDNA) [30].

7.7.2 Genomics Technologies in Treatment of Rare Diseases

Technologies include genetic therapies including gene therapy, gene editing, and RNA-based drugs, showing a lot of promise in managing rare disorders: such treatments work by correcting or compensating for the underlying defect, thus, hopefully curing otherwise untreatable illnesses.

Gene Therapy: It involves the delivery of a functional copy of an abnormal gene so that its original or normal function can be restored. Luxturna, an FDA-approved gene therapy, cures Leber congenital amaurosis, a rare inherited condition causing blindness due to abnormalities in the RPE65 gene [31].

Gene Editing: In the case of CRISPR-Cas9, it means that faulty genes can be targeted and edited exactly. CRISPR has been employed to modify the hematopoietic stem cells in order to treat sickle cell anemia by inducing fetal hemoglobin creation [32].

RNA-Based Therapies: ASOs are being employed to reprogram gene expression in rare diseases. The ASO drug, Spinraza, increases the expression of functional SMN protein to treat SMA [33].

Spinal Muscular Atrophy (SMA): DNA mutations lead to progressive weakening of muscles. The gene therapy Zolgensma transfers a healthy copy of the SMN1 gene into affected newborns, significantly accelerating their motor milestones [34].

Hemophilia A: Delivery of F8 gene by AAV-based gene therapy restored the production of clotting factors in patients suffering from severe Hemophilia A and decreased significantly the number of bleeding episodes [35].

Amyotrophic Lateral Sclerosis (ALS): Genetic studies have identified mutations in the SOD1 gene as the cause of familial ALS. ASO treatments targeting SOD1 mRNA are now in clinical trials and appear to slow disease progression [36].

7.8 Discovery of New Disease Mechanisms

This area has opened a new horizon to further study the pathophysiology of rare diseases. Researchers have established new genetic pathways and causes behind diseases by exploring genomes. For instance, in neurodevelopmental diseases, researches revealed mutations in important genes in the synaptic function process of the brain to understand its formation and connectivity processes.

Genomic research identifies new biomarkers and potential therapeutic targets. Genomic research has also led to the identification of novel biomarkers that can help with diagnosis and treatment. Biomarkers such as particular mutations, gene expression profiles, and epigenetic alterations have become essential tools in rare illness research [37].

Biomarkers for Early Diagnosis: In Gaucher's disease, glucosyl sphingosine levels serve as a biomarker [37].

Figure 7.1 illustrates the role of genomics in oncology, showing how genetic profiling and patient stratification enable precision cancer therapy compared to conventional treatment approaches.

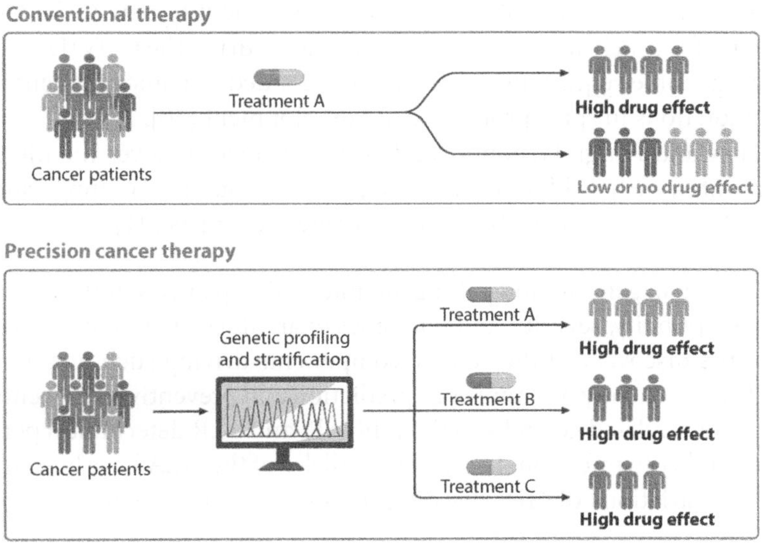

Figure 7.1 Genomics in oncology.

Therapeutic Targets: Genomics has identified druggable targets in diseases like Fabry disease, so that enzyme replacement drugs can be tailored to specific mutations [37].

7.8.1 Genomics in Communicable Diseases

Genomics has brought about a revolution in the field of study and management of communicable diseases by providing unprecedented insights into biology of pathogens, host-pathogen interactions, and processes of disease transmission. Techniques of surveillance, diagnosis, and treatment are designed to be targeted both to the pathogen and the host, exploiting advances in genetical technologies [38].

7.8.1.1 Genomic Research in Infectious Diseases

Genomic research has significantly improved our knowledge of infections by bacteria, viruses, fungi, and parasites. Genomic sequencing enables the identification of the genetic structure of the pathogenic process involved in infection processes, mechanisms of evolution, and the development of resistance.

Bacteria: Genomics have enabled the identification of bacterial genome structures, development of virulence factors, and drug resistance mechanisms. One example includes the detection of mutations responsible for M. TB resistance to medication through genomic studies [39].
Viruses: Viral genomics allows one to quickly identify new variants thus allowing for timely intervention. For instance, during the COVID-19 pandemic, genome sequencing of SARS-CoV-2 aided scientists in identifying vital mutations in spike proteins to aid development [40].
Parasites and Fungi: Genomic studies have helped uncover the life cycle of parasites, such as Plasmodium falciparum (malaria), and fungi, such as Candida auris, thus identifying new therapeutic targets [41].

The human genome and pathogens' interaction plays a significant role in the susceptibility, severity, and outcomes of an illness. Genomics has facilitated the discovery of the genetic components driving such interactions, which subsequently has improved prediction and preventive interventions.

Immune-related genes like HLA, IFNAR, and TLR determine a person's susceptibility to infection. For instance, alleles of the HLA have been implicated in variations of HIV disease progression and resistance to malaria [42].

Pathogen Evasion Mechanisms: Genomics has uncovered how pathogens use host genes to evade immunity. For example, Salmonella can reduce expression of host genes to increase the likelihood of its survival within the cells [43].
Research on the Human Microbiome: Advances in metagenomics have unveiled the significant role that human microbiome plays in altering the outcomes of infectious illness and microbial imbalances often exacerbating diseases [44].

7.8.2 Advances in Genomic Technologies for Infectious Disease Surveillance

Genomic epidemiology uses sequencing technologies to find the spread of infectious diseases, providing accurate and up-to-date data for public health actions.

COVID-19 Pandemic: Genomic sequencing played a critical role in tracing the emergence and spread of SARS-CoV-2 variants like Delta and Omicron. These efforts informed public health measures, vaccine updates, and therapeutic development [45].
Ebola Virus Outbreaks: The sequencing of the virus during the West African Ebola epidemic in 2014 helped trace the evolution of the virus, pinpointing its zoonotic origin and transmission dynamics.
Zika Virus Epidemic: Genomic studies of the Zika virus have elucidated its transmission routes, confirming its link to microcephaly and guiding control measures [46].

Genomic technology has revolutionized vaccine development and diagnostics for infectious diseases:

mRNA Vaccines: The genome sequence of SARS-CoV-2 helped to quickly make mRNA vaccines like Pfizer-BioNTech and Moderna. These vaccines rely on genomic information to encode the spike proteins that induce a powerful immune response [47].
Pathogen-Specific Diagnostics: Sequencing enables the development of highly specific molecular diagnostics. For instance, genetic knowledge about Escherichia coli strains has led to the development of assays that distinguish pathogenic from non-pathogenic variants [48].
Universal Vaccine Targets: Genomics has identified conserved regions in pathogen genomes, opening the potential for vaccinations against diseases such as influenza [49].

7.8.2.1 *Genomic Tools for Targeted Therapies*

Personalized medicine in infectious diseases uses genomic data to customize therapy for specific patients. By evaluating the genetic profiles of both the host and pathogen, doctors can maximize therapy options.

Antiviral Treatments: Host genetic variables, such as CCR5 polymorphisms, influence the use of medications like maraviroc in HIV treatment. Similarly, genomic investigations of hepatitis C virus (HCV) variations influence the choice of direct-acting antivirals [50].
Antibiotic Selection: Pathogen sequencing detects resistance genes, enabling the exact selection of effective antibiotics. For example, sequencing Staphylococcus aureus strains can assist evaluate their susceptibility to methicillin or vancomycin [51].

Antimicrobial resistance is a significant threat to global health. Genomic sequencing plays a key role in tracking and containing AMR through:

Resistance Gene Surveillance: Sequencing has identified resistance genes in bacteria, such as blaNDM-1, carbapenem resistance, allowing for timely intervention to prevent epidemics [52].
Tracking Resistance Evolution: Genomic studies of Neisseria gonorrhoeae have shown the development of resistance to many antibiotics, which will inform updates to treatment guidelines [53].
Predictive Modelling: By combining genomic data with machine learning, researchers can foresee the establishment of resistance and devise preventative treatments [54].

7.8.2.2 *Personalized Treatments According to Pathogen and Host Genetic Profiles*

This integration of pathogen and host genetic information results in very targeted and effective treatments.

Host-Based Medicines: Genomic discoveries about the pathways of immunological responses have enabled the development of host-directed medicines, including drugs interfering with the JAK-STAT pathway in the context of viral infections [55].
Pathogen-Specific Interventions: For example, the research is currently done on the systems of CRISPR-Cas for their application in targeting particular viral genomes to yield treatments for chronic conditions like hepatitis [56].

7.9 Genomics and Noncommunicable Diseases

Noncommunicable illnesses, characterized by long-term progression and lifestyle-related risk factors, are the major cause of death globally. The main categories are:

Cardiovascular Diseases (CVDs): Conditions like coronary artery disease, stroke, and hypertension are influenced by inherited predispositions as well as changeable variables like food and exercise [57].
Diabetes Mellitus: Both Type 1 and Type 2 diabetes have genetic components, with polygenic and monogenic variants contributing to disease vulnerability [58].

Cancer is a disease that results due to multifactorial changes affecting oncogenes and tumor suppressor genes mainly caused by altered genetic makeup resulting from environmental carcinogens like smoke or UV exposures [59].

Genomics has made profound contributions to unraveling the underlying genetics of complex diseases:

The NCDs are polygenic, with the large majority being multi-factorial effects of numerous genes, each carrying a tiny bit of risk. Genome-wide association studies have identified a number of common SNPs involved in NCDs, which include variants within the APOE gene related to Alzheimer's disease and the TCF7L2 gene to Type 2 diabetes [60].

Epigenetics in NCDs: Epigenetic changes such as DNA methylation and histone acetylation regulate gene expression without altering the DNA sequence. Researchers have linked abnormal epigenetic patterns to diseases such as cancer and diabetes [61].

Gene-Environment Interactions: Genomics has elucidated how genetic susceptibility interacts with environmental factors. For instance, certain genetic variations predispose people to obesity only under specific dietary conditions [62].

7.9.1 Precision Medicine in Noncommunicable Diseases

The drugs for NCDs are also made more potent and specific according to the patient's genetic signature through precision medicine.

Cancer: Genomic profiling of tumor guides personalized therapy. For instance, the existence of EGFR gene mutations in non-small cell lung cancer might be treated by tyrosine kinase inhibitors, such as erlotinib. BRCA1/BRCA2 mutation in breast cancer patients dictates whether or not the patient will have PARP drugs [63].
Cardiovascular Diseases: Genomic awareness allows the advancement of treatment options for diseases like hypercholesterolemia and arrhythmia. For instance, mutations in PCSK9 gene led to drugs like evolocumab, a monoclonal antibody Pharmacogenomics dictates the use of anticoagulants like warfarin, where CYP2C9 and VKORC1 mutations determine dose [64].

7.9.2 Risk Detection and Prediction Using Genomics

Genomic techniques enhance the early identification and prevention of noncommunicable diseases, reducing morbidity and mortality through proactive healthcare strategies:

Screening for Genetic Predispositions: Genetic variants associated with diseases such as CVDs or cancer can be identified to initiate preventive actions. For instance, screening for BRCA mutations enables preventive operations on breast and ovarian cancers or enhanced surveillance [65].
Population Genomic Studies: Large-scale initiatives, like the UK Biobank, are providing useful data resources for the discovery of large-effect genetic variants associated with NCDs and population-specific risk predictions [66].

7.9.3 Use of Genomic Screens in Risk Determination

Genomic screens make use of cutting-edge tools such as Polygenic Risk Scores (PRS) to calculate an individual's predisposition to developing NCDs:

7.9.3.1 Cardiovascular Diseases

PRS for coronary artery disease calculates the risk coming from thousands of SNPs, resulting in an overall risk evaluation. Early identification of familial hypercholesterolemia resulting from mutations in the LDLR or APOB genes is possible with genomic screening, thereby avoiding early cardiovascular events [67].

As illustrated in Table 7.2, genomics supports healthcare by enabling early diagnosis, targeted therapies, outbreak tracking, and precision interventions in multiple disease categories.

Table 7.2 Genomic applications in rare and communicable diseases.

Disease category	Role of genomics	Examples	Impact on healthcare
Rare Diseases	Identification of genetic mutations and pathways.	Duchenne muscular dystrophy, cystic fibrosis.	Early diagnosis, personalized treatments, and identification of novel therapeutic targets.
Cancer (NCD)	Identification of driver mutations, biomarker discovery, and targeted therapies.	BRCA1/BRCA2 mutations in breast cancer.	Tailored therapies, reduced side effects, and improved survival rates.
Communicable Diseases	Tracking pathogen evolution, understanding host-pathogen interactions, and vaccine development.	COVID-19, Ebola, Zika virus.	Improved outbreak response, diagnostics, and vaccine efficacy.
Cardiovascular Diseases (NCD)	Identification of genetic risk factors and biomarkers for early detection.	Mutations in LDLR for familial hypercholes-terolemia.	Risk prediction, preventive interventions, and targeted treatments.
Antimicrobial Resistance	Monitoring resistance genes in pathogens and guiding antibiotic use.	Methicillin-resistant Staphylococcus aureus (MRSA).	Improved surveillance and precision in antimicrobial therapy.

7.9.3.2 *Cancer*

Liquid biopsies detect ctDNA, thereby detecting malignancies at an early stage such as colorectal and lung cancer. MSI-H (microsatellite instability-high) marker is a positive predictor for checking responsiveness to immunotherapy in colorectal cancer [68].

7.9.3.3 *Diabetes*

Genomic markers TCF7L2 SNPs can indicate risk of Type 2 diabetes so that preclinical phases can be detected and lifestyle modified before the development of the disease [69].

7.9.4 Biomarkers for Early Diagnosis

Genomic biomarkers, identified through transcriptomics and proteomics, play a role in early detection and disease surveillance.
CVD Biomarkers: Variants of the IL6R gene are associated with inflammation and predict the risk of CVD [70].
Cancer Biomarkers: Genomic studies have identified KRAS and TP53 mutations as potential early markers for colorectal and pancreatic cancers [71].
Diabetes Biomarkers: Genomic studies have connected GLUT2 gene variations to early diagnosis of glucose metabolism problems [72].
Challenges in Genomic Applications for NCDs: Ethical, Social, and Economic Issues.

7.10 Ethical Concerns

The use of genomic data sets forth privacy issues, especially on sensitive health information. Incidental findings, for example, susceptibility to incurable conditions such as Huntington's disease, have brought forward ethical issues [73].

7.10.1 Social Issues

Genomics may provide solutions to the worsening health inequalities. Underserved groups are denied the use of genomic testing and treatment based on their genetic makeup. The misinterpretation of genetic data leads to stigma and discrimination [74].

7.10.2 Economic Issues

The high costs of the diagnostic tests and treatments make it inaccessible to people in large numbers. Limited health funds in low-income countries do not allow them to integrate genetics into their public health systems [75].

7.11 Need for Infrastructure and Accessibility

Successful implementation of genomics in NCDs requires robust infrastructures and balanced access.

Infrastructure: Investment in sequencing facilities, bioinformatics resources, and skilled workforce is a must to adopt genomic medicine [76]. Accessibility: There should be an effort to decrease the cost of genetic testing and make it available in resource-poor settings [76].

Figure 7.2 depicts how genomics interacts with environmental, developmental, and socioeconomic factors in influencing disease susceptibility and progression.

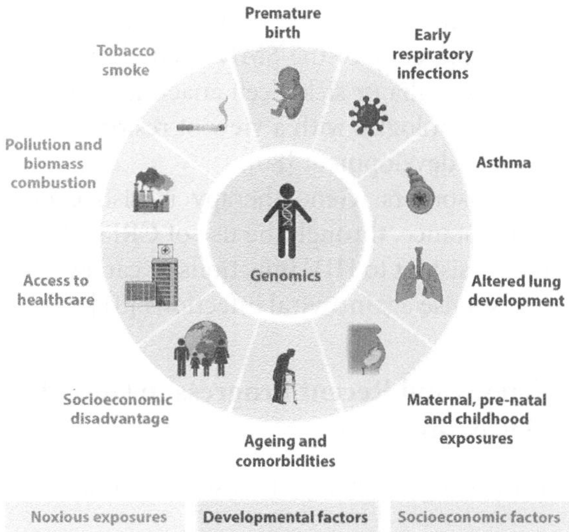

Figure 7.2 Role of genomics in diseases.

7.11.1 Integrating Genomic Data into Healthcare Systems

Integrating genetic data into normal healthcare requires joint efforts.

Data Sharing: International collaborations, such as the Global Alliance for Genomics and Health (GA4GH), promote sharing data with maintaining privacy and security [77].
Electronic Health Records: The inclusion of genomic data in EHRs enables clinicians to make proper decisions from genetic profiles [77].

7.12 Gene Therapy and Future Directions

Gene therapy has evolved medicine into having the possibility to correct genetic faults at their inception. The rise of genomics and biotechnology has allowed this to be so, treating previously untreatable diseases. Gene therapy and its transformative potentials for rare as well as communicable diseases is explored in this section along with emerging trends in genomic medicine and ethical considerations determining its future direction [78].

7.12.1 Gene Therapy for Rare and Infectious Diseases

Gene therapy is the introduction, removal, or alteration of genetic material in a patient's cells to treat or prevent disease. It holds particular promise for:

Most rare diseases are genetic in nature, making them an excellent target for gene therapy. SMA can be treated with Zolgensma, a gene therapy that provides a working copy of the SMN1 gene [79]. The inherited disorders, such as haemophilia or sickle cell anaemia, are now being treated with gene-editing technologies with a view of restoring normal functioning or slowing disease development [80].

Communicable disorders: Gene therapy is also explored for infectious diseases. For instance, through the use of CRISPR, it can change the immune cells to be resistant to HIV infections or can even engineer T cells to increase their response against viral infections [81].

7.12.2 Challenges and Recent Progress in Gene Therapy for Rare Diseases

Promising as the field of gene therapy is, it still encounters many challenges that include:

Delivery mechanisms: Safe delivery of genetic material to target cells is still challenging. Viral vectors, adeno-associated viruses (AAVs), are primarily used but induce immunological response or insertional mutation [82].
Cost and Availability: Many patients cannot afford to access treatments like Zolgensma, which cost millions of dollars. Mass adoption can be achieved only when the production costs are brought down and made affordable [83].

Some breakthroughs include non-viral delivery techniques, improved vector engineering, and regulatory clearances that have expanded the therapeutic space. Examples include Beta-thalassemia, which has been treated with lentiviral vectors. Hereditary blood diseases are being explored in clinical trials using CRISPR-based therapeutics [84].

7.12.3 Gene Editing Technologies: CRISPR-Cas9 and Its Applications

CRISPR-Cas9 represents a revolutionary technology for gene editing, which promises to introduce alterations in DNA sequences at high precision with great efficiency; this has direct medicinal implications. Clinical trials for somatic cell editing have, for example, used genetically modified hematopoietic stem cells in the treatment of diseases such as sickle cell anaemia and leukaemia, achieving success in repairing the HBB gene mutation causing beta-thalassemia [85]. CRISPR applications include both *ex vivo* and *in vivo* approaches: *ex vivo* editing affects cells outside the body, as shown in CAR-T cell therapy for cancer, whereas *in vivo* editing targets illnesses directly within the body, such as Leber congenital amaurosis. Looking ahead, advances like as base editors and prime editing provide increased accuracy, allowing for single-nucleotide repairs or insertions without generating double-stranded DNA breakage [86].

7.12.4 Emerging Trends in Genomic Medicine

AI and ML are revolutionizing genetic medicine by speeding up the process in data analysis, medication development, and customized health care. Algorithms of ML work much faster with large genomic data sets than the usual methods and pinpoint disease-causing genetic variations faster than traditional methods. In drug discovery, AI enables the identification of therapeutic targets, predicts protein structures, and formulates gene editing techniques. Such therapy adaptability according to genetic, clinical, and lifestyle information is seen with personalized medicine due to ML; besides, predictions made by technologies like DeepMind's AlphaFold for the folding of proteins also help drug developers [87]. There are additional revolutions in the healthcare delivery system through the integration of genomics with digital health tools. For example, wearable devices now combine health monitoring and genomic data in an ability to predict illness risk, then to detect at an early stage. Enhancements in telemedicine include remote consultations based on genomic data, thus ensuring more health care in underserved areas. Additionally, incorporation of genomic data into electronic health records facilitates the provision of tailored therapy recommendations and improved predictive disease modeling [88].

7.13 Future Considerations and Challenges

Ethical difficulties in genomic medicine include privacy, equity, informed consent, and societal ramifications, all of which pose challenges to the field's responsible progress. The preservation and sharing of genetic data raise privacy concerns, as well as the potential of unwanted access and exploitation. While initiatives such as GDPR seek to protect data privacy, global implementation remains challenging. Informed consent is important because patients need to know the extent of genetic testing, including unanticipated results and long-term implications [89]. Access to genetic technologies shows disparities, particularly in low-income areas, leading to ethical dilemmas about fairness. In addition, the misuse of genomic information by insurers or employers may lead to discrimination, although various safeguards, including the Genetic Information Nondiscrimination Act (GINA) in the United States, may not be comprehensive enough. Social impacts, such as the idea of "designer babies" and enhancement therapies, have fueled arguments on societal injustice and ethical bounds [90]. Future challenges in gene editing are off-target effects that must be minimized to ensure precision, and ethical issues concerning germline editing, which could alter the human genome since the changes will be heritable. There is a need for strict regulatory oversight to balance innovation, safety, efficacy, and compliance with ethics [91].

7.14 Conclusion

Genomics has emerged as a revolutionary force in modern medicine, providing ground-breaking information about the processes, diagnosis, and treatment of both rare and communicable diseases. In rare diseases, where conventional diagnostic tools frequently fail due to their complexity and rarity, genomics offers a precise tool for identifying underlying genetic abnormalities. Technologies like WES and WGS have uncovered genetic bases of such diseases as Duchenne muscular dystrophy, cystic fibrosis, and Huntington's disease so that the diagnosis can take place early with intervention. Even in communicable illnesses, the understanding of genomics has given a new aspect to pathogen-host interactions; thus, they have played a crucial role in outbreak control. During the COVID-19 pandemic, genomic sequencing followed viral variations and informed vaccine development, demonstrating genomics' great potential in addressing global health challenges.

Genomic innovations will fuel the development of precision medicine, the transformation of the healthcare industry. The effectiveness of such treatments with minimal side effects has been clearly evident in the case of targeted drugs for cancers, cardiovascular disease, and diabetes. Improved detection and prevention techniques due to better genomic risk prediction and biomarker discovery have minimized the burden of disease and optimized patient outcomes. Such benefits characterize a turn from general modalities to those personalized interventions as related to their genetic profiles toward an improvement of therapy efficiency for each patient.

Potential for developing genomic medicine is immense in the future. Precise and efficient genome modification by new technologies like CRISPR-Cas9, primary editing, and AI-driven genomics will create avenues for this work to be further done. Another important work, now becoming routine practice in healthcare systems, is the addition of genomic data into various healthcare institutions, which will fast-track the application of customized therapy on a larger scale. This will make it possible to integrate genetic information into electronic health records, and real-time decision-making and predictive modeling could be achieved. Wearable gadgets that are linked with genomic findings may help individuals control their health more proactively.

Global genomics activities are key to ensuring that everyone can access the benefits. The Human Genome Project has served as a pathway to international cooperation, and such similar programs distributed in underprivileged areas would be democratizing genetic medicine. Therapeutic breakthroughs, like mRNA-based vaccines and CAR-T cell therapy, will mean new arenas to test what medicine can accomplish, especially within the realms of uncommon diseases and cancer. Genomics will also play a part in reducing the rising issue of antibiotic resistance by tracing genetic mutations in pathogens as well as enhancing infectious disease surveillance systems.

However, the road ahead is not without its challenges. To ensure that genomic technologies are used responsibly, ethical problems like privacy and the possibility of genetic prejudice must be addressed. The cost of genomic testing, along with the necessity for robust infrastructure, continues to be a huge obstacle in low- and middle-income nations. It is also further supplemented by not only the need for technological breakthroughs but also educating healthcare workers and public awareness about genetic information being integrated into general healthcare systems.

Global collaboration is the only way the potential of genomics can be realized. Governments, academic institutions, and the commercial sector will have to invest heavily in genetic research and technology. More funding for basic and applied research will accelerate discoveries and their translation into practical healthcare solutions. Developing strong regulatory frameworks will ensure the ethical use of genomic treatments, particularly in gene editing and personalized medicine. International cooperation should focus on capacity building in underprivileged areas so that access to genetic advances is fair. Genetic data should be shared across borders, with tight privacy measures, so that more research can be done and answers can be given more rapidly to public health concerns.

Public awareness and education regarding genomics are equally essential. The best-informed patient makes better-informed decisions about his or her care and becomes a more trustworthy user of genomic medicine. The public, with good education and awareness, can bring forth policies which can encourage equity of access to fair genomics. A focus on education and communication would help many benefit from advances in genetics. This will serve the purpose of bringing health-care innovation to everybody, not a privileged few.

Genomics has introduced medicine to a new era—one where diseases could be understood, diagnosed, and treated at the molecular level. Adding genetics into health care will allow for the possibility of precision medicine, early detection, and targeted medicines. All this has drastically changed the approach to health and disease. Still, it would only be met if long-term investment, awareness in ethics, and global cooperation were made. We can, as a group, apply genetics in enhancing public health, improving life, and contributing to a healthy, fair world.

AI disclosure statement: The author confirms that no artificial intelligence (AI)-generated content or AI-assisted writing tools were used in the preparation of this chapter (including summaries and the abstract). All text, analysis, and conclusions are the result of the author's original work.

References

1. Dive, D. and Content, P., Rare Disease Day 2024–The importance of shining a light on rare diseases.
2. De Cock, K.M., Global Health's Evolution and Search for Identity. *Emerg. Infect. Dis.*, 31, 1, 1, 2025.
3. Chung, C.C.Y., Project, H.K.G., Chu, A.T.W., Chung, B.H.Y., Rare disease emerging as a global public health priority. *Front. Public Health*, 10, 1028545, 2022.
4. Davies, M.T. and Duckett, M.E., The Global Burden of Disease, in: *A Healthcare Students Introduction to Global Health: Low- and Middle Income Countries*, pp. 11–24, Springer, 2024.
5. Nguengang Wakap, S., Lambert, D.M., Olry, A., Rodwell, C., Gueydan, C., Lanneau, V., *et al.*, Estimating cumulative point prevalence of rare diseases: Analysis of the Orphanet database. *Eur. J. Hum. Genet.*, 28, 2, 165–73, 2020.
6. Ropers, H.H. and van Karnebeek, C.D., Rare diseases: human genome research is coming home, in: *Molecular Case Studies*, vol. 8, p. a006210, Cold Spring Harbor Laboratory Press, 2022.
7. Sakai, T. and Morimoto, Y., The history of infectious diseases and medicine. *Pathogens*, 11, 10, 1147, 2022.
8. Kingsmore, S.F., Nofsinger, R., Ellsworth, K., Rapid genomic sequencing for genetic disease diagnosis and therapy in intensive care units: A review. *npj Genomic Med.*, 9, Nature Research; 2024.
9. Frésard, L. and Montgomery, S.B., Diagnosing rare diseases after the exome. *Mol. Case Stud.*, 4, 6, a003392, 2018.
10. Kosorok, M.R. and Laber, E.B., Precision medicine. *Annu. Rev. Stat. Appl.*, 6, 1, 263–86, 2019.
11. Brittain, H.K., Scott, R., Thomas, E., The rise of the genome and personalised medicine. *Clin. Med.*, 17, 6, 545–51, 2017.
12. Stefanicka-Wojtas, D. and Kurpas, D., Personalised Medicine—Implementation to the Healthcare System in Europe (Focus Group Discussions). *J. Pers. Med.*, 13, 3, 2023 Mar 1.
13. Garibyan, L. and Avashia, N., Polymerase chain reaction. *J. Invest. Dermatol.*, 133, 3, 1–4, 2013.
14. Kim, M.J., Yum, M.S., Seo, G.H., Lee, Y., Jang, H.N., Ko, T.S., *et al.*, Clinical application of whole exome sequencing to identify rare but remediable neurologic disorders. *J. Clin. Med.*, 9, 11, 1–14, 2020 Nov 1.
15. Polonis, K., Blommel, J.H., Hughes, A.E.O., Spencer, D., Thompson, J.A., Schroeder, M.C., Innovations in Short-Read Sequencing Technologies and Their Applications to Clinical Genomics. *Clin. Chem.*, 71, 1, 97–108, 2025.
16. Wang, Y., Zhao, Y., Bollas, A., Wang, Y., Au, K.F., Nanopore sequencing technology, bioinformatics and applications. *Nat. Biotechnol.*, 39, 11, 1348–65, 2021.

17. Pereira, R., Oliveira, J., Sousa, M., Bioinformatics and computational tools for next-generation sequencing analysis in clinical genetics. *J. Clin. Med.*, 9, MDPI; 2020.
18. Park, S.H. and Bao, G., CRISPR/Cas9 gene editing for curing sickle cell disease. *Transfus. Apheresis Sci.*, 60, 1, 103060, 2021.
19. Kabata, F. and Thaldar, D., The human genome as the common heritage of humanity. *Front. Genet.*, 14, 1282515, 2023.
20. Iyer, S., Maxson Jones, K., Robinson, J.O., Provenza, N.R., Duncan, D., Lázaro-Muñoz, G., *et al.*, The BRAIN Initiative data-sharing ecosystem: Characteristics, challenges, benefits, and opportunities. *Elife*, 13, e94000, 2024.
21. Vilhekar, R.S. and Rawekar, A., Artificial intelligence in genetics. *Cureus*, 16, 1, 2024.
22. Xie, J., Jin, Y., Shen, M., Chen, L., Zhang, S., A Patient-Centric, Coordinated Care Model for Rare Diseases: The Multidisciplinary Consultation Program at Peking Union Medical College Hospital. *NEJM Catal. Innov. Care Deliv.*, 4, s1, 2023.
23. Stoller, J.K., The challenge of rare diseases. *Chest*, 153, 6, 1309–14, 2018.
24. Willmen, T., Willmen, L., Pankow, A., Ronicke, S., Gabriel, H., Wagner, A.D., Rare diseases: Why is a rapid referral to an expert center so important? *BMC Health Serv. Res.*, 23, 1, 904, 2023.
25. Fernandez-Marmiesse, A., Gouveia, S., Couce, M.L., NGS Technologies as a Turning Point in Rare Disease Research, Diagnosis and Treatment. *Curr. Med. Chem.*, 25, 3, 404–32, 2017 Jul 19.
26. Nowak, K.J. and Davies, K.E., Duchenne muscular dystrophy and dystrophin: Pathogenesis and opportunities for treatment: Third in molecular medicine review series. *EMBO Rep.*, 5, 872–6, 2004.
27. Stoker, T.B., Mason, S.L., Greenland, J.C., Holden, S.T., Santini, H., Barker, R.A., Huntington's disease: Diagnosis and management. *Pract. Neurol.*, 22, 1, 32–41, 2022.
28. Shao, X., Lv, N., Liao, J., Long, J., Xue, R., Ai, N., *et al.*, Copy number variation is highly correlated with differential gene expression: A pan-cancer study. *BMC Med. Genet.*, 20, 1, 2019 Nov 9.
29. Ku, C.S., Polychronakos, C., Tan, E.K., Naidoo, N., Pawitan, Y., Roukos, D.H., *et al.*, A new paradigm emerges from the study of de novo mutations in the context of neurodevelopmental disease. *Mol. Psychiatry*, 18, 2, 141–53, 2013.
30. Bľandová, G. and Repiská, V., Prospective New Technologies of mtDNA Diagnosis, in: *Translational Mitochondrial Medicine*, pp. 57–86, Springer, 2024.
31. Darrow, J.J., Luxturna: FDA documents reveal the value of a costly gene therapy. *Drug Discovery Today*, 24, 949–54, Elsevier Ltd, 2019.
32. Ran, F.A., Hsu, P.D., Wright, J., Agarwala, V., Scott, D.A., Zhang, F., Genome engineering using the CRISPR-Cas9 system. *Nat. Protoc.*, 8, 11, 2281–308, 2013.

33. Zhu, Y., Zhu, L., Wang, X., Jin, H., RNA-based therapeutics: an overview and prospectus. *Cell Death Dis.*, 13, 7, 644, 2022.
34. Zakharova, M., Modern approaches in gene therapy of motor neuron diseases. *Med. Res. Rev.*, 41, 5, 2634–55, 2021.
35. Chernyi, N., Gavrilova, D., Saruhanyan, M., Oloruntimehin, E.S., Karabelsky, A., Bezsonov, E., *et al.*, Recent advances in gene therapy for hemophilia: Projecting the perspectives. *Biomolecules*, 14, 7, 854, 2024.
36. Abati, E., Bresolin, N., Comi, G., Corti, S., Silence superoxide dismutase 1 (SOD1): A promising therapeutic target for amyotrophic lateral sclerosis (ALS). *Expert Opin. Ther. Targets*, 24, 4, 295–310, 2020.
37. Iacobucci, I., Hay Mele, B., Cozzolino, F., Monaco, V., Cimmaruta, C., Monti, M., Andreotti, G., Monticelli, M., Enzyme replacement therapy for FABRY disease: Possible strategies to improve its efficacy. *Int. J. Mol. Sci.*, 24, 5, 4548, 2023 Feb 25.
38. Khoury, M.J. and Holt, K.E., The impact of genomics on precision public health: beyond the pandemic. *Genome Med.*, 13, BioMed Central Ltd, 2021.
39. Punina, N.V., Makridakis, N.M., Remnev, M.A., Topunov, A.F., Whole-genome sequencing targets drug-resistant bacterial infections. *Hum. Genomics*, 9, 1–20, 2015.
40. Dhar, B.C., Diagnostic assay and technology advancement for detecting SARS-CoV-2 infections causing the COVID-19 pandemic. *Anal. Bioanal. Chem.*, 414, 9, 2903–34, 2022.
41. Funmilayo, A.J., Agboola, O., Agboola, S., Oluboyo, B., Egbebi, A., Sijuade, A., *et al.*, The Roles of Genomics and Proteomics in Human Parasitology: Closing the Knowledge Gap. *Infect. Epidemiol. Microbiol.*, 10, 2, 157–74, 2024.
42. Olaniyan, M.F., Oladele, A.A., Tijani, A., Muhibi, M.A., Olaniyan, T.B., Susceptible Genes and Polymorphisms Associated with Communicable and Noncommunicable Diseases. *J. Bio-X Res.*, 7, 0001, 2024.
43. Wang, M., Qazi, I.H., Wang, L., Zhou, G., Han, H., Salmonella virulence and immune escape. *Microorganisms*, 8, 3, 407, 2020.
44. Martín, R., Miquel, S., Langella, P., Bermúdez-Humarán, L.G., The role of metagenomics in understanding the human microbiome in health and disease. *Virulence*, 5, 3, 413–23, 2014.
45. Jolly, B. and Scaria, V., Genomics and infectious diseases: Lessons learnt from the COVID-19 pandemic, in: *Genomics, Populations, and Society*, pp. 183–206, Elsevier, 2025.
46. Nunes, M.L., Carlini, C.R., Marinowic, D., Kalil, F., Fiori, H.H., Scotta, M.C., *et al.*, Microcephaly and Zika virus: A clinical and epidemiological analysis of the current outbreak in Brazil. *J. Pediatr. (Rio J.)*, 92, 3, 230–40, 2016.
47. Xia, X., Detailed dissection and critical evaluation of the Pfizer/BioNTech and Moderna mRNA vaccines. *Vaccines (Basel)*, 9, 7, 734, 2021.
48. Rivas, L., Mellor, G.E., Gobius, K., Fegan, N., *Detection and typing strategies for pathogenic Escherichia coli*, Springer, 2015.

49. Scorza, F.B., Tsvetnitsky, V., Donnelly, J.J., Universal influenza vaccines: Shifting to better vaccines. *Vaccine*, 34, 26, 2926–33, 2016.
50. Kumar, N., Sharma, S., Kumar, R., Tripathi, B.N., Barua, S., Ly, H., *et al.*, Host-directed antiviral therapy. *Clin. Microbiol. Rev.*, 33, 3, 10–1128, 2020.
51. Sanchini, A., Recent developments in phenotypic and molecular diagnostic methods for antimicrobial resistance detection in Staphylococcus aureus: A narrative review. *Diagnostics*, 12, 1, 208, 2022.
52. Caliskan-Aydogan, O. and Alocilja, E.C., A review of carbapenem resistance in Enterobacterales and its detection techniques. *Microorganisms*, 11, 6, 1491, 2023.
53. Unemo, M. and Shafer, W.M., Antimicrobial resistance in Neisseria gonorrhoeae in the 21st century: Past, evolution, and future. *Clin. Microbiol. Rev.*, 27, 3, 587–613, 2014.
54. Prabhod, K.J., The Role of Machine Learning in Genomic Medicine: Advancements in Disease Prediction and Treatment. *J. Deep Learn. Genomic Data Anal.*, 2, 1, 1–52, 2022.
55. Badia, R., Garcia-Vidal, E., Ballana, E., Viral-host dependency factors as therapeutic targets to overcome antiviral drug-resistance: A focus on innate immune modulation. *Front. Virol.*, 2, 935933, 2022.
56. Dubey, A.K., Kumar Gupta, V., Kujawska, M., Orive, G., Kim, N.Y., Li, C.Z., *et al.*, Exploring nano-enabled CRISPR-Cas-powered strategies for efficient diagnostics and treatment of infectious diseases. *J. Nanostruct. Chem.*, 12, 5, 833–64, 2022.
57. Bhatnagar, A., Environmental determinants of cardiovascular disease. *Circ. Res.*, 121, 2, 162–80, 2017.
58. Udler, M.S., Type 2 diabetes: Multiple genes, multiple diseases. *Curr. Diab. Rep.*, 19, 1–9, 2019.
59. Parsa, N., Environmental factors inducing human cancers. *Iran. J. Public Health*, 41, 11, 1, 2012.
60. Khera, A.V., Chaffin, M., Aragam, K.G., Haas, M.E., Roselli, C., Choi, S.H., *et al.*, Genome-wide polygenic scores for common diseases identify individuals with risk equivalent to monogenic mutations. *Nat. Genet.*, 50, 1219–24, Nature Publishing Group, 2018.
61. Khajebishak, Y., Alivand, M., Faghfouri, A.H., Moludi, J., Payahoo, L., The effects of vitamins and dietary pattern on epigenetic modification of non-communicable diseases. *Int. J. Vitam. Nutr. Res.*, 2021.
62. Reddon, H., Guéant, J.L., Meyre, D., The importance of gene–environment interactions in human obesity. *Clin. Sci.*, 130, 18, 1571–97, 2016.
63. Syn, N.L.X., Yong, W.P., Goh, B.C., Lee, S.C., Evolving landscape of tumor molecular profiling for personalized cancer therapy: A comprehensive review. *Expert Opin. Drug Metab. Toxicol.*, 12, 8, 911–22, 2016.
64. Bayona, A., Arrieta, F., Rodríguez-Jiménez, C., Cerrato, F., Rodríguez-Nóvoa, S., Fernández-Lucas, M., *et al.*, Loss-of-function mutation of PCSK9

as a protective factor in the clinical expression of familial hypercholesterolemia: A case report. *Med. (United States)*, 99, 34, 2020 Aug 21.

65. Geanta, M., Boata, A., Cucos, B., Cioroboiu, C., Popescu, R., Genetic Health, in: *A Glimpse at Medicine in the Future*, pp. 79–91, Springer, 2024.
66. Gurdasani, D., Barroso, I., Zeggini, E., Sandhu, M.S., Genomics of disease risk in globally diverse populations. *Nat. Rev. Genet.*, 20, 9, 520–35, 2019.
67. Klarin, D. and Natarajan, P., Clinical utility of polygenic risk scores for coronary artery disease. *Nat. Rev. Cardiol.*, 19, 291–301, Nature Research, 2022.
68. Bettegowda, C., Sausen, M., Leary, R.J., Kinde, I., Wang, Y., Agrawal, N., *et al.*, Detection of circulating tumor DNA in early- and late-stage human malignancies. *Sci. Transl. Med.*, 6, 224, 2014 Feb 19.
69. Canivell Fusté, S., The association of DNA methylation patterns in TCF7L2 and GIPR genes with Type 2 Diabetes, 2014.
70. Georgakis, M.K., Malik, R., Richardson, T.G., Howson, J.M.M., Anderson, C.D., Burgess, S., *et al.*, Associations of genetically predicted IL-6 signaling with cardiovascular disease risk across population subgroups. *BMC Med.*, 20, 1, 245, 2022.
71. Yang, W., Shi, J., Zhou, Y., Liu, T., Zhan, F., Zhang, K., *et al.*, Integrating proteomics and transcriptomics for the identification of potential targets in early colorectal cancer. *Int. J. Oncol.*, 55, 2, 439–50, 2019.
72. Zhou, K., Yee, S.W., Seiser, E.L., Van Leeuwen, N., Tavendale, R., Bennett, A.J., *et al.*, Variation in the glucose transporter gene SLC2A2 is associated with glycemic response to metformin. *Nat. Genet.*, 48, 9, 1055–9, 2016.
73. Ayyalasomayajula, M.M.T., Ethical Considerations in AI Implementation for Neurodegenerative Disease Diagnosis and Treatment, in: *Integrating Neuroimaging, Computational Neuroscience, and Artificial Intelligence*, pp. 105–27, CRC Press, 2025.
74. Khoury, M.J., Bowen, S., Dotson, W.D., Drzymalla, E., Green, R.F., Goldstein, R., *et al.*, Health equity in the implementation of genomics and precision medicine: A public health imperative. *Genet. Med.*, 24, 1630–9, Elsevier B.V., 2022.
75. Albarqouni, L., Arab-Zozani, M., Abukmail, E., Greenwood, H., Pathirana, T., Clark, J., *et al.*, Overdiagnosis and overuse of diagnostic and screening tests in low-income and middle-income countries: A scoping review. *BMJ Glob. Health*, 7, BMJ Publishing Group, 2022.
76. Olono, A., Mitesser, V., Happi, A., Happi, C., Building genomic capacity for precision health in Africa. *Nat. Med.*, 30, 7, 1856–64, 2024.
77. Louie, B., Mork, P., Martin-Sanchez, F., Halevy, A., Tarczy-Hornoch, P., Data integration and genomic medicine. *J. Biomed. Inf.*, 40, 1, 5–16, 2007.
78. Gonçalves, G.A.R. and Paiva R de, M.A., Gene therapy: Advances, challenges and perspectives. *Einstein (Sao Paulo Brazil)*, 15, 369–75, 2017.
79. Jensen, T.L., Gøtzsche, C.R., Woldbye, D.P.D., Current and future prospects for gene therapy for rare genetic diseases affecting the brain and spinal cord. *Front. Mol. Neurosci.*, 14, 695937, 2021.

80. Song, X., Liu, J., Chen, T., Zheng, T., Wang, X., Guo, X., Gene therapy and gene editing strategies in inherited blood disorders. *J. Genet. Genomics*, 2024.
81. Bunnell, B.A. and Morgan, R.A., Gene Therapy for Infectious Diseases. 11, 1998.
82. Słyk, Z., Stachowiak, N., Małecki, M., Recombinant Adeno-Associated Virus vectors for Gene Therapy of the Central Nervous System: Delivery routes and clinical aspects. *Biomedicines*, 12, 7, 1523, 2024.
83. Garrison Jr., L.P., Lo, A.W., Finkel, R.S., Deverka, P.A., A review of economic issues for gene-targeted therapies: Value, affordability, and access. *Am. J. Med. Genet. Part C: Semin. Med. Genet.*, 64–76, Wiley Online Library, 2023.
84. Gorter, R., Retroviral vectors and transposons for stable gene therapy and manufacturing vaccines.
85. O'Keeffe, F.J., Alphonse, A.J., Mendz, G.L., Somatic Genome Editing: Technical Challenges and Ethical Appraisal. *Eur. J. Med. Health Res.*, 2, 3, 239–47, 2024.
86. Volodina, O. and Smirnikhina, S., The Future of Gene Therapy: A Review of *In Vivo* and *Ex Vivo* Delivery Methods for Genome Editing-Based Therapies. *Mol. Biotechnol.*, 1–13, 2024.
87. Quazi, S., Artificial intelligence and machine learning in precision and genomic medicine. *Med. Oncol.*, 39, Springer, 2022.
88. Bombard, Y., Ginsburg, G.S., Sturm, A.C., Zhou, A.Y., Lemke, A.A., Digital health-enabled genomics: Opportunities and challenges. *Am. J. Hum. Genet.*, 109, 1190–98, Cell Press, 2022.
89. Johnson, S.B., Slade, I., Giubilini, A., Graham, M., Rethinking the ethical principles of genomic medicine services. *Eur. J. Hum. Genet.*, 28, 2, 147–54, 2020.
90. Smith, C.E., Fullerton, S.M., Dookeran, K.A., Hampel, H., Tin, A., Maruthur, N.M., *et al.*, Using genetic technologies to reduce, rather than widen, health disparities. *Health Aff.*, 35, 8, 1367–73, 2016.
91. Naeem, M., Majeed, S., Hoque, M.Z., Ahmad, I., Latest Developed Strategies to Minimize the Off-Target Effects in CRISPR-Cas-Mediated Genome Editing. *Cells NLM (Medline)*, 9, 2020.

8

Cancer Genomics: Innovations in Diagnosis, Prognosis and Therapy

Habeebulahi Ajibola Adekilekun[1,2,3]*, Olorunshola Dave Omodamiro[2,3], Ngozi Kalu Achi[2], Habeebat Adekilekun Oyewusi[4] and Racheal Majekodunmi Omodamiro[5,6]

[1]*Department of Pharmacology, Federal University of Health Sciences, Ila-Orangun, Osun State, Nigeria*
[2]*Department of Biochemistry, Michael Okpara University of Agriculture, Umudike, Abia State, Nigeria*
[3]*Department of Pharmacology and Toxicology, Kampala International University, Bushenyi, Western Campus, Ishaka, Uganda*
[4]*Department of Medical Biochemistry, Federal University of Health Sciences, Ila-Orangun, Osun State, Nigeria*
[5]*Department of Pharmaceutical Chemistry and Pharmacognosy, Kampala International University, Bushenyi, Western Campus, Ishaka, Uganda*
[6]*National Root Crops Research Institute, Umuahia, Umudike, Abia State, Nigeria*

Abstract
The advent of cancer genomics has transformed the understanding of tumorigenesis, enabling the development of modern diagnostic, prognostic, and curative strategies. Modern developments on Next-generation sequencing, epigenetic investigation, and immunogenomics have improved the identification and documentation of genetic biomarkers, driver mutations, and tumor-specific vulnerabilities. The need to highlights the recent applications of cancer genomics, including genomically tailored therapy option, targeted therapies, and immunotherapies is also expressed in this research. Implementation of genomics-based medicine face some drawbacks which could be overcome by improving on data analysis complexities, tumor heterogeneity, and regulatory concerns. These would bring promising innovations in patient precision outcomes, personalized medicine, and better understanding of cancer bioscience that involves biomarker prognosis, immunotherapy optimization and liquid biopsy. This review provides a comprehensive

**Corresponding author*: adekilekun.ajibola@kiu.ac.ug

Swetha Kumari Koduru, Sarad Pawar Naik Bukke, Hussaini Bello, Muhamad Mustafa Mundu and Ayush Madan (eds.) Genomics and Precision Healthcare: Innovations, Applications and Challenges, (233–254) © 2026 Scrivener Publishing LLC

assessment of the current status of cancer genomics, its impact on clinical praxis, and future directions for research and translation integrating multi-omics data, AI-driven analytics, and functional genomics to further refine diagnosis, prognosis, and therapy.

Keywords: Cancer genomics, next-generation sequencing, epigenetics, immunogenomics, targeted therapy, immunotherapy, personalized medicine

8.1 Introduction

Cancer genomics has emerged as a rapidly advancing field that has revolutionized our understanding and approach to cancer diagnosis, prognosis, and therapy (Siddiqui, 2024). The ability to acquire comprehensive genomic data through next-generation sequencing has provided unprecedented insights into the molecular profiles of various cancer types, enabling personalized and targeted treatment strategies (Hassan *et al.*, 2022; Siddiqui, 2024). The incorporation of multi-omics (metabolomics, proteomics, and/or genomics) and other advanced bioinformatics tools has enhanced the prospects for precision oncology therapy (de Anda-Jáuregui & Hernández-Lemus, 2020; Meystre *et al.*, 2023). Multinational missions, such as the Cancer Genome Atlas; International Human Cancer Consortium; have played crucial roles in this endeavor by mapping the genomic landscapes of different cancers (Meystre *et al.*, 2023). These efforts have steered to identification of numerous actionable mutations and alterations that could be targeted with specific drug therapies or their downstream pathways can be addressed (de Anda-Jáuregui & Hernández-Lemus, 2020; Meystre *et al.*, 2023). As a result, personalized medicine and targeted therapies have shown promise in improving patient outcomes for some cancer subtypes. However, there is complexity of cancer genomics that also presents with challenges one of which is the advancements in personalized medicine and targeted therapy which has not yet been systematically shown improved patient outcomes across cancer varieties (Meystre *et al.*, 2023). Current developments in genome know-how, following the expression of genomic data have enhanced the merging of discoveries in clinical remedy. Successfully, translating cancer genomics information into diagnostics and therapeutics actions has supported its possibility to agree on personalized cancer remedy. However, the challenges towards converting a genome discovery to a concrete clinical use are many and difficult; therefore there is a need to emphasize the significance of creating the application of genomic innovation in executing its clinical benefit by avoiding major obstacles from the laboratory to the bedside (Meystre *et al.*, 2023).

8.2 Emergence of Cancer Genomics Advances

Cancer is evidently a prominent cause of human death in the world; more so, it is established that cells within the body may undergo mutations (genomic changes) (Nones & Patch, 2020). There have been great strides in exploring the genetic origin of human diseases which has impacted cancer genetics profoundly, where explosion of the molecular profiling and genomic arrangement information has elaborated the complication of malignancies in humans (Mardis, 2021). Emerging cancer genomics advancement has rapidly changed the face of medical care for cancer; therefore, the study of progressions of inherited diseases *via* particular gene mutations with link to different comorbidities, including Parkinson's disease, Alzheimer's disease and cancer, is described as *Genetics*, but *Genomics* is the comprehensive word and comprises the totality of exactly how, i.e., cancer genomics is the totality of intervening genes implicated in cancer progression (Pecorino, 2021). In a cancer cell, many genes might be abnormal in copy number or structure, while some may also be operationally uttered. Insights into the tumor molecular pathogenesis provides new treatment plans while a further understanding of cancer will necessitate fresh computational and statistical methods for the analysis of genetic and signaling networks which govern an individual's tumor activity and condition (Satam *et al.*, 2023).

8.3 Cancer Classification in the Genomic Age

Disease taxonomy is vital to knowledge of development, diagnosis, and treatment. The classification of cancer and subtypes is centered on the observed biological and clinical properties. Cancer nomenclature and terms are predominantly based on location of organ, e.g., "liver cancer" or "breast cancer." Based on this organ-specific-type, the finer subgroups can be designated depending on patients' age, histological grades, cell nature, and occasionally molecular biomarkers, for example, hormonal receptor level in the breast tumor or volatility of microsatellite in colorectal cancer (Satam *et al.*, 2023). Additionally, the new discoveries of somatic dissimilarities in epigenomic, protein, RNA, or DNA features for widespread cancers uncovered new biological diversity of human cancers that can enable the discovery of many previously unidentified types of cancers with different biological features. At the same time, these data's unprecedented volume and complexity presents considerable challenges to biostatisticians, cancer biologists, and clinicians. There are new challenges for current cancer taxonomy and interpretations such as discovery of cancer subtypes with their interpretations, investigation of novel classification with their co-existing classification, intratumor heterogeneity and tumor evolution (Pecorino, 2021).

8.4 The Genomics Role in Personalized Cancer Care

The monogenic approach was used in the past for diagnosis and treatment of cancer but since the accomplishment of human genome, a better therapeutic target discovery by enabling the mining of the human genome by means of advantage of proteogenomics ways has been revealed. This discovery has made novel and favorable cancer targets possible from the genome programming which are used for diagnostics, prognostic evaluation, and therapeutics, which has helped in monitoring response to cancer care (Dalghi, 2023) (Figure 8.1). The cancer disparate character has stalled development in knowing the primary process that leads to irregular cellular progression. International Genome Consortium Projects, and The Cancer Genome Atlas (TCGA), have made enormous development in genome sequencing with vast records of cancer genome. This approach has transformed the understanding of different cancer types with their diagnosis and treatment. Genomics and Proteomics Technologies have generated huge genomic data on clinical tumors, making transformation to cancer research more plausible in prognosis and diagnosis. A whole molecular outlook of the cancer setting is essential to understand the underlying processes for cancer initiation to develop diagnosis and prognosis that eventually will give way to personalized therapy. Remarkably, cancer proteome analysis allows the identification of biomarkers to examine patients on drugs and their resistances during radiation (Santucci, 2023). The proteome and genome of cancer depend on high-throughput techniques like pharmacogenomics data, protein–protein interactions, full proteome, and epigenome. Precision medicine may benefit from the multidimensional research data of omics (genes and proteins) that these methods have produced (Figure 8.1). However, the translation of massive omics data into useful clinical advantages is not well understood due to slow progress in computational approaches. Following the completion of the human genome, several genes have been linked to specific roles in the development of disease, improving the understanding of the disease process and enabling more focused treatments. A cancer subject's genetic mutational setting is incredibly helpful for making an accurate diagnosis and suggesting therapy options.

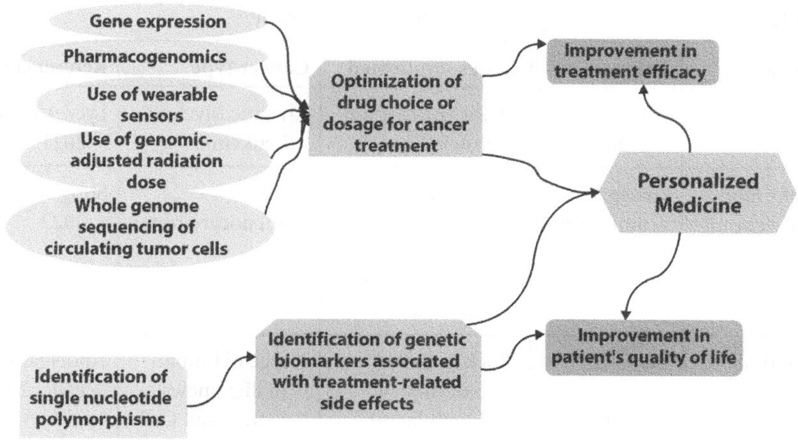

Figure 8.1 Omic contributing to effective personalized cancer care.

Numerous genome sequencing studies have found that specific genetic changes are enclosed in a patient's tumor and that these changes drive carcinogenesis. It is intriguing that a distinct collection of genomic changes in a certain cancer subtype could result in a customized medical treatment that is focused on the precise genetic changes of a single patient. Through Next-Generation Sequencing and bioinformatics research, the genetic and genomic differences among distinct cancer subtypes and/or types may demonstrate a substantial contribution to the advancement of precision medicine. A tumor sample's comprehensive genomic analysis would aid in tracking significant genomic alterations, hence facilitating the creation of practical solutions for precise patient diagnosis and treatment (Zalli *et al.*, 2023).

A thorough analysis of the cancer genome atlas (TCGA) has clearly demonstrated that a vast collection of genetic alterations and abnormalities in a single genome and epigenome could accumulate and start cancerous revolutions. These purported changes are unique to a given tumor and are actually investigated for explicit management, which includes the discovery of special biomarkers (Dalghi, 2023). The mutational form of the EGFR receptor gene, which is known to cause unchecked development and hence suppress apoptotic signals, is a well-known illustration of precision medicine techniques. Genetic Tests, for example, Comparative Genomic Hybridizations (CGH) arrays, Genotyping, complete genome sequencing, and Exome, may offer a better understanding on the exact genetic variants isolated in a biopsy sample. Tumor samples are genetically profiled and their contribution to the important clinical state of the patient could link

Table 8.1 Patient-specific genetic information therapy.

Drug	Enzyme	Genes implicated	Cancer type	References
Imatinib	ABL1 kinase inhibitor	BCR–ABL1 fusion protein	Chronic Myeloid Leukemia	Lyczek *et al.*, 2021
Gefitinib, Crizotinib	kinase inhibitors, such as EGFR	EGFR testing for mutations and ALK rearrangements, ALK fusion	Lung adenocarcinoma	Song *et al.*, 2022
Erlotinib	Tyrosine kinase inhibitor	Block mutant epidermal growth factor receptor (EGFR) activity; HER1	Range of human malignancies: Non-Small Cell Lung Cancer (NSCLC) and Pancreatic Cancer	Abdelgalil *et al.*, 2020
Cetuximab	Inhibit growth of EGFR	KRAS mutational	Colorectal, Head and neck squamous cell carcinoma (HNSCC)	De Pauw *et al.*, 2021

specific genetic variants to target, which is an important step in personalized medicine (Santucci, 2023; Dalghi, 2023) (Table 8.1). Progress is being made, since one of the studies on population-based cancer therapy showed an increase of the five-year survival rate in the United States from 48.9% in the 1970s to 68.5% in 2015 (Santucci, 2023).

8.5 Innovations in Cancer Diagnosis

8.5.1 Next-Generation Sequencing (NGS) Tools

With both therapeutic and research uses, genomic sequencing technologies are evolving quickly. With its unique capacity to resolve RNA and DNA molecules in a high-resolution and economical manner, the introduction of NGS has revolutionized genomics research (Satam *et al.*, 2023). The very of these parallel or NGS methodology to DNA analysis incorporates a number of potential advantages over the conventional approaches like Pyrosequencing, Sanger sequencing and Allele-particular PCR, counting the capacity of NGS to completely arrange vast quantities of qualities in a solitary effort. NGS can at the same time identify insertions, translocations, copy number alterations, deletions, and exome-wide base replacements in all identified cancer-connected genes (Zalli *et al.*, 2023). NGS could also reduce the high cost of creating sequence information and manage an excess of sequence-based methods for examining a genome in many species (Hu *et al.*, 2021). Remarkable advancement has been recorded in the NGS tools *via* contributions from molecular biology and engineering. So also, NGS is developing into a molecular basis thus getting to be a practically used tool in every biomedical study area (Hu *et al.*, 2021; Zalli *et al.*, 2023). This transformative skill has speedily forced genomics developments through various areas. NGS permits prompt sequencing of lots of DNA fragments concurrently, giving inclusive acumens to genetic variations, epigenetic modifications, gene expression outlines, and genome structure. The flexibility of NGS tools has lengthened the space for genomics exploration, in other to simplify studies on erratic genetic diseases, infectious diseases, microbiome analysis, population genetics, and cancer genomics. Additionally, NGS has permitted the advancement of targeted therapies; the precision medicine methods; and improvement of analytical strategies making it an essential tool guiding the diagnosis and treatment in modern clinical exercise (Satam *et al.*, 2023; Zalis *et al.*, 2024). Presently, new sequencing NGS dependent tools are continuously being developed; these include: (1) Transposase Accessible Chromatin Assay with NGS (ATAC-Seq) which is a technique for genome-wide chromatin availability for profiling that uses hyperactive Tn5 transposase; (2) Chromatin Immunoprecipitation Combined with NGS (ChiP-Seq), which is a method using antibodies for identification of specific nucleosomes or proteins to augment for the DNA fragments that are attached to the proteins or nucleosomes. These are considered as vital tools to recognize genome-wide modifications in the regulatory elements in the gene expression such as transcription factor binding sites, insulators, and enhancer (Nones & Patch, 2020).

8.5.2 Genome-Wide Association Studies (GWAS)

Genome-wide association studies have transformed the understanding of the genetic foundations of complex diseases and traits by leveraging on high-throughput genotyping and sequencing technologies to systematically examine the relationship between genetic variants across the entire genome and implicated phenotypes (Lappalainen *et al.*, 2021). Initially, GWAS efforts have yielded a wealth of genetic associations, although the interpretation and translation of its findings into biological vision and clinical applications remains a significant challenge (Uffelmann *et al.*, 2021; He & Gai, 2023). Recent advances in genome analysis and bioinformatics have opened new paths to address these challenges. Network-based approaches that integrate GWAS data with gene interaction networks have shown promise in prioritizing disease-associated genes and elucidating their functional roles (He & Gai, 2023). Additionally, fast growth of the whole-genome sequencing data has enabled more comprehensive cataloging of genetic variation and improved functional interpretation of both common and rare variants (He & Gai, 2023).

8.5.3 Epigenetic Markers for Cancer Diagnosis/Epigenetic Alterations

Emergent of cancer prevalence and mortality globally requires development of true biomarkers to ascertain diagnosis, prognostication, detection, and observations. Urologic parts, lung, colorectal, and breast are the most common cancer types with major developments in their characterization, which do not still tailored towards the biomarkers responsive in practice. The epigenetic alterations are considered to be advanced cancer biomarkers due to their stability, reversibility, accessibility, and frequency in body fluids; therefore they are used as potential assay for development that helps in managing the patient (Frank, 2024). "Epigenetic" could be defined as all inherited alterations in chromatin structure and gene expression which are not coded in DNA sequence (Constâncio *et al.*, 2025). Epigenetic modifications, including histone modifications, and DNA methylation act an important function in the development of various types of cancer; this field has gained significant consideration recently, because researchers now explore the underlying mechanisms by which epigenetic changes can affect the initiation and progression of cancerous cells. DNA methylation, in particular, has been generally studied as a likely target for management of cancer diagnosis (Constâncio *et al.*, 2025).

8.5.4 Liquid Biopsy and Circulating Tumor DNA (ctDNA) Investigation

Liquid biopsy, a revolutionary diagnostic and non-invasive approach, has gained momentous traction in the area of oncology, offering an alternative to traditional tissue biopsies. Unlike traditional tissue biopsies, liquid biopsy involves the analysis of natural fluids, such as blood, to detect and characterize ctDNA – genetic material introduced into the bloodstream by tumors (Casanova-Sala *et al.*, 2024). These circulating tumor DNA (ctDNA) molecules exhibit unique genetic and epigenetic signatures that can provide valuable data about the underlying tumor biology, including its genetic profile, heterogeneity, and evolution (Casanova-Sala *et al.*, 2024). Advances in molecular and genomic techniques are growing the choice of possible applications for circulating tumor DNA (ctDNA), in a "liquid biopsy" for cancer. Cell-free DNA (cfDNA) is released primarily into the bloodstream by cell death, though active secretion might have a function (Casanova-Sala *et al.*, 2024). Since detection of fetal cfDNA in maternal fluid, cfDNA analysis has been promptly executed in preparation for noninvasive prenatal analysis. More so, to identify ctDNA in persons or presymptomatic individuals, before diagnosing cancer, insinuates the likelihood of use of ctDNA as biomarker for early cancer screening. Observing numerous mutations in parallel can boost the accuracy of ctDNA discovery, thereby making the assessment of the clonal advancement of patients' disease and identification of resistance mutations before observing clinical progression. ctDNA analysis is now moving from the research set only into clinical use. The European Medicines Agency (EMA) and the U.S. Food and Drug Administration have also permitted ctDNA diagnostic tests for particular symptoms in the lack of tumor tissue evaluation (Cescon *et al.*, 2020).

8.6 Diagnosis of Cancer with Subtypes Using Genomics

Significant innovations in genomic technologies have unfolded new procedures for the accurate diagnosis with characterization of cancer subtypes, paving the means for more effective personalized treatments. One of the key challenges in cancer diagnosis and management has been the high grade of heterogeneity within different cancer types because both at the inter-tumor and intra-tumor levels, it poses significant challenges for detection, treatment, and monitoring of disease recurrence (Patel *et al.*, 2020; Chhillar & Singh, 2023). The incorporation of "multi-omics" and "non-omics" data has appeared as a promising tactic to overcome cancer diagnosis challenges and this has helped in cancer detection at early stages. Various diagnostic and treatment methods used to determine the kinds of cancer cells and their treatment sensitivity include positron emission tomography, magnetic resonance imaging (MRI), computed tomography, mass spectrometry and genomic sequencing, polymerase chain reaction (Wisniewska *et al.*, 2024). Artificial Intelligence (AI) or Machine Learning is exceptionally capable of handling huge amounts of multifaceted non-linear files (multi-omics and non-omics) that are generated from cancer researches and treatments that are capable of improving oncologic care (Patel *et al.*, 2020). Conventional remedies in oncology such as radiotherapy and chemotherapy remain the backbone of treatment, but in many circumstances, a targeted approach should be considered to avoid vulnerability of the patient to drug resistance (Menyhárt & Győrffy, 2021; Chhillar & Singh, 2023; Khan, 2024).

8.7 Advances in Cancer Prognosis

Cancer is a multifaceted and complex disease having been a focus of extensive research and medical advancements. The scientific community recently took remarkable strides in understanding the underlying molecular mechanisms, development of innovative diagnosis and therapeutic approaches in cancer (Nayak *et al.*, 2021). The rapid advancements in genomic, proteomic, metabolomic, and epigenomic technologies have revolutionized the understanding of cancer biology, enabling researchers to uncover novel actionable mutations and delineate the genetic basis of various cancer types (Nayak *et al.*, 2021).

Integration of big data, bioinformatics, and artificial intelligence has been a major progress in cancer research. Multinational missions, for example, the International Human Cancer Consortium (IHCC) and the Cancer Genome Atlas, have generated vast amounts of molecular and phenotypic data, which are now being leveraged to advance a comprehensive understanding of cancer hallmarks and to develop personalized treatment strategies (Hassan *et al.*, 2022). The application of advanced computational analyses, including the incorporation of non-omics and multi-omics data, has the potential to solve accuracy challenges in detection, categorization, and monitoring of cancer. AI-driven analysis of the complex and heterogeneous data can offer a holistic understanding of the mechanism, novel diagnosis and prognosis markers, which enable accurate management beyond the traditional disease stratification arrangements (Patel *et al.*, 2020; Hassan *et al.*, 2022). There is a significant variation in the level of toxicity that patients suffer for each dose of radiation, which is linked to variations in the underlying causes of each person's normal tissue radiosensitivity. The lymphocyte apoptosis evaluation is a good prospect, and presently under soon-to-be authentication. Genetic analysis and individualized prescriptions could lessen toxicity experienced by patients undergoing radiosensitive, thus, allows isotoxic dose rise for local regulator increase in radio-resistant individuals for better outcomes. Gene expression reporting provides prognostic data that cannot be got from histologic or clinical analysis (Hassan *et al.*, 2022; Gabbia & Martin, 2023).

8.8 Precision Medicine and Targeted Therapies

Genomic information is progressively being integrated into clinical cancer attention through the provision of clinically pertinent information concerning somatic point mutations, translocations, copy number alterations, and gene fusions. Large-scale sequencing explorations have developed the knowledge of the genomic setting in cancer, thus contributing to the formation sequence of alterations that are used for supporting cancer prognosis, diagnosis, and management (Bombard *et al.*, 2013; Chakravarty & Solit, 2021; Zalli *et al.*, 2023). Current clinical trials are assessing the use of agent working at molecular level in genomically based divisions of cancers across tumor histologies (Siddiqui, 2024). Better thoughts about patient-to-patient genomic variations in breast cancer are evolving, making the possibility of directing the right treatment to the exact patient at the accurate time, which is considered "precision cancer medicine" (Zalli *et al.*, 2023; Siddiqui, 2024). PARPi (PARP inhibitors), is a cancer remedy aiming at poly (ADP-ribose) polymerase exploiting synthetic lethality as a genetic model. Tumors rising in patients carrying germline mutations in any of *BRCA2* or *BRCA1* are sensitive to PARPi because they possess a particular kind of defect DNA restoration. PARPi displays encouraging action in many common cancers that share the modification defect *via* resistance to PARPi rises in disease (Siddiqui, 2024).

8.9 Immunogenomics and Cancer Therapy

Immunogenomics, the intersection of immunology and genomics, has appeared to be a potential field in cancer therapy, offering novel insights and approaches harnessing the influence of the immune system in the fight against this formidable disease. The immune system, with its complex network of cells and signaling pathways, has the remarkable possibility to identify and eliminate cancer cells, making it a striking target for therapy (Waldman *et al.*, 2020). Immunotherapy has recently gained significant traction, aiming to encourage the immune system's natural energy to detect and extinguish cancer cells. Immunotherapy has shown outstanding success in certain cancer types, with the potential to induce durable responses and improve patient outcomes, even in cases where traditional therapies have failed (Quitt *et al.*, 2021). The development of immunotherapies has been heavily influenced by the growing understanding of the molecular mechanisms that govern the interface between cancer cells and the body immune system. Researchers have identified key immune barriers, such as CTLA-4 and PD-1, which are often hijacked by cancer cells to elude immune recognition and destruction. By targeting these checkpoints, immunotherapies can reinvigorate the immune system, enabling it to support a more operational anti-tumor response. Furthermore, the immunogenomics field has provided valuable insights into the epigenetic and genetic factors that shape the immune reaction to cancer (Waldman *et al.*, 2020; Quitt *et al.*, 2021).

8.9.1 Neoantigen-Based Therapies

Cancer immunotherapies use medicines to stimulate patients' personal immunity against tumor cells. An encouraging tactic for cancer immunotherapy includes the introduction of personalized vaccines that are considered to activate *de novo* T cell reactions against neoantigens; these are targets of specific individuals, in order to intensify and widen the endogenous selection of tumor-explicit T cells, and this has called for the need for regulatory concerns in tumor immunotherapies, especially for vaccines, antibody-based therapies, and adoptive cell therapy (Blass & Ott, 2021). Neoantigens are newly discovered antigens that are generated by tumor cells as a result of several tumor-specific alterations, including integrated viral open reading frames, dysregulated RNA splicing, disordered post-translational modification, and genomic mutation. The development of NGS and bioinformatics is probably responsible for the quick identification and extrapolation of tumor-specific neoantigens. In contrast, tumor-specific neoantigens provide novel targets for tailored or customized cancer immunotherapies, because tumor-associated antigens are highly immunogenic. In conclusion, investigating the main mechanism of the neoantigen-induced anti-tumor immune response and the reformation process of neoantigen-based immunotherapies would advance the development and further discovery of cancer therapies (Xie *et al.*, 2023; Ho *et al.*, 2023).

8.9.2 Cancer-Immune-Tumor Microenvironment

Tumor Microenvironment (TME) is a congregation for a multifaceted system of heterogeneous stromal cells with corresponding or contradictory functions relying on the main signals within the environment. Mutual paracrine interactions between cancer cells with other cells within the tumor stroma usually restructure the TME in other to promote tumor (Tallima & El-Ridi, 2022). Recently, the paradigm in cancer treatment has shifted to immunotherapy, such as immune checkpoint inhibitors (ICI) that have been seen to show excellent clinical results. It is paramount to explore the features of TME that could influence the effectiveness of immunotherapy, which is a process whereby cancer matures and at the same time escaping invades from the body immune system contained in the TME (Kim and Cho, 2022). Tumors cells evade attacks from body immunity *via* different mechanisms such as obstructing the immune system, induction of T cell collapse and limiting antigen identification. Also, tumors accumulate particular metabolites and signal features within the TME or starve the immune cells as a way of evade or inhibit the immune system (Kim and Cho, 2022).

8.9.3 Cancer Genome-Derived Vaccine Development

Vaccines are developed using contributions from proteomics, microarrays, immunoinformatics, and high-throughput immunology analyses with better management of available data in genome databases to a convenient size (Figure 8.2). The three forces involved in vaccines development are computational immunology, microbial genome sequencing, and fresh vaccine

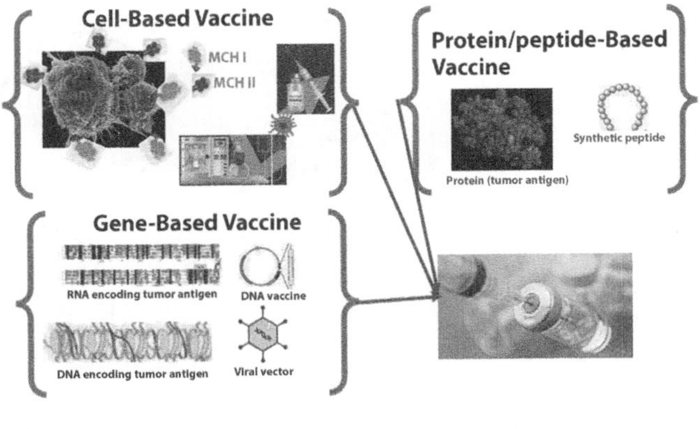

Figure 8.2 Schematic description of anticancer vaccine development.

technologies, thus shifting genome excavating for vaccines into a leading aspect of immunology research (Joon *et al.*, 2022). Although it is challenging to identify the antigens that effectively trigger the immune response to cancer cells, it is also possible that capable immune responses against cancer are limited to the neoplasm's "immunome." The components of the immunome are identified by the use of high-throughput immunology testing, microarrays, bioinformatics sequence analysis techniques, and epitope mapping techniques in order to develop new vaccines. Even up to clinical trials, the immunome-derived vaccine era has been much improved by the quick development of immunoinformatics techniques and immunological assays (Penson *et al.*, 2020).

8.9.4 Adoptive T-Cell Treatment and CAR-T Cells

Adoptive T-cell treatment has appeared as a favorable methodology in the area of cancer immunotherapy, featuring the advantage of harnessing the influence of the patient's immunity to invade malignant cells. The tumor microenvironment (TME) is a complex ecosystem consisting of cancer cells, immune cells, fibroblasts, blood vessels, and extracellular matrix. In this environment, cancer cells develop mechanisms to evade immune surveillance and promote immune suppression. Understanding these mechanisms is crucial for developing effective cancer immunotherapies (Figure 8.3) (Yu *et al.*, 2024). Chimeric antigen receptor T-cell therapy is one of the

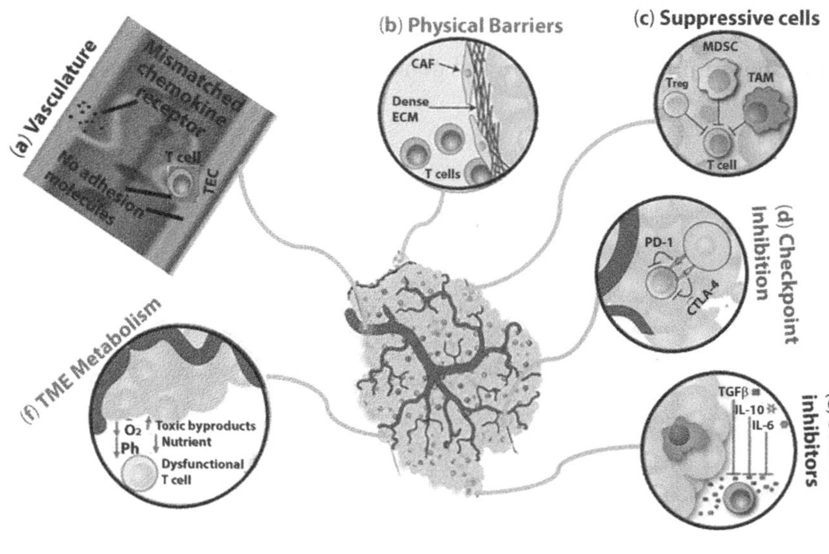

Figure 8.3 Processes of immune suppression in the tumor microenvironment.

new strategies involving engineering the patient's T cells that will identify and target particular antigens which are exposed on the surface of cancer cells. T cells extracted directly from the patient (autologous approach) or from healthy donors (allogeneic method) may be used in T-cell treatment (López-Cantillo *et al.*, 2022). Adoptive T-cell therapy, which uses either autologous or allogeneic T cells to deliver therapeutic immunity, is also used in autologous situations. In autologous situations, patient-derived T cells are modified to express TCR or chimeric antigen receptor (CAR) transgenes that are reactive to tumor-associated antigens (López-Cantillo *et al.*, 2022). Furthermore, the development of CAR-T cell treatments for the treatment of hematological cancers, such as lymphoma and B-cell leukemia, has sparked significant interest in the field of oncology (Titov *et al.*, 2021; López-Cantillo *et al.*, 2022).

8.10 Challenges and Limitations in Cancer Genomics

Cancer genomics, the study of the genomic landscape in oncology area, has been a rapidly advancing field with significant effects for the understanding and management of this complex disease. However, the field is not without its challenges and restrictions, which need to be attended to for realization of the concept of precision medicine in cancer. Cancer drug therapy is progressing from the former pregenomic cytotoxic age to the fresh postgenomic era. Some of the challenges remain, such as the idea of individualized combinatorial genome-based remedy, by using drugs that target a significant molecular aberration in cancer (Hassan *et al.*, 2022; Siddiqui, 2024). Another primary challenge is the heterogeneity of the disease because cancer cells can exhibit vast genetic diversity within a single tumor, with various subclones harboring distinct genomic modifications (Monberg, 2022). The intra-tumor heterogeneity can obscure the interpretation of genomic information and obstruct the identification of driver mutations that are dire for cancer progression (Mongerg, 2022). Innovative machine learning methods are eligible for recognizing the too complex patterns in the enormous datasets but these perspective technical survey analyses in Machine Learning (ML), and Deep Learning (DL) methodologies for genome analysis also have their limitations (Abdelazim *et al.*, 2020). Conclusively, the challenges include the data analysis and interpretation complexities, standardization and regulatory issues, tumor heterogeneity and resistance to therapy and most importantly ethical considerations protecting genetic privacy and informed consent.

8.11 Future Perspectives

Despite better understanding of cancer, much demands further exploration, especially the process through which genomic and non-genomic features recruit carcinogenesis, motivate cell invasion with metastasis, and thereby empower cancer to grow. Technology advancements aid the exploration of whole genomes, involving different tumors across different populations globally, so that mutation signatures peculiar with specific tumor types can be identified. There are issues bordering the implementations of vital genomic outcomes in clinical situations with disparate resource availability and which could affect the regulations and practices that will benefit the populace (Ginsburg *et al.*, 2021). How to address the high level of genetic and epigenetic heterogeneity in cancer caused by different biological factors (metabolic processes, genetic and epigenetic variations) and environmental impacts that can raise the risk of cancer formation depending on the circumstances is one of the main challenges facing the TCGA Project (Zalli *et al.*, 2023). Development of sequencing tools and bioinformatics will improve the sensitivity, accuracy and costs of diagnosing somatic mutations. The large-scale missions should also be used to organize cancer genome sequencing to aid the production and distribution of high-quality constant genetic information for more opportunities in studies and applications in cancer pathogenesis (Hassan *et al.*, 2022). A significant future challenge that must be solved to bridge the gap between research and clinical application is the translation of large oncogenomic data to clinical usage in personalized medicine agendas. According to Siddiqui (2024), narrowing the gap will require translational customized trials, which will undoubtedly lead to a significant shift in the standard clinical procedures in oncology.

8.12 Conclusion

The multi-gene disease known as cancer is caused by mutations and epigenetic changes that are connected to the activation of complex signaling pathways. The use of biomarkers for early cancer detection, staging, and treatment personalization has improved patient care. Biomarker development will undoubtedly need to be addressed due to a number of important issues, including tumor heterogeneity, which is a very dynamic aspect of the extrinsic and intrinsic determinants of radio- and chemo- resistance, as well as the adaptability and diversity of cancer stem cells (CSC).

The advent of proteomic and genomic tools with advancement of bioinformatic approaches permits the analysis of various genetic molecules at the same time making the detection of new tumor autographs, which are useful for early disease detection, progression monitoring, and proper treatment choice, supporting the method of individualized cancer treatment (Hassan *et al.*, 2022; Siddiqui, 2024). More precise diagnostic instruments, predictive biomarkers, and tailored treatments have been made possible by recent advancements in genomic technologies, bioinformatics, and precision medicine. Personalized, genomics-driven strategies that customize therapy to the unique molecular makeup of each malignancy are ultimately the way of the future for cancer treatment (Siddiqui, 2024). Furthermore, advancements in next-generation sequencing techniques are facilitating the methodical examination of entire cancer genomes, raising awareness of the genetic heterogeneity and the history of somatic mutations.

AI disclosure statement: The author acknowledges the use of AI-based tools for minor language editing or grammatical correction. However, all intellectual contributions, analyses, and conclusions presented in this chapter are entirely the author's own.

References

Abdelazim, M.A., Nasr, M.M., Ead, W.M., A survey on classification analysis for cancer genomics: Limitations and novel opportunity in the era of cancer classification and Target Therapies. *Ann. Trop. Med. Public Health*, *23*, 24, 2020.

Blass, E. and Ott, P.A., Advances in the development of personalized neoantigen-based therapeutic cancer vaccines. *Nat. Rev. Clin. Oncol.*, *18*, 4, 215–229, 2021, https://doi.org/10.1038/s41571-020-00460-2.

Casanova-Salas, I., Aguilar, D., Cordoba-Terreros, S., Agundez, L., Brandariz, J., Herranz, N., Mateo, J., Circulating tumor extracellular vesicles to monitor metastatic prostate cancer genomics and transcriptomic evolution. *Cancer Cell*, *42*, 7, 1301–1312, 2024.

Chakravarty, D. and Solit, D.B., Clinical cancer genomic profiling. *Nat. Rev. Genet.*, *22*, 8, 483–501, 2021, https://doi.org/10.1038/s41576-021-00338-8.

Charmsaz, S., Prencipe, M., Kiely, M., Pidgeon, G.P., Collins, D.M., Innovative technologies changing cancer treatment. *Cancers (Basel)*, 10, 6, 208, Jun 19 2018, doi: 10.3390/cancers10060208.

de Anda-Jáuregui, G. and Hernández-Lemus, E., Computational oncology in the multi-omics era: State of the art. *Front. Oncol.*, *10*, 423, 2020, https://doi.org/10.3389/fonc.2020.00423.

Gabbia, D. and De Martin, S., Tumor mutational burden for predicting prognosis and therapy outcome of hepatocellular carcinoma. *Int. J. Mol. Sci.*, *24*, 4, 3441, 2023, https://doi.org/10.3390/ijms24043441.

Ginsburg, O., Ashton-Prolla, P., Cantor, A., Mariosa, D., Brennan, P., The role of genomics in global cancer prevention. *Nat. Rev. Clin. Oncol.*, *18*, 2, 116–128, 2021, https://doi.org/10.1038/s41571-020-0428-5.

Ho, S.Y., Chang, C.M., Liao, H.N., Chou, W.H., Guo, C.L., Yen, Y., Chang, W.C., Current trends in neoantigen-based cancer vaccines. *Pharmaceuticals, 16*, 3, 392, 2023, https://doi.org/10.3390/ph16030392. https://doi.org/10.1016/j.jgg.2011.02.003.

Hu, T., Chitnis, N., Monos, D., Dinh, A., Next-generation sequencing technologies: An overview. *Hum. Immunol.*, 2021, doi: 10.1016/j.humimm.2021.02.012.

Khan, R.U., Innovations in Cancer Diagnosis and Challenges. *Pak. J. Health Sci.*, *5*, 08, 01, 2024, https://doi.org/10.54393/pjhs.v5i08.2187.

Kim, S.K. and Cho, S.W., The evasion mechanisms of cancer immunity and drug intervention in the tumor microenvironment. *Front. Pharmacol.*, *13*, 868695, 2022, https://doi.org/10.3389/fphar.2022.868695.

Kumar, K.R., Cowley, M.J., Davis, R.L., Next-generation sequencing and emerging technologies, in: *Seminars in Thrombosis and Hemostasis*, Thieme Medical Publishers, 2024.

Lappalainen, T. and MacArthur, D.G., From variant to function in human disease genetics. *Science*, *373*, 6562, 1464–1468, 2021, https://doi.org/10.1126/science.abi8207.

Menyhárt, O. and Győrffy, B., Multi-omics approaches in cancer research with applications in tumor subtyping, prognosis, and diagnosis. *Comput. Struct. Biotechnol. J.*, *19*, 949–960, 2021.

Meystre, S.M., Gouripeddi, R., Alekseyenko, A.V., Molecular, Genetic, and Other Omics Data, in: *Clinical Research Informatics*, pp. 309–328, Springer International Publishing, Cham, 2023, doi.org/10.1007/978-3-031-27173-1_16.

Monberg, M., Leveraging Single Cell Technologies for the Characterization and Treatment of Refractory Pancreatic Cancer. https://www.researchgate.net/publication/361717657_Leveraging_single_cell_technologies_for_the_characterization_and_treatment_of_refractory_pancreatic_cancer, 2022.

Nones, K. and Patch, A.M., The impact of next generation sequencing in cancer research. *Cancers*, *12*, 10, 2928, 2020, https://doi.org/10.3390/cancers12102928.

Patel, S.K., George, B., Rai, V., Artificial intelligence to decode cancer mechanism: Beyond patient stratification for precision oncology. *Front. Pharmacol.*, *11*, 1177, 2020, https://doi.org/10.3389/fphar.2020.01177.

Penson, A., Camacho, N., Zheng, Y., Varghese, A.M., Al-Ahmadie, H., Razavi, P., Berger, M.F., Development of genome-derived tumor type prediction to inform clinical cancer care. *JAMA Oncol.*, *6*, 1, 84–91, 2020, doi: 10.1001/jamaoncol.2019.3985.

Quitt, O., Luo, S., Meyer, M., Xie, Z., Golsaz-Shirazi, F., Loffredo-Verde, E., Protzer, U., T-cell engager antibodies enable T cells to control HBV infection and to target HBsAg-positive hepatoma in mice. *J. Hepatol.*, *75*, 5, 1058–1071, 2021, https://doi.org/10.1016/j.jhep.2021.06.022.

Satam, H., Joshi, K., Mangrolia, U., Waghoo, S., Zaidi, G., Rawool, S., Thakare, R.P., *et al.*, Next-generation sequencing technology: Current trends and advancements. *Biology*, 12, 7, 997, 2023, https://doi.org/10.3390/biology12070997.

Waldman, A.D., Fritz, J.M., Lenardo, M.J., A guide to cancer immunotherapy: From T cell basic science to clinical practice. *Nat. Rev. Immunol.*, *20*, 11, 651–668, 2020, https://doi.org/10.1038/s41577-020-0306-5.

Wisniewska, K., Marku, E., Ugurbas, M.V., Hartmane, I., Shukurova, M., Innovations in cancer diagnosis and treatment: prospects and challenges. *Healthc. Low-resour. Settings*, 2024.

Xie, N., Shen, G., Gao, W., Huang, Z., Huang, C., Fu, L., Neoantigens: Promising targets for cancer therapy. *Signal Transduction Targeted Ther.*, *8*, 1, 9, 2023, https://doi.org/10.1038/s41392-022-01270-x.

Yu, L., Fu, Y., Miao, R., Cao, J., Zhang, F., Liu, L., Ou, M., Regulatory T Cells and Their Derived Cell Pharmaceuticals as Emerging Therapeutics against Autoimmune Diseases. *Adv. Funct. Mater.*, 2405133, 2024, https://doi.org/10.1002/adfm.202405133.

Zalis, M., Viana Veloso, G.G., Aguiar Jr., P.N., Gimenes, N., Reis, M.X., Matsas, S., Ferreira, C.G., Next-generation sequencing impact on cancer care: Applications, challenges, and future directions. *Front. Genet.*, *15*, 1420190, 2024, https://doi.org/10.3389/fgene.2024.1420190.

Zalli, D., Mai, Z., Ferati, E., Ramaj, A., Bregu, R., Pranjol, M.Z.I., Advancing Precision Medicine: Integrating Next-Generation Sequencing and Tumor Markers for Early Cancer Detection and Personalized Treatment, in: *Handbook of Cancer and Immunology*, pp. 1–31, 2023, https://doi.org/10.1007/978-3-030-80962-1_413-1.

Siddiqui, F., The Role of Genomic Medicine in Cancer Treatment and Diagnosis. *Front. Biotechnol. Genet.*, *1*, 02, 166–183, 2024.

Hassan, M., Awan, F.M., Naz, A., deAndrés-Galiana, E.J., Alvarez, O., Cernea, A., Fernández-Brillet, L., Fernández-Martínez, J.L., Kloczkowski, A., Innovations in genomics and big data analytics for personalized medicine and health care: A review. *Int. J. Mol. Sci.*, 23, 9, 4645, 2022, https://doi.org/10.3390/ijms23094645.

Titov, A., Zmievskaya, E., Ganeeva, I., Valiullina, A., Petukhov, A., Rakhmatullina, A., Bulatov, E., Adoptive immunotherapy beyond CAR T-cells. *Cancers*, *13*, 4, 743, 2021, https://doi.org/10.3390/cancers13040743.

López-Cantillo, G., Urueña, C., Camacho, B.A., Ramírez-Segura, C., CAR-T cell performance: How to improve their persistence? *Front. Immunol.*, *13*, 878209, 2022, https://doi.org/10.3389/fimmu.2022.878209.

Joon, S., Singla, R.K., Shen, B., Vaccines and immunoinformatics for vaccine design, in: *Translational Informatics: Prevention and Treatment of Viral Infections*, pp. 95–110, Springer Nature Singapore, Singapore, 2022, https://doi.org/10.1007/978-981-16-8969-7_5.

Tallima, H. and El Ridi, R., The Stroma and Tumor Microenvironment Are Not the Most Instrumental Players in Tumor Cells Immune Evasion, Growth, and Resistance to Therapy. https://www.preprints.org/manuscript/202206.0316, 2022.

Nayak, J., Favorskaya, M.N., Jain, S., Naik, B., Mishra, M., *Advanced Machine Learning Approaches in Cancer Prognosis*, Springer International Publishing, 2021.

Chhillar, I. and Singh, A., An Insight into Machine Learning Techniques for Cancer Detection. *J. Inst. Eng. India Ser. B*, 104, 963–985, 2023, https://doi.org/10.1007/s40031-023-00896-x.

Cescon, D.W., Bratman, S.V., Chan, S.M., Siu, L.L., Circulating tumor DNA and liquid biopsy in oncology. *Nat. Cancer*, *1*, 3, 276–290, 2020, https://doi.org/10.1038/s43018-020-0043-5.

Constâncio, V., Lobo, J., Sequeira, J.P., Henrique, R., Jerónimo, C., Prostate cancer epigenetics—from pathophysiology to clinical application. *Nat. Rev. Urol.*, 1–23, 2025, https://doi.org/10.1038/s41585-024-00991-8.

Uffelmann, E., Huang, Q.Q., Munung, N.S., De Vries, J., Okada, Y., Martin, A.R., Posthuma, D., Genome-wide association studies. *Nat. Rev. Methods Primers*, *1*, 1, 59, 2021, https://doi.org/10.1038/s43586-021-00056-9.

He, J. and Gai, J., Genome-Wide Association Studies (GWAS), in: *Plant Genotyping. Methods in Molecular Biology*, vol. 2638, Y. Shavrukov (Ed.), Humana, New York, NY, 2023, https://doi.org/10.1007/978-1-0716-3024-2_9.

Mardis, E.R., The emergence of cancer genomics in diagnosis and precision medicine. *Nat. Cancer*, *2*, 12, 1263–1264, 2021, https://doi.org/10.1038/s43018-021-00305-6.

Pecorino, L., *Molecular biology of cancer: mechanisms, targets, and therapeutics*, Oxford University Press, 2021.

Dalghi, E., Bridging Genomics and Oncology: Molecular Approaches to Personalized Medicine. *Int. J. Biolife Sci. (IJBLS)*, *2*, 2, 281–290, 2023, https://doi.org/10.22034/ijbls.2023.185500.

Santucci, C., Progress in cancer incidence, mortality, and survival. https://hdl.handle.net/2434/951503, 2023.

Abdelgalil, A.A., Al-Kahtani, H.M., Al-Jenoobi, F.I., Erlotinib, in: *Profiles of Drug Substances, Excipients and Related Methodology*, vol. 45, pp. 93–117, Academic Press, 2020, https://doi.org/10.1016/bs.podrm.2019.10.004.

Song, P., Zhou, J., Wu, K., Wang, W., Gu, S., The underlying mechanism involved in gefitinib resistance and corresponding experiment validation in lung cancer. *Mediators Inflammation, 2023*, 1, 9658912, 2023, https://doi.org/10.1155/2023/9658912.

Lyczek, A., Berger, B.T., Rangwala, A.M., Paung, Y., Tom, J., Philipose, H., Seeliger, M.A., Mutation in Abl kinase with altered drug-binding kinetics indicates a novel mechanism of imatinib resistance. *Proc. Natl. Acad. Sci., 118*, 46, e2111451118, 2021, https://doi.org/10.1073/pnas.2111451118.

9

Genomic Research in Infectious Diseases: Impact and Advances

Ananda Kumar Chettupalli[1], Aman Sisodia[2], Rajendra Kumar Jadi[3]*, Akiful Haq[3] and Nagarani Kagithala[2]

[1]*Department of Pharmaceutical Technology, School of Health and Medical Sciences, Adamas Knowledge City, Adamas University, Kolkata, West Bengal, India*
[2]*Department of Pharmaceutical Sciences, School of Pharmacy, Galgotias University, Greater Noida, Uttar Pradesh, India*
[3]*Department of Pharmaceutics, Center for Nanomedicine, School of Pharmacy, Anurag University, Venkatapur, Ghatkesar, Medchal-Malkajgiri, Hyderabad, Telangana, India*

Abstract

Over the last decade, microbiology and infectious disease research have witnessed the most significant transformation since Pasteur. Most infections' blueprints have finally been uncovered thanks to genomic sequencing. The high pace of viral and microbial genomes research, which is mostly driven by the development of genomics technology, provides new data-generating platforms as well as a revolutionary knowledge foundation for better understanding illnesses and infections. Systematic molecular biology studies using genomes information and technologies have aided in the elucidation of virulence and pathogenicity mechanisms, whilst genomics-based medical genetic investigations have been utilized to improve understanding of pathogen susceptibility. Genomics, transcriptomics, metabolomics, structural genomics, proteomics, and immunomics are being used to improve target discovery, develop new vaccines and medications, and forecast their effects in patients. Furthermore, human genomics and associated research are shedding light on components of host biology that are crucial in infectious disease. The introduction of high-throughput technologies such as whole-genome sequencing (WGS), metagenomics, and powerful bioinformatics tools has greatly improved our ability to detect new diseases, track disease outbreaks, and trace transmission networks. These developments have not only increased our ability to respond to emerging and reemerging infections,

**Corresponding author*: rajendra.rajaji@gmail.com

Swetha Kumari Koduru, Sarad Pawar Naik Bukke, Hussaini Bello, Muhamad Mustafa Mundu and Ayush Madan (eds.) Genomics and Precision Healthcare: Innovations, Applications and Challenges, (255–294) © 2026 Scrivener Publishing LLC

but have also accelerated the development of targeted diagnostics, vaccines, and treatments. Unfortunately, the great transformation of basic science research and the resulting profusion of promising data, particularly in the arena of human health and disease, has not been accompanied by a corresponding growth in the practical application of this information. On the contrary, the number of new potential medications in development has continuously decreased, indicating the existence of bottlenecks that limit the conversion of promising research into medically applicable therapeutic or diagnostic applications. The article discusses how genomic technologies can alter infectious disease research, mentioning applications such as antimicrobial resistance surveillance, vaccine development, and pandemic preparedness. Innovative methods including CRISPR-based diagnostics, pan-genomic studies, and artificial intelligence will be used to improve genomic research.

***Keywords*:** Genomic sequencing, antimicrobial resistance, precision medicine, vaccine development, metagenomics, global health challenges

9.1 Introduction

9.1.1 The Evolution of Infectious Disease Research

Infectious disease research has evolved tremendously over time, from simple observation and primitive treatment procedures to the intricate molecular biology techniques used today. Traditionally, understanding infections was dependent on recognizing microbes as disease-causing agents. Early pioneers were Louis Pasteur and Robert Koch, who scientifically founded microbiology and developed the germ theory of germs and disease. However, most of our understanding of pathogen molecular processes and host interactions remained obscure until genomics emerged [1].

With modern molecular biology and sequencing tools, there is tremendous change in the research world of infectious disease. Genomic sequencing has placed the field at a new trajectory in which the identification of pathogens not only proceeds with an unparalleled level of precision but their very genomic blueprints, the fine details of pathogenicity, resistance mechanisms, and virulence are deciphered. This new research development has unlocked the possibilities for new diagnostics, therapeutic strategies, and preventive approaches that will contribute to the understanding of pathogen and host biology [2].

9.1.2 Genomics: A Game-Changer in Understanding Infections

Since the advent of genomics, the way research, diagnosis, and treatment are conducted for infectious diseases has completely changed. Genomics is a study of the whole DNA sequence of an organism, which includes studying the structure, functions, evolution, and mapping. In pathogen genome sequencing, it was found that there existed genetic makeup of bacteria, viruses, and fungi, giving researchers a very powerful tool to study how these pathogens caused disease [3].

Earlier, infections were primarily studied by growing microbes, analysing their activity under the predefined conditions, and knowing their basic biological properties. Now, genomics provides the possibility of going into details concerning genetic factors affecting survival, virulence, and drug resistance of pathogens. Next-generation sequencing has been augmented by developing tools in bioinformatics that have made identification of pathogens much faster and have also provided information about their evolutionary history, mutation rates, and mode of transmission [4].

The real strength of genomics is actually its capacity to provide better information for more tailored diagnoses, vaccinations, and therapeutic options. This will enable the tracking of outbreaks in real time with pinpoint accuracy down to the microbial strains and enhance our understanding of infections' abilities to adapt to their host environments. This has speeded the development of personalized medicine, where the patient genomic profile can direct a more specific therapy strategy relative to susceptibility and immune response [5].

9.1.3 Scope and Significance of the Chapter

This chapter will look at the revolutionary value of genomic research in infectious disease management, considering both gains and challenges it invites to this rapidly developing field. It will examine how technological advances in genomics, such as whole-genome sequencing, metagenomics, and CRISPR-based approaches, have shaped the identification of pathogen, conceived pathogenicity, and devised new therapeutic methods [6].

Aside from summarizing the basic genomics technologies, this chapter discusses the broad areas of applications where genomics is used, as well as how its products are used in antimicrobial resistance (AMR) surveillance, vaccine development, and pandemic preparation. Further covered is the integration of genomics with exciting new technologies such as artificial intelligence and multi-omics techniques to improve the accuracy of predictions of infectious disease cause, diagnostic capabilities, and therapeutic possibilities [7].

Finally, the chapter will conclude by discussing current challenges impeding translation of genomic findings into clinical practice and proposed solutions to address these issues. In this regard, this chapter aims to highlight genomics' critical impact on the future of global health while enhancing patient care around the world by providing a comprehensive overview of the status of genomic research in infectious diseases [8].

9.2 Foundations of Genomic Technologies

9.2.1 Historical Overview of Genomics in Microbiology

One of the most important developments in biomedical science has been the integration of genetics into microbiology. The field of genomics originated in the mid-twentieth century, when scientists discovered DNA as the genetic material. Early microbiological research was primarily limited to culture-based techniques and phenotypic characterizations. However, the major breakthrough occurred in 1977 when Frederick Sanger developed DNA sequencing techniques, allowing for the first time thorough investigation of microbial genomes [9].

The first complete sequence of a bacterial genome, published in 1995 for the bacterium Haemophilus influenzae, marked a great milestone in the field of microbial genomics. It opened avenues for the full sequencing of all major viruses with new insights in their genomic organization, diversity, and potential for pathogenicity. The change was accelerated by human genome sequencing in 2003, which highlighted the enormous potential of genomics not only in microbiology but across all biological sciences [10].

Infectious illness research has progressed beyond the detection of pathogens to include an examination of the genetic structure and mechanisms of pathogen virulence, resistance, and evolution. Genomics has become a significant tool for discovery, epidemiology, diagnostics, treatments, and preventive methods as sequencing technologies have advanced and costs have decreased. Currently, genomics is critical to our understanding of infectious illnesses because it enables real-time epidemic surveillance, the detection of new pathogens, and insights into host-pathogen interactions [11].

9.2.2 Genomic Sequencing Platforms

Innovative sequencing platforms have driven significant progress in genomic technology. Next-Generation Sequencing (NGS) is the foundation of current genomics, replacing previous Sanger sequencing methods due to its high throughput, accuracy, and low cost. NGS has transformed microbial genomics by allowing the sequencing of whole microbial genomes in a fraction of the time and expense required previously [12].

Some of the most popular NGS platforms are as follows:

9.2.2.1 *Illumina Sequencing*

Illumina sequencing utilizes sequencing by synthesis (SBS) technology. It is commonly used for whole-genome sequencing, metagenomics, transcriptomics, and large-scale genomic research. It includes high precision, scalability, and cost-effectiveness for big sequencing programs [13].

9.2.2.2 *Ion-Torrent Sequencing*

Based on semiconductor technology, hydrogen ions generated during DNA polymerization are detected and analysed. It is mostly utilized for targeted sequencing and small to medium-sized sequencing projects. It has quick turnaround and is portable and easy to use [14].

9.2.2.3 *PacBio Sequencing*

This uses Single Molecule, Real-Time (SMRT) sequencing technology, long-read sequencing and complicated genome assembly. The major advantages are long reads allow for the entire assembly of difficult-to-sequence genomic areas, particularly repetitive sequences and structural changes [15].

9.2.2.4 *Oxford Nanopore Sequencing*

It enables real-time sequencing of lengthy DNA segments as they pass through a protein nanopore. Field-based applications include quick genomic surveillance for pathogens, which is especially beneficial in the event of an outbreak. Advantages include portability and readiness, real-time sequencing, produces lengthy reads [16].

9.3 Overview of Bioinformatics Tools for Genomic Data Analysis

The genetic data explosion caused by sequencing platforms highlighted the significance of bioinformatics tools for analysing, interpreting, and storing data. Bioinformatics refers to the computational methods and techniques used to analyse and visualize biological data, particularly genetic sequences. NGS produces complex and abundant data, necessitating more powerful computational pipelines, and these technologies have received a lot of attention to meet this [17].

9.3.1 BLAST (Basic Local Alignment Search Tool)

BLAST is used to identify local regions of similarity among sequences. Researchers can search for putative homologs by comparing DNA, RNA, or protein sequences to databases. The Basic Local Alignment Search Tool is useful for aligning sequences, identifying diseases, and comparing microbial genomes to reference databases [18].

9.3.2 Galaxy

Galaxy is a web-based platform for creating, running, and sharing bioinformatics workflows. Microbial genomics include sequencing assembly, annotation, and analysis. It provides a user-friendly interface for non-programmers [19].

9.3.3 Bowtie and Burrows-Wheeler Aligner (BWA)

BWA is the tool for aligning short DNA readings to a reference genome. This program is used to analyse high-throughput sequencing data, including variant detection, mutation finding, and read alignment in microbial genomes [20].

9.3.4 Cufflinks

Cufflinks is a tool for RNA-Seq analysis that assembles and quantifies transcripts. Applications: It has been used in transcriptomics research to assess gene expression and identify differentially expressed genes as a result of infection [21].

9.3.5 VEP (Variant Effect Predictor)

VEP predicts the functional impact of genetic variations in genomes. It is used to assess the impact of mutations on pathogens and hosts in order to better understand host virulence, resistance, and susceptibility [22].

9.3.6 IGV (Integrative Genomics Viewer)

IGV is a visualization tool for discovering genomic data and aligning sequencing results to reference genomes. Used in large-scale genomic datasets to facilitate interpretation of sequence variants, structural variances, and gene expression levels [23].

9.3.7 QIIME (Quantitative Insights into Microbial Ecology)

QIIME is a bioinformatics pipeline for microbiome analysis using 16S rRNA gene sequencing and metagenomics data. Therefore, it is crucial for studying the microbiome, recognizing microbial diversity, and understanding microbial populations in infectious disorders [24].

9.3.8 DeepVarant

DeepVarant is a deep learning-based variant caller for NGS data. It is based on deep learning that improves accuracy in the identification of complex genomes to some extent [25].

9.4 Technological Advances in Genomic Research

9.4.1 Whole Genome Sequencing (WGS)

The ability to completely sequence a pathogen's genome has revolutionized the study of infectious diseases. This method dramatically increases the detection and characterization of pathogens, so that strains otherwise very closely related can be distinguished and the pathogen's history can be followed. WGS is particularly useful in outbreak investigations because it can help trace transmission chains, identify sources of infection, and carry out targeted control measures. For instance, the sequencing of the genome is vital in tracking the emergence of new variants in the COVID-19 pandemic, their impact on transmissibility, and effectiveness of vaccination. Mutational changes associated with AMR can also be assessed using WGS. The scientific community can discover treatments and the spread of resistant bacteria by recognizing genetic changes connected to resistance mechanisms. This capacity assesses the genetic variation within diseases and could be used in identifying conserved regions, which should be great sites for developing vaccines [26].

9.4.2 Metagenomics

Metagenomics is an approach that is revolutionary in infectious diseases, where sequences of DNA are achieved directly from the complex microbial environment without requiring culture. It has proven quite useful in microbiome research, novel pathogen detection, and the dynamic interactions of microbes in their many habitats; including the human body, soil, and wastewater. It significantly offers the ability to detect diseases that can't be cultivated using traditional microbiological procedures. For example, researchers have used metagenomics in identifying earlier recognized viruses and bacteria causing respiratory and gastrointestinal diseases. In the second place, AMR surveillance critically applies this concept, mainly in detecting resistance genes found in microbial populations. Applied in environmental monitoring, for example, wastewater-based epidemiology, metagenomics permits the early detection of new disease risks, thus offering proactive avenues through which public health interventions may be implemented [27].

9.4.3 Transcriptomics and Proteomics

Transcriptomics and proteomics are two vital tools for identifying gene expression and protein function in infectious diseases. Transcriptomics, primarily with the aid of RNA sequencing, has enabled the study of expression patterns of genes in pathogens under various conditions. It gives insights regarding the pathogen's response to environmental stimuli and host immunological defence mechanisms. Hence, critical regulatory mechanisms associated with virulence and pathogenicity may be identified. However, proteomics identifies the proteins expressed by pathogens, thus providing a deeper understanding of their biological roles. These approaches are extremely useful for researching host-pathogen interactions because they disclose how pathogens alter host cellular processes to establish infection. Furthermore, transcriptome and proteomic analysis aid in the identification of biomarkers for early illness detection and personalized therapy techniques. Analysing the differences in gene and protein expression between infected and healthy persons can lead to the development of targeted medicines for improving treatment outcomes [28].

9.4.4 CRISPR and Genome Editing Technologies

CRISPR has provided the holy grail of medical microbiology and infectious disease research, opening new avenues for pathogen surveillance and genetic changes to facilitate therapy development. CRISPR-Cas systems are increasingly being used for rapid and accurate pathogen detection, with diagnostic platforms such as SHERLOCK and DETECTR providing very sensitive and specific identifications of viral and bacterial infections. Apart from diagnostics, another key product that has enabled the detection of new infectious diseases early on and the reinforcement of timely public health interventions has been developed. Some of the other applications of CRISPR include editing microbial genomes to further the understanding of mechanisms of disease and targeted therapeutics. For instance, methods based on CRISPR are being used for developing bacteriophages selectively killing antibiotic-resistant bacteria as an alternative to traditional antibiotics. Furthermore, genetic editing of pathogens with the help of CRISPR technology has assisted in the creation of attenuated vaccine strains, thus making vaccinations both more effective and safer. Synthetic biology coupled with CRISPR may prove to be a promising approach to the development of customized probiotics that can outcompete harmful bacteria and provide new avenues for the treatment of infectious diseases [29].

9.5 Understanding Pathogenesis through Genomics

9.5.1 Virulence Factors Identification

Genomic research has greatly enhanced our ability to identify and define virulence components in pathogenic bacteria. It has allowed the investigation of whole-genome sequences, which reveal particular genes that allow a virus to penetrate host tissues, evade immune responses, and cause disease. One of the most powerful applications of genomics in this field is gene annotation, which enables scientists to predict gene function based on sequence similarities to known virulence factors. The study of the comparison between the pathogenic and non-pathogenic strains has proved useful for determining virulence-associated genes crucial to their life cycles. Genome-based studies have established toxin-, adhesin- and secretory-mechanism-coding genes for various pathogens, such as Escherichia coli and Salmonella, that have caused infection in many patients. This virulence-variable mapping proves crucial to gaining an insight into mechanisms of diseases, thus implying a necessity to carry out targeted therapy and vaccinations [30].

9.5.2 Molecular Mechanisms of Pathogenicity

Pathogens have varied mechanisms of molecular strategies that aid in the establishment of infection, evasion of the immune system, and adaptation to their host environment. Such strategies include various tactics discovered by genomic studies that further identified genes implicated in immune evasion, host cell manipulation, and stress tolerance. The microbes Mycobacterium TB and Staphylococcus aureus have been engineered such that they evade the host's immune system either through proteins decorating the surface, disrupting phagocytosis or acting as host-molecule mimics. Further, genomic studies have provided answers to why some viruses like HIV and SARS-CoV-2, escape neutralizing antibodies, causing protracted infections, and thus are hard to immunize with suitable vaccines. Beyond that, pathogens need to adapt themselves constantly to environmental challenges like antibiotic exposure, temperature variations, and oxidative stress. Genomic studies have unravelled regulatory networks controlling gene expression in response to such obstacles. Such findings allow the understanding of how bacteria construct antibiotic resistance and viruses enhance their replication efficiency under host settings. This genetic modulation not only furthers our comprehension of the biological aspects of a pathogen, but also points to novel pharmacological targets, as well as disease intervention approaches [31].

9.6 Genomics in Disease Detection and Diagnosis

9.6.1 Next-Generation Diagnostics

The integration of genomic technologies in disease diagnostics improves the speed, accuracy, and sensitivity of detecting pathogens. Next-generation diagnostics has utilized advanced genomics techniques including CRISPR-based platforms and genomic markers in improving early identification and surveillance of illness. Among the CRISPR-based diagnostic tools are SHERLOCK, which is specific, high-sensitivity enzymatic reporter UnLOCKing, and DETECTR, both being sensitive and quick. These detect specific pathogen genetic sequences and deliver results in minutes, without requiring advanced laboratory facilities. Such techniques have been applied successfully in resource-constrained situations to detect infectious diseases such as COVID-19, TB, and the Zika virus [32].

Outside of CRISPR, genomic markers are also important for the early detection and surveillance of any illness. Once specific genetic signatures for infectious agents are established, highly specific molecular diagnostics can be designed, including PCR-based assays and NGS-driven diagnostic panels. In addition to early diagnosis, genetic markers help to track pathogen evolution, antibiotic resistance, and treatment efficacy. Clinically, genomic-based techniques have already transformed how pathogen strains may be easily recognized and discriminated, hence increasing the speed and efficacy of therapies during infectious disease epidemics [33].

9.6.2 Genomic Epidemiology

Genomic epidemiology is a relatively young subject that uses WGS and other genomics methods to study the distribution and evolution of infectious illnesses. Comparing pathogen genomes at different time intervals and locales allows researchers to reconstruct transmission patterns, detect mutation hotspots, and potentially predict future outbreaks. This has proven to be extremely useful in addressing global health emergencies, particularly for monitoring emerging and reemerging infectious illnesses [34].

One of the most illuminating examples of genetic epidemiology was the detection of SARS-CoV-2 mutations during the COVID-19 epidemic. Sequencing viral genomes from multiple geographical regions enabled the identification and categorization of variants of concern based on genetic changes and the resulting implications for transmission and vaccination efficacy. This genomic information influenced public health initiatives, directed vaccination updates, and led to the implementation of specific containment measures in certain cases [35].

Outside of the COVID-19 application, genomics-based surveillance has greatly facilitated the identification of outbreaks for some bacterial diseases such as Salmonella and E. coli. Some more applications of Genomics in research of infectious disease are shown in Table 9.1. Furthermore, genomic epidemiology has played a huge role in comprehensively understanding the antimicrobial resistance for Mycobacterium TB and Neisseria gonorrhoeae; in return, better surveillance and intervention tactics are followed [36].

Incorporating genetic data into the traditional methodologies of epidemiology will help the public health authority develop better preventive and control plans against diseases. Further development in sequencing technologies and tools in bioinformatics will widen the scope of genomic epidemiology as a foundational practice in the new infectious disease surveillance and control of outbreaks [37].

Table 9.1 Key genomic technologies and their applications in infectious disease research.

Genomic technology	**Applications**	**Impact on infectious disease research**
Whole Genome Sequencing (WGS)	Pathogen identification, epidemiological surveillance	Enables the discovery of new pathogens, tracking outbreaks, understanding mutations and strain variations
Metagenomics	Pathogen discovery in complex microbial environments, environmental monitoring	Identifies pathogens without culturing, detects unknown microbes, enables comprehensive microbiome analysis
Transcriptomics	Gene expression profiling in pathogens and hosts	Helps understand pathogen behaviour and host responses, aids in identifying virulence factors
Proteomics	Protein expression analysis in pathogens and infected hosts	Provides insights into protein functions, host-pathogen interactions, and immune responses
CRISPR/Cas9 Technology	Gene editing, pathogen genome surveillance, diagnostic applications	Allows precise genome editing for studying pathogens, creates novel diagnostics, and can be used in therapeutic applications
Bioinformatics	Data analysis, pathogen genome assembly, molecular modelling	Speeds up genome sequence analysis, aids in drug design, and identifies biomarkers for diagnostics
Artificial Intelligence	Analysing large genomic datasets, predicting disease outcomes and treatments	Improves precision medicine, accelerates drug discovery, and enhances predictive epidemiological models

9.7 Human Genomics and Infectious Disease Susceptibility

9.7.1 Genetic Predisposition to Infections

Human genetic variation determines an individual's susceptibility to infectious diseases. Some genetic variables influence how successfully the human immune system fights pathogens, hence determining illness severity and development. For example, mutations in the SLC11A1 (formerly known as NRAMP1) gene are linked to tuberculosis susceptibility because they alter macrophage activity. In addition, genetic variations of the Duffy antigen receptor for chemokines gene give resistance to Plasmodium vivax malaria by preventing the parasite from penetrating red blood cells. GWAS have identified a variety of loci associated with increased or decreased susceptibility to numerous infectious diseases, highlighting the importance of human genomics in understanding host-pathogen interactions. These findings pave the way for the development of tailored therapies such as genetic screening in at-risk populations and precision medicine approaches to infectious disease management [38].

9.7.2 Immune System Genomics

Genomics has revealed that there is a great deal of genetic variation among the genes that comprise the immune system, indicating how one would react to pathogens. Humans' HLA systems are highly genetically diverse. It influences susceptibility to viruses, bacteria, and parasites by regulating antigens' ability to deliver their peptides. Certain HLA alleles have been associated with a range of HIV clinical outcomes, most notably slower progression to disease for some carriers of the B57 and B27 alleles. Increased immune vigilance against viral peptides underlies this altered phenotype. Polymorphisms in cytokine genes, including IL-6 and TNF-α, impact the level of inflammation observed in disorders such as sepsis and COVID-19 [39].

Immune genomics predicts susceptibility to infection, as well as patient outcomes and treatment responses. For instance, alterations in the gene IFNL3, which was known as IL28B, can drive response to interferon-based therapy in treatment of the hepatitis C virus. Individualized therapeutic regimens are now pursued based on such genetic variations, determining the level of illness as well as doing utmost with available therapies for the patient [40].

9.7.3 Pharmacogenomics in Infectious Diseases

Pharmacogenomics is the study of how changes in genes affect pharmacological responses. This has revolutionized the treatment of infectious diseases, making individualized medicine possible. Examples include variations in genetic makeup in drug-metabolizing enzymes, transporters, and receptors affecting the efficacy and toxicity of antibiotics. For example, CYP2B6 gene mutations alter the metabolism of the major antiviral HIV drug efavirenz, and hence, patients receive different concentrations of the drug, which may have an impact on the side effects experienced by a particular population. Similarly, NAT2 gene mutations alter the outcomes of isoniazid, a key antibiotic used to treat tuberculosis, thereby changing its therapeutic outcomes and risk of toxicity [41].

Genomic profiling-based tailored therapy allows patients to have the most effective and safest therapies for infectious diseases. Pharmacogenomics in the clinical practice also enables healthcare providers to change the prescription dosages, reduce adverse responses to drugs, and improve successful treatment outcomes. Infectious disease pharmacogenomics is a very exciting topic that offers new opportunities for maximum antimicrobial therapy, suppression of drug resistance, and survival with precision medicine approaches [42].

9.8 Genomic Approaches to Vaccine Development

9.8.1 Genomic Data for Vaccine Design

Genomic technologies have changed the course of vaccine development by finding certain antigens that elicit robust immune responses. This is achieved by genomics-based analysis of pathogen genomes where highly immunogenic and conserved proteins can be found as vaccine candidates. By this way, vaccine design takes less time, and the vaccine is more effective based on the pathogen-specific components which are essential for infection and progression [43].

COVID-19 vaccines developed within a short time period is an example of genomics-based vaccine development. Since the complete genome of SARS-CoV-2 had been sequenced, scientists discovered the S protein as the ideal antigen that would confer protective immunization. This leads to the discovery of several vaccinations. There are three categories of vaccines: mRNA-based, viral vector-based, and protein subunit-based. It not only shortened the timetable for vaccine manufacture, but it also permitted *in-situ* tracking of viral changes, hastening the generation of new vaccine formulations to circumvent emergent variations. It exemplifies the potential of genetic data to assist vaccine research and pandemic preparedness [44].

9.8.2 Next-Generation Vaccines

Genomic understanding has paved the way for next-generation vaccines, such as mRNA and DNA-based platforms. These new vaccines do not use inactivated or attenuated viruses but rather genetic instructions that tell cells to produce antigenic proteins, leading to an immune response. This change in the field of immunizations is shown in mRNA vaccines of Pfizer-BioNTech COVID-19, Moderna, and others such as the genetic sequences encoding for the SARS-CoV2 spike protein leading to the exact immune response desired by the human body. Then there are the DNA vaccines as in the ZyCoV-D COVID-19 vaccine that brings about immunity *via* the use of plasmid DNA. Each of these approaches has many more advantages such as rapid development scalability and the adaptation to new disease [45].

Although highly successful, genomic-based vaccinations face additional challenges such as stability and delivery systems, as well as global access. mRNA vaccines must be stored at extremely low temperatures, which makes their distribution challenging in resource-poor settings. Additionally, genomic surveillance must continuously monitor pathogen evolution to ensure vaccine effectiveness against newly emerging strains. The present-day solution of such challenges by developing improved formulations, enhanced delivery methods, and equitable distribution tactics will pave the way for the future of genomics-driven vaccinology. Simply put, the advent of genetic data use in vaccine design ushered in a new age of precision immunology. Starting from antigen discovery to next-generation vaccine platforms, genetic information has shaped the battle against infectious diseases by enabling quicker, more efficient, and agile immunization techniques [46].

9.9 Antimicrobial Resistance (AMR) and Genomics

9.9.1 Genomic Surveillance of AMR

Recently, AMR has emerged as one of the most severe global health concerns, and the standard antibiotics are becoming less and less effective.

Genomic technologies can rapidly identify the resistance genes present in pathogens. This changed the whole landscape of AMR surveillance. With the advent of WGS and metagenomics, scientists can now trace the history of resistance in different bacteria through mutations and mobile genetic elements that give resistance [47].

Genomic surveillance offers public health organizations real-time information into the spread of AMR, enabling them to build effective containment strategies. For example, WGS has been used to monitor Mycobacterium tuberculosis strains resistant to rifampin and isoniazid, and it has thus enabled the early detection of multidrug-resistant tuberculosis. Such efforts have also isolated plasmid-borne resistance genes, among them blaNDM-1 gene and mcr-1, which bestow carbapenem resistance and colistin resistance to Escherichia coli and Klebsiella pneumoniae, respectively. Integration of genomic information with classical epidemiological analyses predicts emerging resist types and shapes infection control practice [48].

9.9.2 Targeted Antibiotics and Resistance Mechanisms

Genomes is also assisting in the fight against AMR by designing tailored antibiotics; with comparative genomes, scientists can identify pathogen-specific vulnerabilities and then build medicines that precisely target bacterial resistance pathways. This strategy improves therapeutic efficacy while minimizing collateral damage to beneficial bacteria. One of the major contributions of genomics to AMR research is the discovery of novel resistance mechanisms. Scientists have identified previously undiscovered genetic alterations and adaptive methods that allow germs to evade drugs by studying resistant bacterial genomes. For example, genomic investigations have discovered efflux pump genes in Pseudomonas aeruginosa causing multidrug resistance and mutations in PBPs in Staphylococcus aureus conferring MRSA [49].

A CRISPR-based genome editing device is the second approach under research to fight AMR. The CRISPR-Cas systems are being researched for selective targeting and removal of antibiotic resistance genes from bacterial populations, which might give rise to innovative medicines restoring the potency of antibiotics while leaving the normal microbiota intact [50].

9.10 Integrating Genomics with Traditional Epidemiological Approaches

9.10.1 Combining Genomic Tools with Classical Epidemiology for Enhanced Disease Tracking

With this integration of genomic tools into classical epidemiology, we are transforming the way in which we trace and control infectious diseases. Conventional epidemiology, which is mostly based on clinical case histories, history, and spatial distribution, has historically understood disease patterns. It is effective, but its inability to offer detailed information at the molecular level limits its application [51].

9.10.2 The Role of Genomics in Understanding Disease Transmission Dynamics

Genomic research becomes a very potent lens in trying to understand mechanisms of spread of infectious illnesses. This is an extension of the classical epidemiological approach, which always tries to know how diseases travel through populations from case reports to contact tracing, and so forth. Genomics provides another dimension by giving molecular markers that would be used for tracking the origin of the source and the pathogen route. For instance, genomic sequencing of bacteria or viruses from different patients may indicate genetic similarities in strains, which would be evidence of whether an outbreak is caused by a single source or multiple sources. This enables realist models of disease spread and allows health officials to determine whether a disease is evolving or spreading through human-to-human contact, animal vectors, or environmental reservoirs. Molecular data from genomics can be used to further improve public health responses, determine areas and populations at higher risk, and predict the future course of an outbreak [52].

9.10.3 Benefits of Integrating Genomics into Field-Based Epidemiological Studies

It is likely that the integration of genetics into field-based epidemiological studies will speed up and enhance the accuracy of outbreak investigations. Traditionally, field epidemiologists have relied on data collection through surveys, interviews, and clinical testing. Genomic technologies allow for faster and more accurate pathogen detection in the field, especially in remote or resource-limited environments where other diagnostic approaches are delayed or ineffective. For instance, genomics is helpful in identifying new or newly discovered infections as well as new versions or mutations that may be much more destructive than others. Genomic technology enables the identification of AMR patterns in real time, which essentially determines the treatment procedures. The use of genomes into the field surveys for epidemiologists offers a way into how well sicknesses are actually transmitted, which would boost their containment efforts over a certain span of time. This is critically important in controlling new infectious diseases, where real-time and exact data can indeed be the difference between life and death [53].

9.11 Pandemic Preparedness and Genomic Research

9.11.1 Real-Time Genomic Surveillance During Pandemics

These days high speed of sequencing and analysing the genome of a pathogen give genomic surveillance in real time. The researcher now has viral and bacterial genome sequencing, enabling how mutations take place, tracking modes of transmission, and devising targeted therapies. It has lately been observed with Ebola, Zika, and COVID-19, wherein the monitoring of genetic trends in diseases and disease-causing organisms formed part of surveillance where public health interventions guided this kind of exercise [54].

In the COVID-19 pandemic, genomie sequences quickly detected and traced new variants such as Alpha, Delta, and Omicron. Data from GISAID enabled freely available global information exchange among researchers to discover mutations that may impact vaccine efficacy or disease severity. The same approach was applied in the understanding of Ebola by genetic research on the evolution of the virus, which informed containment techniques and vaccine development. These cases demonstrate how genetic technology accelerates our understanding of pathogens during a health crisis and therefore, better interventions are implemented [55].

9.11.2 Predictive Genomics in Preventing Disease Outbreaks

Other than real-time surveillance, predictive genomics offers an active approach to preventing and controlling future epidemics of disease. Scientists can predict which viruses or bacteria are likely to infect humans by analysing pathogen genomes and evolutionary patterns. This is particularly important for zoonotic infections; such diseases create a large percentage of new infectious diseases. Understanding zoonotic viruses like SARS-CoV-2, H5N1 influenza and Nipah virus have substantially relied on genomics data that describe the viral transmission process. Table 9.2 summarizes the major genomic applications in infectious disease management. Understanding reservoir viruses in animals clarifies whether this infection would very efficiently transfer barriers between species toward causing outbreaks within humans. Since metagenomics can identify unidentified or emergent viruses from the wildlife sources ahead of infection events in human patients, an avenue for earlier control measures emerges as

Table 9.2 Genomic applications in infectious disease management.

Application area	**Genomic approach**	**Examples**
Vaccine Development	Genomic sequencing to identify antigens, mRNA vaccine design	COVID-19 mRNA vaccines (Pfizer/BioNTech, Moderna), vaccine candidates for Zika, Ebola
Antimicrobial Resistance (AMR)	Genomic surveillance of resistance genes, pathogen strain profiling	Tracking AMR in bacteria (e.g., MRSA, E. coli), identifying novel resistance mechanisms
Pandemic Preparedness	Genomic surveillance, genomic epidemiology, pathogen prediction	Genomic surveillance, genomic epidemiology, pathogen prediction
Diagnostics	CRISPR-based diagnostics, genomic biomarkers for early detection	CRISPR-based diagnostics for rapid pathogen detection (e.g., SHERLOCK), genomic testing for HIV, TB
Host-Pathogen Interactions	Human genomic analysis to identify susceptibility genes	Identifying genetic variations in humans that affect susceptibility to diseases like malaria and tuberculosis
Personalized Medicine	Pharmacogenomics to guide treatment decisions	Tailoring treatments based on patient's genomic data (e.g., HIV drug resistance testing, tuberculosis susceptibility)

well. Predictive genomics is also valuable in the production of preventive countermeasures like universal vaccinations and broad-spectrum antivirals. Through huge genomic datasets, machine learning and AI models can identify genetic markers associated with high-risk diseases, thus enabling better outbreak predictions and preparedness [56].

9.12 Challenges in Translating Genomic Discoveries into Clinical Applications

9.12.1 Regulatory and Ethical Hurdles

One of the major regulatory and ethical challenges is genetic discovery into therapy. The data from genetics, its complexity, and the probable implications for a patient's treatment require rules ensuring accuracy, dependability, and safety. Although the FDA and EMA have frameworks for genomics-based diagnostics and medicines, the approval process is lengthy and resource-intensive. Ethical issues hinder the clinical application of genomics. Patient consent, data privacy, and equal access to genetic medicine are some of the issues that need to be addressed appropriately. The collection and storage of genomic data raise issues of potential exploitation, discrimination, and data security vulnerabilities. Informed consent is particularly important since patients need to know precisely how their genomic data will be applied in research and therapeutic settings. Besides that, inequality in access to genetic technology between affluent and impoverished areas is an ethical issue regarding universal health equality worldwide [57].

9.12.2 Challenges in Drug and Vaccine Development

Although the genetic information is abundant, converting the outcomes into useful medicines and vaccines remains a difficult and complex process. An important problem is the "translational gap," wherein promising genetic results are unable to be translated into clinical applications. This can be ascribed in part to the difficulty of identifying druggable targets from genetic data, as well as the high expense of drug development itself. The COVID-19 pandemic demonstrated the potential of genomics to accelerate the development of vaccinations, such as mRNA vaccines. However, regulatory constraints and lengthy clinical studies frequently cause the approval process for new genomic-based medicines to take a long time. The issue here is to standardize genomic data interpretation while also ensuring reproducibility in the clinical settings [58].

9.12.3 Interdisciplinary Collaboration Needs

This will require seamless coordination among genetic researchers, doctors, bioinformaticians, and public health professionals. The genetics of infectious illnesses is extremely complicated, necessitating a multidisciplinary approach that combines molecular biology, epidemiology, data science, and healthcare policy. Interdisciplinary interactions would bridge gaps in research application through knowledge transfer and the alignment of research objectives with clinical demand. This element fortifies collaboration between academia, pharmaceutical businesses, and regulatory bodies for quicker diagnostics and treatments. In addition, genomic monitoring is even more helpful in infectious diseases through global cooperation and data exchanges such as the Global Pathogen Monitoring Network [59].

9.13 Innovative Genomic Approaches on the Horizon

9.13.1 Artificial Intelligence and Genomics

The integration of artificial intelligence and machine learning into genomics has revolutionized the study, diagnosis, and treatment of infectious diseases. The amazing speed and accuracy with which AI-driven algorithms scan massive volumes of genetic data lead to the discovery of patterns that traditional bioinformatics methods would not be able to identify. Figure 9.1 illustrates the genomic workflow used in infectious disease research.

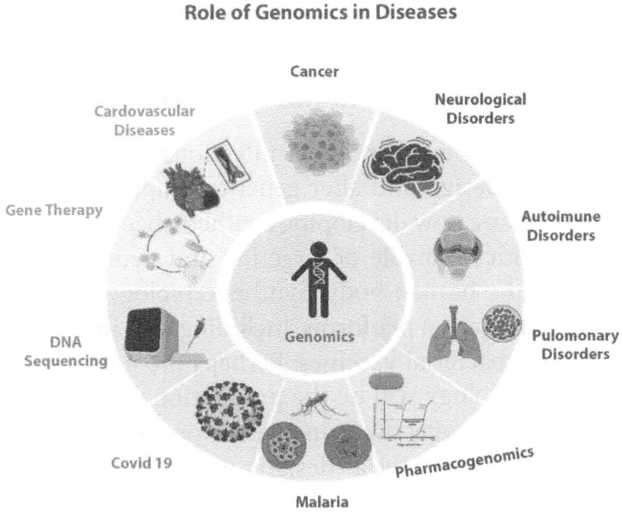

Figure 9.1 Genomics and diseases.

This allows for rapid identification of disease-causing mutations, antimicrobial resistance markers, and potential targets for treatment. Another benefit of AI is the improvement in predictive genomics. It allows us to predict outbreaks of diseases, track the evolution of pathogens, and predict treatment efficacy. Deep learning models are now being developed on genomic sequences and correlated with patient outcomes to enable precision therapy in infectious diseases. AlphaFold, for example, is using AI to revolutionize protein structure prediction, accelerate drug development, and design vaccines. With the growing capabilities of computers, AI and genomics will likely dominate future control of infectious diseases [60].

9.13.2 Pan-Genomics and Multi-Omics Approaches

Pan-genomics is the study of the complete genome of a species, giving more detailed insight into the genetic diversity of pathogens. Comparing the core and accessory genomes between strains helps researchers find conserved virulence factors, drug resistance genes, and vaccination targets. This approach is useful in studying the development and adaptation of bacteria and viruses. The integration of genomes, transcriptomics, proteomics, and metabolomics allows for a complete understanding of the biology of the pathogen as well as host responses. The combination of these datasets may provide new biomarkers for the early detection of illness, analyse the interaction between hosts and pathogens at various levels, and provide new treatment strategies that are more efficient. This drives multi-omics science forward into personalized medicine through the adaptation of treatment based on the individual's genomic and molecular profile [61].

9.13.3 Synthetic Biology and Infectious Disease Management

Synthetic biology is a new frontier in infectious disease research. It has made it possible to design and alter genetic components for therapeutic purposes. Scientists are now developing synthetic bacteria that can be used as live biotherapeutics, capable of targeting and neutralizing pathogenic pathogens within the human body. Synthetic biology is also used in the creation of new vaccination platforms, including self-amplifying RNA vaccines and synthetic bacterial vaccines. Perhaps the most promising application of synthetic biology will be the production of synthetic pathogens to study diseases more effectively in the lab or controlled environment, for instance. Virus and bacterial designs with specified mutations may give researchers a much clearer view of what pathogenicity is all about and provide ways to break open therapeutic windows. Another potential application of synthetic biology could be the engineering of immunity, such as constructing personalized immune receptors, which might lead to new therapies for infectious disorders [62].

9.14 The Impact of Genomics on "One Health" Approaches to Infectious Diseases

9.14.1 Understanding the Interconnectedness of Human, Animal, and Environmental Health through Genomics

The "One Health" approach recognizes that the health of humans, animals, and the environment are inextricably linked, and that treating health issues in one domain has repercussions for the others. Genomics has played an important role in understanding this interconnectedness, particularly in the context of infectious diseases. Studies in humans, animals, and the environment that use genomics technology to characterize pathogens are sure to unravel new information regarding how disease emergence and dissemination take place. Therefore, for effective monitoring and tracking, genomics would unveil the origin and sources of zoonotic diseases that affect a wide range of animal species, and the species to which these pathogens may transmit to humans. An integrated understanding would help build more holistic, coordinated efforts in disease prevention, surveillance, and control across all domains of health [63].

9.14.2 Genomic Surveillance in Animal Reservoirs and Environmental Sources of Infectious Diseases

Animal reservoirs and environmental sources are important for transmitting and establishing infectious diseases, hence holding a prime position in genetic surveillance. Outbreaks may occur in humans through pathogen spread from animal populations or environmental conditions. Wildlife, livestock, and pets come into the realm of animal genomic surveillance, and such insights into the genetic diversity of the diseases that are transmitted in animals can draw in the possible human infection reservoirs. Examples include genetic analysis that traced the SARS-CoV-2 virus to its animal hosts, which helped public health interventions and global responses during the pandemic. More general use of genomic technologies has been made in ecosystems that harbour infectious agents, such as waterborne pathogens or antimicrobial-resistant bacteria in agricultural settings. This is how genomic surveillance can be used to incorporate into early warning systems the ability to prevent cross-species transmission and outbreaks from becoming a problem before they reach human populations [64].

9.14.3 Advancing "One Health" Strategies with Genomic Data to Prevent Cross-Species Transmission

It also provides focal points for advancing "One Health" methods, Figure 9.2 depicts various one heath approaches. Such as reducing the spread of infectious diseases from one species to another. Combining human, animal, and environmental genomes data would potentially assist in anticipating which hot regions are most vulnerable to disease spillover. For example, it could identify the genetic markers responsible for the pathogen's multi-species infection potential, allowing public health organizations to target interventions more effectively. Furthermore, genomic sequencing of pathogens can track the evolution and mutation of these pathogens in real time, providing valuable information about how diseases cross species and if they grow more virulent or resistant. It can also be used to guide vaccine plans, surveillance programs, and policy decisions that will aid in the prevention of disease transmission from animals to people. Genomics-based approaches contribute to a more anticipatory and integrated approach to global health, in which efforts to prevent and control infectious diseases

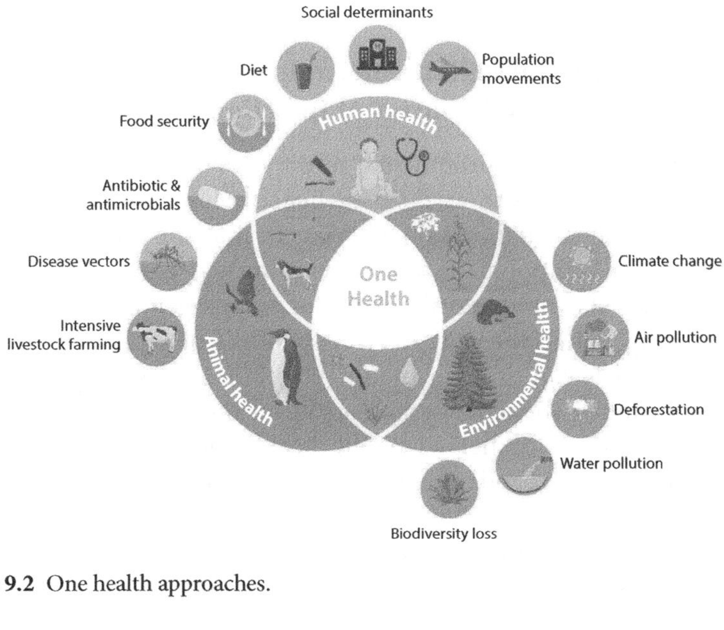

Figure 9.2 One health approaches.

consider the larger ecological context as well as the relationships between human, animal, and environmental health [65].

9.15 The Role of Microbiome Genomics in Infectious Disease Prevention

9.15.1 Investigating the Human Microbiome's Role in Infection Resistance and Susceptibility

The human microbiome, which is made up of trillions of bacteria that live in and on the human body, plays an important role in immune system modulation and infection resistance. Genomic research has revealed how the variety and composition of the microbiome affect susceptibility to a wide range of infectious diseases, including gastrointestinal infections, respiratory and skin problems. Recent research has employed genome sequencing methods to identify not just specific microbial species, but also the genes from those species that protect or predispose a person to certain infections. For instance, several good gut bacteria produce immune stimulation through enhanced antimicrobial peptide synthesis or the direct prevention of colonization by pathogens. Dysbiosis, defined as an imbalance in the microbiome, leads to increased risks of infections and autoimmune disorders as well as chronic diseases, making genomics crucial in elucidating these complex interactions and providing insights into the importance of maintaining a healthy microbiome as one measure to prevent these diseases [66].

9.15.2 Potential of Microbiome-Based Therapies and Vaccines in Infectious Disease Control

Microbiome-based therapies are increasing in popularity today for the prevention and treatment of infectious illnesses. Advances in the field of genomics related to the microbiome open new areas of investigation about the possibility of manipulating the human microbiome toward better host defence against pathogens. A potential mechanism may be employing probiotics or prebiotics or FMT to re-colonize gut or other likely compromised body area with a homeostatic and balanced microbial presence. Hence, the sequence of microbiome data may enable a precise elucidation of exactly which strains or communities of bacteria must be reinstated to help reconstitute immunity or avoid infection. How vaccination efficacy interfaces with the microbiomes becomes ever more transparent, and yet the relationship between the microbiota and vaccination-induced immunity is an area that genetic technologies are still exploring. Such knowledge will likely be helpful in developing microbiome-based vaccines or medicines that enhance the immune response of vaccines or inhibit infection by promoting health-promoting bacteria communities [67].

9.15.3 Genomic Approaches to Understanding the Relationship between Gut Microbiota and Infection

The gut microbiota is one of the most intensively studied microbiomes in the prevention of infectious diseases, and genetic research has significantly increased our knowledge of this relationship. Genomic technologies, including 16S rRNA sequencing, metagenomics, and shotgun sequencing, have enabled scientists to explore the genetic profiles of microbes in the gut. This study discovered that some of the intestinal microbiota synthesize and excrete specific metabolites of short-chain fatty acids that have antimicrobial potential, or collaborate with immune cells in a synergistic way to enhance their protective function against pathogens invading the body. More importantly, genomics perspectives indicated that changes in the gut microbial community can predispose individuals to illnesses caused by Clostridium difficile, Salmonella, and influenza. These avenues are leading toward tailored drugs to manage the gut microbiota with a view toward the prevention and treatment of disease. One study team is, for instance, exploring the possibility of modifying certain probiotics so as to manufacture specific chemicals for defensive purposes and therapeutic diets intended at balancing the microbiome and raising the levels of infection resistance. These represent exciting frontiers in infectious disease prevention, where treatments directed toward the maintenance or reconstitution of a healthy microbiome may well form an essential instrument in global health strategy [68].

9.16 Personalized Medicine in Infectious Disease Treatment

9.16.1 The Impact of Genomic Sequencing on Individualized Therapeutic Strategies

The whole genetic sequencing process has revolutionized the area of personalized medicine, especially in the treatment of infections. This method permits the sequencing of pathogen genomes as well as the formulation of targeted therapy tactics that are designed exactly to the individual's genetic characteristics. A clinician can administer the best kind of treatment by identifying the genetically unique pathogen from mutations of antibiotic resistance, thus avoiding all the guesswork that frequently goes with previous techniques. Similarly, the genomic sequencing of the genetic profile of the patient helps detect probable drug interactions, predispositions for side effects, and changes in drug metabolism. Such information will ensure that a specific drug is not only effective in treating the infection but also as harmless as possible to the patient. More frequent use of genetic sequencing in the therapeutic field will enhance the effectiveness of treatment, reduce the likelihood of complications, and reduce the incidence of relapse due to the availability of precision medicine drugs tailored to specific patients [69].

9.16.2 Tailoring Drug Treatments Based on Genetic Profiles of Pathogens and Patients

The most interesting feature of personalized medicine is its potential to adjust medicines as per the genetic compositions of the pathogens and patients. In infectious diseases, this two-way approach ensures that the right set of drugs is administered in the right dosage both for eradicating the pathogen effectively and for safety to the patient. For instance, through genomics in bacterial infections, individual strains and levels of resistance of bacteria are identified, ensuring that the right medicine is administered. That is a more successful method because the spread of antimicrobial resistance will be limited inasmuch as the use of broad-spectrum antibiotics is minimal. The comprehension of the variability will help the patients predict the response to the particular medicine. For instance, genetic variations affecting drug metabolism—antiretrovirals, antivirals, among others—might result in doctors adjusting prescription doses, thereby optimizing safety and effectiveness. The overall sum of such pathogen as well as patient-specific genetic data places personalized medicine in a firm position to deal with infectious diseases [70].

9.16.3 Challenges and Opportunities in Implementing Personalized Medicine for Infectious Diseases

Although personalized medicine has tremendous potential for treating infectious diseases, several challenges need to be addressed before it can be applied on a large scale. One of the most important challenges to genome sequencing is cost and access. It is still prohibitively expensive despite the drop in cost, limiting its use in the majority of the world. The intricacy of genetic data will also make it difficult to incorporate it into routine clinical practice because interpreting genomic data needs strong bioinformatics skills. Finally, there have been worries that data privacy has been jeopardized by keeping and sending genetic information without adequate consent for the possible use of such genetic information. Despite the benefits that genome sequencing could provide, it is not a solution to all of these issues. In most cases, it is hard to foresee all probable mutations of pathogens, such as viruses. Nonetheless, the benefits of customized medicine for improving therapeutic outcomes, lowering AMR, and increasing patient safety are apparent. Personalizing medicine is a fantastic possibility for infectious disease therapy, thanks to advances in genomic technologies, more effective interpretation and application of genetic data, and more global collaboration [71].

9.17 Global Collaboration in Genomic Research

9.17.1 International Genomic Initiatives

Along with GISRS, the Human Microbiome Project and the Global Genetic Medicine Collaborative are working to develop large-scale databases of genetic sequences and health data. Such platforms aid in the identification of genetic markers for disease susceptibility, antibiotic resistance, and vaccination efficacy. These worldwide collaborations propel global efforts toward precision medicine by sharing genetic data and research findings, allowing genomic advances to benefit all nations, regardless of resource availability [72].

9.17.2 Data Sharing and Global Health Equity

International collaboration has been a significant factor in the progress of genetic research, but ensuring fair access to genomic technology is a very serious issue. Low- and middle-income countries face several issues in conducting high-tech genetic research due to poor infrastructure, finance, and personnel. Such unequal distribution of resources may also create an imbalance in the involvement of countries in international genetic databases and research programs, thus depriving them of equitable benefits [73].

Cross-border data sharing will have to be necessary in order to bridge these gaps. Organizations such as GA4GH are building rules and frameworks for the ethical use, transparent treatment, and secure exchange of genomic information. These initiatives help to bridge gaps between high and low resource settings, ensuring genomic research contributes to global health equity. These countries can then become active participants in global health efforts, thereby reaping benefits to the people all over the world, by providing training, infrastructure, and financing for genomic research in disadvantageous areas [74].

9.18 Addressing Public Concerns Around Genomic Research in Infectious Diseases

9.18.1 Public Perception of Genomic Research

Public consciousness about genomic research, especially in infectious diseases, is one of the most persuasive factors influencing the feasibility and introduction into health-care practices. Although genetic technologies have the potential to revolutionize the diagnostic, preventive, and curative features of illness, the public is unaware of this advancement. Many individuals who are not aware of genetic research have shown a mix of curiosity, hope, and fear. The genomics tool seems to provide a new cure for infectious diseases, but this opens the possibility of unintended consequences of genetic manipulation—ethical dilemmas about genetic privacy and designer viruses. All this requires straightforward presentation of the merits and limitations of genetic research by scientists, healthcare professionals, and politicians. A better sense of what genetic research can do for health improvement could be created through educational quality programs, public forums, and outreach campaigns [75].

9.18.2 Addressing Fears Surrounding Gene Editing and the "Unknown" Consequences of Modifying Pathogens or Human Genomes

The development of gene editing technologies, for instance, CRISPR-Cas9, has been deemed a revolutionary weapon against infectious illnesses. It raises new issues on unknown "consequences" that can result from tampering with viruses or even the human genome. The greatest source of public apprehension emanates from perceived unpredictability in altering pathogens through genetic modification. In some cases, some of these concerns question whether gene-edited organisms may unintentionally introduce new diseases, not to mention escape the laboratory setting and thus bring on unforeseen threats to public health. Analogously, whether for therapeutic or enhancement purposes—like greater resistance to disease—worry about human genome editing has led to ethical concerns about "designer babies" and unintended genetic effects for subsequent generations. Addressing worries and re-emphasizing stringent safety precautions, as well as the ethical considerations that guide genomic research through transparent and responsible science communication is particularly important. With such an amount of regulation in place, there will be enough surveillance to ensure that experiments are being conducted appropriately in genes. A few potential applications of gene editing technologies include the ability to battle infectious diseases and improve public health [76].

9.18.3 Misinformation and Genomic Research

Misinformation about genomic research can impact public opinion in a very important way and hinder the acceptance of beneficial technology. False or exaggerated statements about genetic research spread like wildfire, largely fueled by social media, spectacular media coverage, and conspiracy theories, which instil widespread anxiety and mistrust. For example, erroneous assertions about gene editing resulting in the creation of hazardous or "designer" creatures tend to outweigh any actual benefits from this technology. Misinformation can only be combated through an active approach to accurate information, which should be available and clear to all. An engagement system with media, particularly social platforms, will assist in correcting misconceptions, addressing concerns, and educating people about the real ramifications of genetic research. Collaboration with respectable scientists, public health groups, and educational institutions will ensure that the public receives reliable information, thereby dispelling myths and building a more scientifically literate society [77].

9.18.4 Building Public Trust in Genomic Research

Building and maintaining public trust is critical to make genetic research effective in reducing infectious illnesses. However, this type of trust is primarily dependent on how well researchers, policymakers, and healthcare professionals communicate the nature of the genome's risks and promises. Community engagement is one of the most effective ways to increase confidence. Scientists can show commitment to the public's welfare and shared decision-making by allowing local communities, among other patient advocacy groups, to listen to discussions on the goals of genetic research and methodological approaches as well as the ethical issues. The fact that genetic research improves societal health would be manifested through the development of communication tactics by giving paramount importance to safety measures. Further, discussions on the ethical, legal, and social implications of genetic discoveries will be allowed to flow, and the public will be encouraged to join in the debate and make informed decisions. Finally, transparency, accountability, and an emphasis on the health and well-being of individuals and communities will help to foster public trust [78].

9.19 Exploring Genomic Approaches in Tropical and Neglected Diseases

9.19.1 Advancements in Genomic Research for Tropical Infectious Diseases

Genomic research has significantly expanded knowledge in the realm of tropical infectious diseases, which garner far less attention than the big diseases of influenza and tuberculosis. However, the genomics revolution has helped uncover genetic markers influencing pathogen virulence and transmission in diseases like malaria, dengue, and Zika virus. Advances in WGS and transcriptomics have facilitated the mapping of the genetic make-up of tropical infections. Thus, understanding better how these organisms interact with their hosts and how they evade the immune responses provides crucial genomic information for the development of targeted medicines, improving diagnostics, and enhancing the production of more effective vaccines. For instance, genomic surveillance is increasingly becoming a tool for monitoring evolutionary changes in resistances among pathogenic microorganisms, including antibiotic-resistant Plasmodium, a malaria parasite whose control is heavily influenced by treatment outcomes [79].

9.19.2 Genomic Innovations for Understanding and Addressing Neglected Diseases

Neglected tropical diseases, which affect millions of people in resource-limited situations, have been grossly underfunded and understudied. Genomic breakthroughs now allow researchers to investigate the genetic basis of such disorders in ways previously inconceivable. Pathogen and host genome sequencing offers novel biomarkers for early diagnosis and prognosis. Research on the genome also sheds light on the genetic pathways that enable pathogens to stay in the human body and evade therapy, as happens in diseases such as leprosy and Chagas disease. This may bring the possibility of developing drugs and vaccines that attack the specific genetic characteristics of NTDs. Using genomes gives us a chance now to begin moving from trying to control these diseases to discovering the permanent end to their existence [80].

9.20 Ethical, Legal, and Social Implications of Genomic Research

9.20.1 Genetic Privacy and the Protection of Personal Data in Research

Some of the most ethical concerns in the area of genomics are presented by the genetic privacy problem. Though genetic information is increasingly utilized to explain disease, it remains very important to ensure that a person's private genetic information should not be accessed or misused without authorization. Genetic data entails sensitive information relating to an individual's health condition, susceptibility, and family background. Research institutes and healthcare providers must follow strict privacy rules in the EU, known as the General Data Protection Regulation or GDPR, and in the United States, known as the Health Insurance Portability and Accountability Act or HIPAA. Informed permission is a fundamental condition for privacy. Individuals must know how their genetic information will be used, kept, and shared. Nevertheless, questions still persist about how de-identification should be done, or how genetic data can be made not to re-identify and shared in genomic data to the greater good of research with respect to protecting individual privacy [81].

9.20.2 Ethical Challenges in Genomic Editing and Pathogen Modification

The rapid development of genomic editing technologies such as CRISPR has raised ethical arguments regarding the prospect of modifying human, animal, or pathogen genomes. On the one hand, these technologies provide unprecedented prospects for the prevention and treatment of illness; however, they raise worries about unintended consequences, particularly germline gene editing, which can be passed on to future generations. The alteration of viruses and bacteria for modification in terms of pathogenicity and resistance mechanisms raises several ethical issues regarding the potential creation of more virulent or resistant strains. To boot, the alteration of diseases creates a biosecurity issue as they may be exploited for bioterrorism or merely disseminated accidentally. International ethical principles and laws are essential in ensuring that these technologies use are responsibly put across so that the public health benefits are weighed against the risks such technologies carry [82].

9.20.3 Addressing Public Concerns Around Genomic Research in Infectious Diseases

A mix of interest, fear, and ignorance shapes the public's opinion of genetic research. It represents promising advances in understanding infectious diseases and the creation of targeted therapeutics but raises significant problems. It has been linked to dread, namely the fear of altering human genomes and the prospect of gene editing infectious organisms. Misinformation, disseminated mainly over social media networks, fuels concerns, creating suspense and resistance around new technologies. Scientists, researchers, and political leaders must start a candid communication with the society on the usefulness and risks in new technologies as a way to enhance public acceptance of genomic study. The available precautions, as illustrated by stringently set regulations and oversight are reassuring. It will also raise public confidence when this is reflected on community involvement in decision-making and the execution of genetic research following social values and needs [83].

9.21 Conclusion

Infectious disease research can be said, without doubt, to have certainly changed its direction over the past ten years over the landscape; genomics will be deciphered to explain all previously unknown mechanisms of infectious disease, pathogenicity, as well as virulence. Each of these discovered genomics translated into novel markers, further-advanced diagnostics are more precise while efficient treatment, and vaccination tactic is developed as well. Genomic technologies utilize whole-genome sequencing, metagenomics, and CRISPR-based methods, enabling the monitoring and tracking of infectious diseases, prediction of potential outbreaks, and forecasting of how the disease will react to a given therapy. The genomic revolution goes far beyond the limitations of the conventional approach by increasing our knowledge about pathogens. This also leads to more precise, data-driven approaches in combating diseases.

There is no greater promise for genomics than in combating emerging and re-emerging infections. As diseases begin their new life cycles, fast genome sequencing allows for prompt detection and investigation, providing critical information for containment and treatment. Genomic surveillance technologies have lately been critical in monitoring diseases such as COVID-19, Ebola, and Zika, allowing for prompt responses and informed decision-making in public health. In addition, it can accelerate the development of vaccines and treatments that can be easily adapted to emerging strains of viruses, leading to better preparedness for pandemics. Detection is also enabled by changes in AMR-related genetic mutations, allowing for very targeted therapies while reducing the overall risk posed by resistant infections worldwide. As new pathogens emerge, genomics will be crucial in keeping ahead of infectious illnesses and maintaining the resilience of global health systems.

Genomic research is important in defining the way forward about infectious disease management. It has always been an ongoing affair and necessitates investment in genomics not only to understand pathogen biology but also in the creation of the next generation of tools and tactics for tackling infectious diseases. For instance, the governments, research institutions, and the commercial sector need to come together to increase funding for genetic research, infrastructure, and the development of accessible technologies. This investment will not only make us better responsive to the challenges posed by current infectious diseases but also prepare us for unforeseen health dangers in the future. In addition, global cooperation and equal access to genomic research will ensure that all sections of society benefit from the fruits of this study and thus reduce health inequalities in the future. As a consequence, the increased use of genetics in public health practice and clinical delivery will define future efforts toward long-term, science-based global health goals that help create a better, safer society.

AI disclosure statement: The author acknowledges the use of AI-based tools for minor language editing or grammatical correction. However, all intellectual contributions, analyses, and conclusions presented in this chapter are entirely the author's own.

References

1. Hooper, K., Supporting the evolution of infectious disease research. *Dis. Models Mech.*, 17, 9, 2024.
2. Speers, D.J., Clinical Applications of Molecular Biology for Infectious Diseases. *Clin. Biochem. Rev.*, 27, 2006.
3. Burton, H., Jackson, C., Abubakar, I., The impact of genomics on public health practice. *Br. Med. Bull.*, 112, 37–46, Oxford University Press, 2014.
4. Goyal, P.A., Bankar, N.J., Mishra, V.H., Borkar, S.K., Makade, J.G., Revolutionizing Medical Microbiology: How Molecular and Genomic Approaches Are Changing Diagnostic Techniques. *Cureus*, 2023 Oct 16.
5. Gardy, J., Loman, N.J., Rambaut, A., Real-time digital pathogen surveillance - the time is now. *Genome Biol.*, 16, BioMed Central Ltd., 2015.
6. Boyce, A.M. and Garibaldi, B.T., Genomics and high-consequence infectious diseases: A scoping review of emerging science and potential ethical issues. *Health Secur.*, 17, 62–8, Mary Ann Liebert Inc., 2019.
7. Bianconi, I., Aschbacher, R., Pagani, E., Current Uses and Future Perspectives of Genomic Technologies in Clinical Microbiology. *Antibiotics*, 12, Multidisciplinary Digital Publishing Institute (MDPI), 2023.
8. Phillips, K.A., Liang, S.Y., Van Bebber, S., Afable-Munsuz, A., Elkin, E., Haas, J., *et al.*, Challenges to the Translation of Genomic Information into Clinical Practice and Health Policy: Utilization, Preferences, and Economic Value. *Curr. Opin. Mol. Ther.*, 10, 3, 260–6, 2008 Jun.
9. Kobras, C.M., Fenton, A.K., Sheppard, S.K., Next-generation microbiology: from comparative genomics to gene function. *Genome Biol.*, 22, BioMed Central Ltd, 2021.
10. Fleischmann, R.D., Adams, M.D., White, O., Clayton, R.A., Kirkness, E.F., Kerlavage, A.R., *et al.*, Whole-genome random sequencing and assembly of Haemophilus influenzae Rd. *Science (1979)*, 269, 5223, 496–512, 1995.
11. Catalano, S., Battelli, F., Traore, Z.I., Raghwani, J., Faust, C.L., Standley, C.J., Pathogen genomics and One Health: A scoping review of current practices in zoonotic disease research. *IJID One Health*, 100031, 2024.
12. Satam, H., Joshi, K., Mangrolia, U., Waghoo, S., Zaidi, G., Rawool, S., *et al.*, Next-Generation Sequencing Technology: Current Trends and Advancements. *Biology*, 12, Multidisciplinary Digital Publishing Institute (MDPI), 2023.
13. Illumina. Technology Spotlight: Illumina ® Sequencing.
14. Merriman, B., R&D Team IT, Rothberg, J.M., Progress in ion torrent semiconductor chip-based sequencing. *Electrophoresis*, 33, 23, 3397–417, 2012.
15. Rhoads, A. and Au, K.F., PacBio sequencing and its applications. *Genom. Proteom. Bioinform.*, 13, 5, 278–89, 2015.
16. Lu, H., Giordano, F., Ning, Z., Oxford Nanopore MinION sequencing and genome assembly. *Genom. Proteom. Bioinform.*, 14, 5, 265–79, 2016.

17. Clark, A.J. and Lillard, J.W., A Comprehensive Review of Bioinformatics Tools for Genomic Biomarker Discovery Driving Precision Oncology. *Genes*, 15, Multidisciplinary Digital Publishing Institute (MDPI), 2024.
18. Altschul, S.F., Gish, W., Miller, W., Myers, E.W., Lipman, D.J., Basic local alignment search tool. *J. Mol. Biol.*, 215, 3, 403–10, 1990.
19. Thiel, W.H., Galaxy workflows for web-based bioinformatics analysis of aptamer high-throughput sequencing data. *Mol. Ther. Nucleic Acids*, 5, 2016.
20. Huang, L., Popic, V., Batzoglou, S., Short read alignment with populations of genomes. *Bioinformatics*, 29, 13, i361–70, 2013.
21. Sareshma, S.D. and Subha, B., Assessing Host-Pathogen Interaction Networks *via* RNA-Seq Profiling: A Systems Biology Approach, in: *Applications of RNA-Seq in Biology and Medicine*, p. 39, 2021.
22. Schikora-Tamarit, MA. and Gabaldón, T., Using genomics to understand the mechanisms of virulence and drug resistance in fungal pathogens. *Biochem. Soc. Trans.*, 50, 3, 1259–68, 2022.
23. Zhang, E., *Visualizing the human genome in virtual reality*, McGill University (Canada), 2019.
24. El-Tohamy, M., Badiea, M.E.A., Azmy, A., Muhammed, R., QIIME: Quantitative Insights into Microbial Ecology. https://qiime.org/1.4.0/.
25. Junjun, R., Zhengqian, Z., Ying, W., Jialiang, W., Yongzhuang, L., A comprehensive review of deep learning-based variant calling methods. *Brief. Funct. Genom.*, elae003, 2024.
26. Rodino, K.G. and Simner, P.J., Status check: Next-generation sequencing for infectious-disease diagnostics. *J. Clin. Invest.*, 134, 4, 2024 Feb 15.
27. d'Humières, C., Salmona, M., Dellière, S., Leo, S., Rodriguez, C., Angebault, C., *et al.*, The Potential Role of Clinical Metagenomics in Infectious Diseases: Therapeutic Perspectives. *Drugs*, 81, 13, 1453–66, 2021 Sep 1.
28. van den Esker, M.H. and Koets, A.P., Application of transcriptomics to enhance early diagnostics of mycobacterial infections, with an emphasis on Mycobacterium avium ssp. paratuberculosis. *Vet. Sci.*, 6, MDPI Multidisciplinary Digital Publishing Institute, 2019.
29. Huang, T., Zhang, R., Li, J., CRISPR-Cas-based techniques for pathogen detection: Retrospect, recent advances, and future perspectives. *J. Adv. Res.*, 50, 69–82, 2023.
30. Peterson, J.W., Bacterial pathogenesis, in: *Medical Microbiology*, 4th ed., 1996.
31. Alberts, B., Johnson, A., Lewis, J., Raff, M., Roberts, K., Walter, P., Cell biology of infection, in: *Molecular Biology of the Cell*, 4th ed., Garland Science, 2002.
32. Hassan, Y.M., Mohamed, A.S., Hassan, Y.M., El-Sayed, W.M., Recent developments and future directions in point-of-care next-generation CRISPR-based rapid diagnosis. *Clin. Exp. Med.*, 25, 1, 1–15, 2025.

33. Pronyk, P.M., de Alwis, R., Rockett, R., Basile, K., Boucher, Y.F., Pang, V., *et al.*, Advancing pathogen genomics in resource-limited settings. *Cell Genomics*, 2023.
34. Hill, V., Ruis, C., Bajaj, S., Pybus, O.G., Kraemer, M.U.G., Progress and challenges in virus genomic epidemiology. *Trends Parasitol.*, 37, 12, 1038–49, 2021.
35. Flores-Alanis, A., Cruz-Rangel, A., Rodríguez-Gómez, F., González, J., Torres-Guerrero, C.A., Delgado, G., *et al.*, Molecular epidemiology surveillance of SARS-CoV-2: Mutations and genetic diversity one year after emerging. *Pathogens*, 10, 2, 1–14, 2021 Feb 1.
36. Uelze, L., Becker, N., Borowiak, M., Busch, U., Dangel, A., Deneke, C., *et al.*, Toward an Integrated Genome-Based Surveillance of Salmonella enterica in Germany. *Front. Microbiol.*, 12, 2021 Feb 10.
37. Duggal, P., Ladd-Acosta, C., Ray, D., Beaty, T.H., The evolving field of genetic epidemiology: From familial aggregation to genomic sequencing. *Am. J. Epidemiol.*, 188, 2069–77, Oxford University Press, 2019.
38. Kwok, A.J., Mentzer, A., Knight, J.C., Host genetics and infectious disease: new tools, insights and translational opportunities. *Nat. Rev. Genet.*, 22, 137–53, Nature Research, 2021.
39. Medhasi, S. and Chantratita, N., Human Leukocyte Antigen (HLA) System: Genetics and Association with Bacterial and Viral Infections. *J. Immunol. Res.*, 2022, Hindawi Limited, 2022.
40. Lenardo, M., Lo, B., Lucas, C.L., Genomics of Immune Diseases and New Therapies*. *Annu. Rev. Immunol.*, 34, 121–49, Annual Reviews Inc., 2016.
41. Mandlik, V., Kabra, R., Singh, S., System pharmacogenomics application in infectious diseases. *Briefings Funct. Genomics*, 16, 274–80, Oxford University Press, 2017.
42. Mochizuki, T., Ikegami, M., Akiyama, T., Factors predictive of second-line chemotherapy in soft tissue sarcoma: An analysis of the National Genomic Profiling Database. *Cancer Sci.*, 115, 2, 575–88, 2024 Feb 1.
43. Rinaudo, C.D., Telford, J.L., Rappuoli, R., Seib, K.L., Vaccinology in the genome era. *J. Clin. Invest.*, 119, 2515–25, 2009.
44. Du, L., He, Y., Zhou, Y., Liu, S., Zheng, B.J., Jiang, S., The spike protein of SARS-CoV - A target for vaccine and therapeutic development. *Nat. Rev. Microbiol.*, 7, 226–36, 2009.
45. Lu, B., Lim, J.M., Yu, B., Song, S., Neeli, P., Sobhani, N., *et al.*, The next-generation DNA vaccine platforms and delivery systems: Advances, challenges and prospects. *Front. Immunol.*, 15, Frontiers Media SA, 2024.
46. Cheng, F., Wang, Y., Bai, Y., Liang, Z., Mao, Q., Liu, D., *et al.*, Research Advances on the Stability of mRNA Vaccines. *Viruses*, 15, MDPI, 2023.
47. Salam, M.A., Al-Amin, M.Y., Salam, M.T., Pawar, J.S., Akhter, N., Rabaan, A.A., *et al.*, Antimicrobial Resistance: A Growing Serious Threat for Global Public Health. *Healthcare (Switzerland)*, 11, Multidisciplinary Digital Publishing Institute (MDPI), 2023.

48. Bogaerts, B., Delcourt, T., Soetaert, K., Boarbi, S., Ceyssens, P.J., Winand, R., *et al.*, A bioinformatics whole-genome sequencing workflow for clinical Mycobacterium tuberculosis complex isolate analysis, validated using a reference collection extensively characterized with conventional methods and *in silico* approaches. *J. Clin. Microbiol.*, 59, 6, 2021 Jun 1.
49. Papp, M. and Solymosi, N., Review and Comparison of Antimicrobial Resistance Gene Databases. *Antibiotics*, 11, MDPI, 2022.
50. Mayorga-Ramos, A., Zúñiga-Miranda, J., Carrera-Pacheco, S.E., Barba-Ostria, C., Guamán, L.P., CRISPR-Cas-Based Antimicrobials: Design, Challenges, and Bacterial Mechanisms of Resistance. *ACS Infect. Dis.*, 9, 1283–302, American Chemical Society, 2023.
51. Chalker, V.J., Smith, A., Al-Shahib, A., Botchway, S., Macdonald, E., Daniel, R., *et al.*, Integration of genomic and other epidemiologic data to investigate and control a cross-institutional outbreak of Streptococcus pyogenes. *Emerg. Infect. Dis.*, 22, 6, 973–80, 2016 Jun 1.
52. Traynor, B.J., The era of genomic epidemiology. *Neuroepidemiology*, 33, 276–9, 2009.
53. Reintjes, R. and Zanuzdana, A., Outbreak Investigations. *Mod. Infect. Dis. Epidemiol.*, 28, 159–76, 2009 Jul, doi: 10.1007/978-0-387-93835-6_9.
54. Harcourt, J., Tamin, A., Lu, X., Kamili, S., Sakthivel, S.K., Murray, J., *et al.*, Isolation and characterization of SARS-CoV-2 from the first US COVID-19 patient. bioRxiv [Internet]. 2020 Mar 7; Available from: http://www.ncbi.nlm.nih.gov/pubmed/32511316.
55. Andre, M., Lau, L.S., Pokharel, M.D., Ramelow, J., Owens, F., Souchak, J., *et al.*, From Alpha to Omicron: How Different Variants of Concern of the SARS-Coronavirus-2 Impacted the World. *Biology*, 12, Multidisciplinary Digital Publishing Institute (MDPI), 2023.
56. Carlson, C.J., Farrell, M.J., Grange, Z., Han, B.A., Mollentze, N., Phelan, A.L., *et al.*, The future of zoonotic risk prediction. *Philos. Trans. R. Soc. B: Biol. Sci.*, 376, Royal Society Publishing, 2021.
57. Scarpa, F. and Casu, M., Genomics and Bioinformatics in One Health: Transdisciplinary Approaches for Health Promotion and Disease Prevention. *Int. J. Environ. Res. Public Health*, 21, 10, 2024 Oct 1.
58. Chin, L. and Gray, J.W., Translating insights from the cancer genome into clinical practice. *Nature*, 452, 553–63, Nature Publishing Group, 2008.
59. Liu, P., Lyndon, A., Holl, J.L., Johnson, J., Bilimoria, K.Y., Stey, A.M., Barriers and facilitators to interdisciplinary communication during consultations: A qualitative study. *BMJ Open*, 11, 9, 2021 Sep 2.
60. Sarantopoulos, A., Mastori Kourmpani, C., Yokarasa, A.L., Makamanzi, C., Antoniou, P., Spernovasilis, N., *et al.*, Artificial Intelligence in Infectious Disease Clinical Practice: An Overview of Gaps, Opportunities, and Limitations. *Trop. Med. Infect. Dis.*, 9, Multidisciplinary Digital Publishing Institute (MDPI), 2024.

61. Wang, T., Antonacci-Fulton, L., Howe, K., Lawson, H.A., Lucas, J.K., Phillippy, A.M., *et al.*, The Human Pangenome Project: A global resource to map genomic diversity. *Nature*, 604, 437–46, NLM (Medline), 2022.
62. Ozdemir, T., Fedorec, A.J.H., Danino, T., Barnes, C.P., Synthetic Biology and Engineered Live Biotherapeutics: Toward Increasing System Complexity. *Cell Syst.*, 7, 5–16, Cell Press, 2018.
63. Chakraborty, T. and Barbuddhe, S.B., Enabling one health solutions through genomics. *Indian J. Med. Res.*, 153, 273–9, Wolters Kluwer Medknow Publications, 2021.
64. Hohenlohe, P.A., Funk, W.C., Rajora, O.P., Population genomics for wildlife conservation and management. *Mol. Ecol.*, 30, 1, 62–82, 2021 Jan 1.
65. Chen, K.T., Emerging Infectious Diseases and One Health: Implication for Public Health. *Int. J. Environ. Res. Public Health*, 19, MDPI, 2022.
66. Burr, A.H.P., Bhattacharjee, A., Hand, T.W., Nutritional Modulation of the Microbiome and Immune Response. *J. Immunol.*, 205, 6, 1479–87, 2020 Sep 15.
67. Ranallo, R.T., Mcdonald, L.C., Halpin, A.L., Hiltke, T., Young, V.B., The state of microbiome science at the intersection of infectious diseases and antimicrobial resistance. *J. Infect. Dis.*, 223, S187–93, 2021 Jun 15.
68. Piccioni, A., Rosa, F., Manca, F., Pignataro, G., Zanza, C., Savioli, G., *et al.*, Gut Microbiota and Clostridium difficile: What We Know and the New Frontiers. *Int. J. Mol. Sci.*, 23, MDPI, 2022.
69. Popova, L. and Carabetta, V.J., The use of next-generation sequencing in personalized medicine.
70. Bissonnette, L. and Bergeron, M.G., Infectious disease management through point-of-care personalized medicine molecular diagnostic technologies. *J. Pers. Med.*, 2, 50–70, MDPI AG, 2012.
71. Goetz, L.H. and Schork, N.J., Personalized medicine: Motivation, challenges, and progress. *Fertil. Steril.*, 109, 952–63, Elsevier Inc., 2018.
72. Turnbaugh, P.J., Ley, R.E., Hamady, M., Fraser-Liggett, C.M., Knight, R., Gordon, J.I., The Human Microbiome Project. *Nature*, 449, 804–10, Nature Publishing Group, 2007.
73. Tekola-Ayele, F. and Rotimi, C.N., Translational Genomics in Low- and Middle-Income Countries: Opportunities and Challenges. *Public Health Genomics*, 18, 4, 242–7, 2015 Jul 25.
74. Rehm, H.L., Page, A.J.H., Smith, L., Adams, J.B., Alterovitz, G., Babb, L.J., *et al.*, GA4GH: International policies and standards for data sharing across genomic research and healthcare. *Cell Genomics*, 1, 2, 2021.
75. Walker, A., Boyce, A., Duggal, P., Thio, C.L., Geller, G., Genomics and Infectious Diseases: Expert Perspectives on Public Health Considerations regarding Actionability and Privacy. *Ethics Hum. Res.*, 42, 3, 30–40, 2020 May 1.
76. Hawsawi, Y.M., Shams, A., Theyab, A., Siddiqui, J., Barnawee, M., Abdali, W.A., *et al.*, The State-of-the-Art of Gene Editing and its Application to Viral

Infections and Diseases Including COVID-19. *Front. Cell. Infect. Microbiol.*, 12, Frontiers Media S.A., 2022.
77. Do Nascimento, I.J.B., Pizarro, A.B., Almeida, J.M., Azzopardi-Muscat, N., Gonçalves, M.A., Björklund, M., *et al.*, Infodemics and health misinformation: A systematic review of reviews. *Bull. World Health Organ.*, 100, 544–61, World Health Organization, 2022.
78. Middleton, A., Adams, A., Aidid, H., Atutornu, J., Boraschi, D., Borra, J., *et al.*, public engagement with genomics. *Wellcome Open Res.*, 8, 310, 2023 Jul 18.
79. Croft, S.L., Neglected tropical diseases in the genomics era: Re-evaluating the impact of new drugs and mass drug administration. *Genome Biol.*, 17, 1, 2016 Mar 14.
80. Melo, G.B.T., Angulo-Tuesta, A., da Silva, E.N., Santos T da, S., Uchimura, L.Y.T., Obara, M.T., Evolution of research funding for neglected tropical diseases in Brazil, 2004–2020. *PLoS Negl.Trop. Dis.*, 17, 3, 2023.
81. Sorani, M.D., Yue, J.K., Sharma, S., Manley, G.T., Ferguson, A.R., Cooper, S.R., *et al.*, Genetic Data Sharing and Privacy. *Neuroinformatics*, 13, Humana Press Inc., 2015.
82. Davis, D.J., Goutham, S., Yeddula, R., CRISPR Advancements for Human Health.
83. Public Attitudes Toward Genetics & Genomics Research Literature and Polling Review Report.

10

Genomics of Prenatal and Paediatric Diseases: Early Detection and Intervention

Priyanka[1]* and Ayush Madan[2]

[1]*Graduate Institute of Biomedical Sciences, China Medical University, Taichung City, Taiwan*
[2]*Department of Biotechnology, School of Research and Technology, People's University, Bhanpur, Bhopal, Madhya Pradesh, India*

Abstract

Advancements in Genomic technologies have revolutionized the detection and management of prenatal and paediatric diseases, providing critical insights into genetic risk factors well before the onset of clinical symptoms. Recent innovations in genomic sequencing, including whole-genome sequencing and Non-invasive prenatal testing, have enabled the early identification of genetic predispositions to a variety of conditions, such as congenital heart disease, metabolic disorders, and rare genetic syndromes. Early diagnosis through these techniques offers a potential for specialized interventions that significantly reduces the long-term morbidity and mortality correlated with many genetic disorders.

However, the incorporation of genomics with clinical practice remains complex, demanding a balance between clinical efficacy and its ethical considerations. While prenatal and paediatric genomic screening promises to increase preventive care, challenges like the evaluation of incidental findings, counselling for parents or guardians and the accessibility of these technologies in diverse healthcare system should abide. Furthermore, the clinical and medical application of genomic data in early intrusion, such as personalized or targeted treatment plans or preventive measures, remains an area of working research, with a focus on improving clinical outcomes for children by targeted therapies and precision medicine.

This chapter explores the scientific advancements driving the genomic medicine in early identification and intervention for prenatal and paediatric diseases. Besides this, up to 25% of genomic results may reveal variants of uncertain

**Corresponding author*: Zoologypriyankasaini@gmail.com

Swetha Kumari Koduru, Sarad Pawar Naik Bukke, Hussaini Bello, Muhamad Mustafa Mundu and Ayush Madan (eds.) Genomics and Precision Healthcare: Innovations, Applications and Challenges, (295–350) © 2026 Scrivener Publishing LLC

significance by complicating its clinical decision-making and counselling. It explores the benefits and issues of incorporating genomic sequencing into regular basis clinical practice and consider it as future directions for genomic medicine, including the expansion of cost-effective screening programs and the expansion of genomic counselling frameworks to ensure the equity and accessibility for the long term. As genomics continues to evolve, its potential to establish the future of paediatric care upholds the promise for more effective, precise, accurate and feasible treatments that could substantially improve patient outcomes and disease prevention across populations without bias.

Keywords: Prenatal, paediatric diseases, genomic sequencing, genomic screening, targeted therapies

10.1 Introduction

10.1.1 Overview of Genomics in Medicine

Genomics is the study of the complete set of DNA that contains all of its genes in an individual or other organism. By mapping the human genome scientists have unlocked the potential to understand diseases at the most fundamental level [2]. Genomics goes beyond its classical genetics by not only identifying specific genes associated with a disease but also exploring how genes engage with each other and the surrounding environment [3]. This intricate interplay is crucial for understanding the complexity of human health and disease. Development in genomic technologies, like next-generation sequencing (NGS) and Bioinformatics, have enabled rapid, high-throughput analysis of DNA and RNA from basic to high level. These innovations have dramatically reduced the time and cost required for sequencing by making genomic analysis more accessible for easy approach in each and every field [4]. Genomics plays an integral role in several domains of medicine, including oncology, cardiology, infectious diseases, and rare genetic disorders like fragile X syndrome, Klinefelter syndrome, Marfan syndrome, etc. By identifying genetic variants associated with these conditions healthcare providers can offer personalized treatments by improving diagnostic accuracy and predict disease susceptibility [5].

One of the most groundbreaking contributions of genomics is the advent of precision medicine, a medical model that tailors treatment to individual genetic profiles by achieving a targeted approach in different ways. Precision medicine has shifted the paradigm from one-size-fits-all approaches to interventions that consider an organism's unique genetic, environmental, and daily routine factors and target and make every individual fit for treatment. For instance, in oncology, genomic profiling of tumors has facilitated the development of targeted therapies, like EGFR inhibitors for non-small cell lung cancer (NSCLC) and HER2-targeted drugs for breast cancer [1]. These treatments have significantly improved outcomes for individuals with specific genetic mutations. Beyond the desired unique treatment, genomics also impacts disease prevention. The validation of the genetic risk factors enables proactive measures to mitigate a disease onset. For example, genetic testing for BRCA1 and BRCA2 mutations in individuals with an inheritance history of breast or ovarian cancer allows for early intervention through surveillance, prophylactic surgeries, or chemoprevention. Additionally, the wide population for genomic screening programs have begun to emerge, aiming to identify individuals at risk of common diseases like diabetes and cardiovascular conditions, for example.

The integration of genomics with healthcare also raises “Ethical, Legal, and Social Implications” (ELSI). Cases with genetic privacy, data security and the potential for genetic discrimination must be addressed to ensure the responsible use of genomic information. Policies and guidelines are continually being developed to balance innovation with ethical considerations that foster trust among patients and the public. The field of genomics in medicine is full of challenges. Despite significant progress, the clinical implementation of genomics faces challenges such as the interpretation of Variants of unknown significance (VUS), the unavailability of diversity in genomic research and the need for robust infrastructure to support data analysis and its storage [42]. The efforts to overcome these leading challenges includes global initiatives such as the Human Variome Project and The All of Us Research Program which primarily aim to expand the diversity of the genomic datasets and also increase the translation of the genomic discoveries into clinical/medical practice.

Genomics has transformed medicine and treatment by providing insights into the genetic basis of health and disease by enabling personalized interventions and paving the way for predictive and preventive healthcare infrastructure future. As recent researches continue to expand understanding of the genome, the potential for genomics to reshape medicine is infinite; it makes a way for achieving targeted/unique treatment for patients.

10.1.2 Importance of Early Detection and Intervention

Early detection and intercession are the foundation of effective healthcare; they significantly influence results in a wide range of diseases [25]. The principle behind early detection is completely straightforward. Identifying a disease in the initial stages allows on time and often less invasive treatment, thereby improving prognosis and quality of life [41]. Genomics is a powerful tool in the domain of initial detection by enabling the identification of molecular changes that precede clinical symptoms. In oncology, early detection is particularly elucidating, as cancer is often asymptomatic in its initial stages and by the time symptoms appear, it may have progressed to later stages with a less treatable state [31]. Genomic biomarkers such as circulating tumor DNA (ctDNA), i.e., a small strand of DNA that comes from tumor cells or cancer cells in the blood and microRNAs, i.e., a small RNA molecule that actually controls gene expression by binding to mRNA (messenger RNA), have been instrumental in the development of non-invasive diagnostic methods, such as liquid biopsies. These techniques allow, as shown in Table 10.1, for the early detection of cancer stages and cancer types by analyses of the disease progression and understanding of treatment response. The detection of the targeted mutations in ctDNA can identify early-stage lung or colorectal cancer by enabling the interventions that significantly improve survival rates. Genomics also plays a pivotal role in the early diagnosis of genetic disorders. For conditions like Cystic fibrosis, Duchenne muscular dystrophy and Spinal muscular atrophy, early identification through newborn screening programs allows for the prompt initiation of treatments that can mitigate disease progression and improve quality of life [11]. Advancements in genomic sequencing have expanded the scope of newborn screening by enabling the detection of a wider range of conditions. This progress underscores the importance of combining genomics with public health strategies that enhance the early identification efforts for long-term outcomes.

In infectious diseases, an early detection that is mainly facilitated by the genomics is critical for effective containment and its treatment. Genomic sequencing of the pathogens enables the rapid identification of its infectious agents that is achieved by monitoring the antimicrobial resistance and tracking of its outbreaks. For instance, during the COVID-19 pandemic, genomic surveillance played an important role in identifying variants of "SARS-CoV-2" virus by guiding public health responses and informing the vaccine development for leading good healthcare infrastructure. This demonstrates the far-reaching impact of genomics in managing infectious diseases and preventing widespread transmission. Cardiovascular diseases (CVDs) are a leading cause of morbidity and mortality worldwide that

Table 10.1 Comparison of early detection techniques.

Technique	Principle	Advantages	Applications
Liquid Biopsy	Detection of circulating tumor DNA (ctDNA)	Continuous, non-invasive monitoring that provides real-time data and insights.	Early cancer detection, treatment response assessment
Newborn Screening	Detection of genetic disorders	Timely and proactive intervention leads to significantly better outcomes.	Cystic fibrosis, phenylketonuria
Genomic Risk Scores	Analysis of multiple genetic markers	Assessment of individual susceptibility to different diseases. This majorly relates to an evaluating genetic, surrounding environmental and daily routine factors that can contribute to a personal risk of developing these specific health conditions.	Cardiovascular disease, diabetes
Pathogen Genomic Surveillance	Sequencing of pathogen genomes	Rapid detection and its effective containment measures.	COVID-19 variant detection

also benefit from the early detection enabled by genomics [1]. Genetic risk scores are generally derived by Genome-Wide Association Studies (GWAS) that can identify individuals at high risk of developing conditions like coronary artery disease or atrial fibrillation. This information allows us better implementation of preventive measures like lifestyle modifications as per need and pharmacological interventions that reduce the chances of adverse outcomes. Moreover, the validation of rare genetic variants associated with the inherited cardiomyopathies or arrhythmias enables early intervention and family screening or detection, preventing sudden cardiac events [28]. The importance of early detection extends to mental health preparation as well. Genomic studies have uncovered genetic variants associated with psychiatric conditions like Schizophrenia, Bipolar disorder and depressive disorder, for example, major depressive leading disorder, persistent depressive disorder, etc. [13, 26]. Although the predictive value of these findings is still moving forward, as they hold the promise for identifying individuals at risk and tailored or unique interventions just to prevent the onset or worsening of the mental health conditions. Personalized medicine has the potential to transform healthcare but faces challenges such as integrating genetic information into medical practice that advanced data analytics for interpreting that patients data and ensuring equitable access to needed treatments. Regulatory frameworks need to adapt a new therapies and there are concerns about patients privacy and ethics regarding genetic data use [18]. For overcoming these challenges is crucial for enhancing the patient result's outcomes. These include the need for exclusive clinical validation of the genomic biomarkers by the integration of genomic data into electronic health records and the development of cost-effective screening programs that make it more accessible for needed people. Ethical considerations such as the psychological impact of the identifying asymptomatic individuals at risk for untreatable conditions that must also be addressed as requirement [41]. Public education and genetic counselling are the essential components of any early detection initiative by ensuring that individuals understand the implications of the genomic findings and can decisions about their healthcare.

Early detection and intervention are fundamental to improving health benefits and well-being, and genomics has significantly enhanced our ability to achieve these goals. That, we can achieve by identifying molecular changes that introduce the clinical manifestations; genomics enables continuously targeting the interventions that can prevent or mitigate disease in large populations. As technologies and methodologies continue to evolve, the role of genomics in early detection will only grow by offering new opportunities to transform healthcare and improve everyone's lives.

10.2 Advancements in Genomic Technologies

Genomic technologies have undergone remarkable advancements over the last few decades, that actually revolutionizing our understanding of the genetic underpinnings of health, dieases and its treatment [25]. These outstanding improvement have been faciliated the Precision/targeted medicine that improved diagnostic capanilities and enabled an early detection of the diseases. Among the most prominent advancement techniques are whole-genome sequencing (WGS) and non -invasive pre natal testing (NIPT) which have redefined the landscape of genomic medicine and its scope into clinical practice.

10.2.1 Whole-Genome Sequencing (WGS)

Whole-genome sequencing is a laboratory method for analysing an entirety of an organism's DNA that encompasses both coding and non-coding regions of the genome. In other words, this is a laboratory method that is used to learn an exact order of all of the building blocks, i.e., nucleotides that make up an organism's genome as a complete set of DNA. This entails sequencing all of an individual's chromosomal DNA as well as DNA contained in the mitochondria [16]. Unlike traditional methods, such as targeted gene panels or exome sequencing, whole-genome sequencing provides an unbiased view of the entire genome, which makes it a powerful tool for identifying genetic variations related to diseases.

- **Advancements in WGS Technology**

The development of next-generation sequencing (NGS) or Massive parallel sequencing or Second-generation sequencing, a technology that sequences a DNA or RNA to determine the order of nucleotides, platforms have significantly enhanced the speed, accuracy and cost-effectiveness of WGS. Early sequencing methods, such as Sanger sequencing or Chain Termination Method—a laboratory method that determines the order of the bases in a DNA segment—were labour-intensive and costly, limiting their application to small-scale studies. In contrast, NGS technologies, like Illumina's HiSeq and NovaSeq systems, can sequence billions of DNA fragments simultaneously, enabling the completion of a human genome sequence in days rather than years [9, 10].

Emerging technologies, such as nanopore sequencing, a sequencing that offers a label-free, electrical, Single strand molecule of DNA sequence methods, obviating the need for amplification or labelling by detecting a direct electrical Signal, have further advanced WGS that is done by measuring "Electronic conductance across a membrane" [47]. Oxford Nanopore's devices allow for real-time sequencing with long-read capabilities by improving the detection of structural variants and complex genomic rearrangements [20]. These advancements have expanded the utility of WGS in research and clinical settings.

- **Applications of WGS in Medicine**

WGS has numerous applications in medical genetics, oncology, infectious diseases, and pharmacogenomics, as mentioned in Table 10.2 as follows:

1. **Rare Genetic Disorders:** WGS is an invaluable tool in diagnosing of rare genetic diseases, particularly where the conventional diagnostic approaches have failed many times. For example, in patients with Dravet syndrome or Severe myoclonic epilepsy of infancy (SMEI), a rare inheritance disorder that causes seizures and developmental delays in infants or children, WGS has successfully identified novel mutations as SCN1A. This led to a successful path for precise diagnosis and tailored or needed treatment plans [24].
2. **Cancer Genomics:** In the principality of oncology, WGS has completely changed the way initially we understand and treat cancer by enabling researchers and clinicians to identify somatic mutations, a change or mutation in DNA that occurs only in cells of the body of organism or in non-germline cells except the reproductive cells or germline cells of an individual. As well as, Copy number variations and complex structural rearrangements within the cells [2]. These insights have been pivotal in the development of targeted or unique needed therapies such as ALK inhibitors for lung cancer patients with ALK rearrangements.
3. **Infectious Diseases:** Over a period of a few decades, the application of WGS has extended to the field of infectious disease by enabling the identification and characterization of pathogens. This technique has been important in supporting outbreak investigations and monitoring trends in antimicrobial resistance. Examples are Ebola and COVID-19 pandemics. WGS helps to understand the evolution of the viruses in both cases and also tracks their transmission pattern in a large population in real time [21, 22]. That makes it an important part for responding in real time, thereby informing public health responses and containment strategies.
4. **Pharmacogenomics:** In the field of Pharmacogenomics, we studied how genes affect a person's or individual's response to drugs that helps us to understand improving health by ensuring patients receive the right drugs at the right quantity or dose. By identifying specific genetic variants that

Table 10.2 Key applications of WGS in medicine.

Medical field	Application	Examples
Rare Genetic Disorders	Accurate diagnosis and tailored treatment plans.	Identification of SCN1A mutations in Dravet syndrome
Cancer Genomics	Identification of genetic mutations for the purpose of developing targeted therapies.	ALK rearrangement detection for lung cancer
Infectious Diseases	The detailed analysis of pathogens and the monitoring of outbreaks.	COVID-19 viral evolution monitoring
Pharmacogenomics	Personalized drug therapy tailored to a unique inheritance or genetic, surrounding environmental and routine of each individual.	CYP2C19 testing for clopidogrel response

influence drug metabolism, efficacy, and leading toxicity, clinicians can make more precise and required selection of medication. Example: CYP2C19 and CYP2D6 Variations; these genes can significantly alter a patient's response to medications such as Clopidogrel that is a blood thinner or antidepressants, respectively. This personalized and targeted approach to medication management not only optimizes the therapeutic results but also minimizes adverse drug reactions [38].

- **Challenges and Future Directions**

Despite its massive potential and advancing medical advantages, the widespread adoption of WGS still faces challenges in implementation in different clinical settings. Some primary challenges include the interpretation of variants of uncertain significance (VUS), data storage and analysis requirements and the need for robust clinical validation that is essential to ensure that findings from WGS can be reliably used in patient care. To mitigate these issues with coming challenges, advanced machine learning algorithms and international genomic databases, such as ClinVar (archives and aggregates information about relationships among variation and human health) and gnomAD (Genome Aggregation Database—a resource developed by an international coalition of investigators), are being developed to improve variant annotation and interpretation, which provides a large-scale population genomic data [16, 41].

Looking ahead, the integration of WGS into routine clinical practice looks promising as costs continue to decrease and computational tools advance [25]. The advent of multi-omics approaches combining WGS with transcriptomics, proteomics, and metabolomics holds promise for a deeper understanding of complex diseases. These tools aim to enhance the accuracy of variant annotation and interpretation, ultimately facilitating better clinical decisions and treatments.

10.2.2 Non-Invasive Prenatal Testing (NIPT)

Non-invasive prenatal testing (NIPT) is a blood screening test for pregnant women that has emerged as a groundbreaking application in genomic technologies that offers a safe and effective method for assessing foetal genetic health [35]. By analysing cell-free foetal DNA (cffDNA) circulating in the maternal bloodstream [37], NIPT provides valuable information about the chromosomal abnormalities and other genetic conditions without creating any risks to the foetus or mother [51].

- **Advancements in NIPT Technology**

The approach of high-throughput sequencing and bioinformatics has launched NIPT from a niche diagnostic tool to a widely accepted and adopted clinical practice. Initially focused on detecting common aneuploidies, such as Trisomy 21 (Down syndrome), Trisomy 18 (Edwards syndrome), and Trisomy 13 (Patau syndrome), NIPT has been broadening its scope to include [49, 51]:

1. **Microdeletions and Duplications:** Advanced knowledge and intelligence now enable the detection of the sub-chromosomal abnormalities in a lifetime, such as 22q11.2 deletion syndrome (DiGeorge syndrome), a condition present since birth that can seed a range of lifelong problems, including heart defects and self-learning difficulties.
2. **Single-Gene Disorders:** Leading NIPT platforms can screen out for monogenic conditions that include cystic fibrosis

and sickle cell anaemia. Cystic Fibrosis is caused by a capable gene that affects the movement of salt (NaCl) and Water (H2O) in and out of cells. In a long way, this along with recurrent infections can result in a build-up of thick, Sticky mucus in the body's tube and passageways, mostly around the lungs and digestive system. In Sickle Cell anaemia or Sickle cell disease, the shape of RBCs (Red Blood Cells) gets elongated while usually it is round and flexible to move easily and carry oxygen through blood vessels.

3. **Foetal Whole-Genome Sequencing:** Cutting-edge research has demonstrated that feasibility of sequencing the entire foetal genome from the maternal plasma is offering out of the ordinary insights into the genetic health of foetus.

- **Clinical Applications of NIPT**

NIPT has completely changed prenatal care by offering initial and accurate risk assessment for the genetic conditions. Key applications for this assessment include:

1. **Screening for Aneuploidies:** NIPT has shown high sensitivity and specificity for detecting the Trisomies. For example, studies have reported a detection rate of almost 99% for Down Syndrome or Trisomy 21, which significantly reduced the need for invasive procedures like amniocentesis, as it has many risks such as infection in the uterus, a few chances of miscarriage, abdominal discomfort, etc.
2. **Identifying RhD Status:** For RhD-negative mothers, NIPT can identify the foetal RhD status, by guiding the administration of "prophylactic anti-D immunoglobulin" only when it is so necessary. That is generally done by a blood test that checks for the presence of the RhD antigen on RBCs. People with the RhD antigen are "RhD positive", while without it is "RhD negative".
3. **Twin Pregnancies and IVF:** NIPT is really effective for identifying the zygosity and aneuploidies in the case of twin pregnancies, as well as in pregnancies resulting from assisted reproductive technologies (ART). ART pregnancy results from fertility treatments that involve manipulating eggs or embryos, that is generally used to treat or handle infertility.

4. **Maternal Health Insights:** NIPT has the major potential to detect maternal genetic conditions and malignancies like cffDNA analysis; it is a non-invasive prenatal test that screens for fetal genetic abnormalities as it is performed on the mother's blood to get analysis of Fetal DNA, and occasionally reveals unexpected findings such as chromosomal abonormalities/aberrations [15].

- **Ethical and Social Considerations**

The unexpected widespread adoption of NIPT raises important ethical and social questions. These include concerns usually about informed consent, the potential for discrimination against individuals with disabilities and the risk of overdiagnosis leading to unnecessary anxiety and mental stress. Ensuring equitable access to NIPT and providing comprehensive genetic counselling are most essential for addressing these upcoming pronounced challenges [18].

- **Future Directions in NIPT**

As genomic technologies continue to advance the future of NIPT is likely to include the following:

1. **Expanded Genetic Screening:** Daily basis NIPT may encompass a wider range of conditions that includes polygenic risk scores, an estimate of an organism's genetic liability to a trait or disease that is calculated according to their genotype profile and relevant genome-wide association study (GWAS) data, for complex or different traits.
2. **Integration with Other Biomarkers:** Having NIPT with proteomics, i.e., a study of structure and function of proteins in living organisms or metabolomics, i.e., a study of metabolites or small molecules in a biological system, that could enhance its diagnostic accuracy and predictive value.
3. **Global Accessibility:** Efforts to reduce costs and simplify workflows will be critical for making NIPT accessible in low-resource settings, majorly addressing disparities in prenatal care. Making health a primary and foremost criterion for both foetus and mother is the most important aspect for upcoming efforts in the medical field, where having low cost and the best equipment and facilities is the most important need of the time.

NIPT represents a paradigm shift in prenatal medicine and care by offering a non-invasive, accurate and comprehensive method for assessing foetal genetic health. Ongoing advancements in the sequencing technologies and bioinformatics promise to further enhance its capabilities, making it an essential tool in modern obstetrics [35].

The last few decades have witnessed a large explosion in genomic technologies that have revolutionized the field of medicine. Advances in the sequencing technologies, computational tools and bioinformatics have enabled researchers to uncover the complex genetic information at an extraordinary scale and speed. These advancements have had profound implications for disease diagnosis, prognosis and therapeutic interventions. In this section, we will explore the key breakthroughs in genomic technologies, focusing on their applications and impact on medical science for both foetus and mother.

- **High-Throughput Sequencing Technologies**

Next-Generation Sequencing (NGS) has emerged as a fundamental principle of modern genomics [34]. Unlike traditional and many-steps-oriented Sanger sequencing, which processes DNA one fragment at a time as it takes a long time for every result, NGS allows the parallel sequencing of millions of DNA fragments with good accuracy and precision while taking much less time than traditional methods [6]. This high-throughput capability has dramatically reduced the costs while increasing speed, precision and accuracy.

- **Applications:**

  - **Oncology**: NGS is mostly used to identify the mutations in cancer-associated genes, including oncogenes, tumor suppressor genes and DNA repair genes such as BRCA1 (breast cancer 1) and BRCA2 (breast cancer 2), i.e., tumor suppressor genes that help to prevent breast, ovarian, prostate and pancreatic cancer, enabling targeted or unique therapies.
  - **Rare Diseases**: Whole-exome sequencing (WES) is more focussed to detect Rare variants and mutations in protein-coding regions, and whole-genome sequencing (WGS) is more focussed to detect the rare and common genetic variations, including structural variants. These methods have been instrumental in diagnosing rare genetic disorders where conventional methods fail.
  - **Infectious Diseases**: Pathogen genomics is a study of the genetic material of disease-causing microorganisms. Using

NGS has accelerated the identification of novel pathogens and their resistance mechanisms as exemplified by the rapid genomic sequencing of SARS-CoV-2 during the COVID-19 pandemic [25].

Key Developments:

- **Third-Generation Sequencing**: Technologies like PacBio's Single-Molecule Real-Time (SMRT) sequencing and Oxford Nanopore sequencing Technology (ONT) have offered\longer read lengths as compared to previous sequencing, and are essential for resolving the complex genomic regions.
- **Single-Cell Genomics**: Single-cell RNA sequencing (scRNA-seq) provides insights into gene expression profiles at the cellular level from individual cells that elucidate heterogeneity within tissues to understand how cells function and interact [3].

10.2.3 Genetic Screening Tools for Paediatric Diseases

Early diagnosis of genetic conditions in children is crucial for timely intervention and improved clinical outcomes [34]. The development of advanced genetic screening tools and methods has significantly enhanced our ability to detect paediatric diseases at early stages [11].

1. Neonatal and Newborn Screening:
Newborn screening programs are among the earliest applications of genomic technologies in the field of paediatrics. They are sometimes also referred to as "Heel Prick test". Tandem or random mass spectrometry and NGS-based assays have expanded the number of detectable conditions in a few decades, including [11, 34]:

- **Metabolic Disorders**: Phenylketonuria (PKU) is an inherited or genetic disorder that shows increased levels of a substance called "phenylalanine" in the blood, and Maple Syrup Urine Disease (MSUD) that actually repels the body to breaking down certain amino acids.
- **Endocrine Disorders**: Congenital hypothyroidism, in which the thyroid gland does not produce enough thyroid hormone at birth. This eventually leads to growth problems, intellectual disability, etc.

- **Haematological Conditions**: Sickle cell disease or Sickle cell anemia and Thalassemia is an inherited blood-related disorder that causes the body to have less haemoglobin (Hb) than normal levels, as it is more important to carry oxygen in the body.

2. Carrier Screening:
Preconception and prenatal carrier screening validate couples at risk of passing on genetic disorders. Technologies like WGS and targeted gene panels have enabled the detection of over 400 recessive conditions, including:

- **Cystic Fibrosis**: Cystic Fibrosis is caused by a faulty gene that affects the movement of salt (NaCl) and Water (H2O) in and out of cells. In the long term, it can result in a formation of thick, Sticky mucus in the body's tube and passageways, mostly around the lungs and digestive system [6].
- **Tay-Sachs Disease**: It is a rare, inherited disease that causes a formation of fatty acids that damages the brain and typically results in muscle control loss, genetic change that causes disease and results in a deficiency of enzyme beta-hexosaminidase A.
- **Spinal Muscular Atrophy (SMA)**: SMA is a genetic disease that causes progressive weakness and wasting; it affects the motor neurons in the spinal cord that controls movement. It has four types [34].

■ **Case Studies:**

- **Duchenne Muscular Dystrophy (DMD)**: DMD is a genetic disorder that is characterized by progressive muscle degeneration and weakness due to the alternations of a protein known as "Dystrophin" that helps keep muscle cells intact and enclosed. The advances in NGS have allowed for the identification of exon deletions and duplications in the DMD gene for guiding the therapy.
- **Severe Combined Immunodeficiency (SCID)**: SCID is a rare genetic disorder that is a reason for life-threatening problems with the immune system. Genomic screening

has facilitated early detection and successful bone marrow transplantation in affected infants.

- **Integration with AI:**

Machine learning algorithms are now integrated with genomic data to predict the disease onset and progression along with time. For example, AI-driven platforms can analyse the genetic variants to assess the risk of autism spectrum disorders [13].

10.2.4 Others

1. Pharmacogenomics:

Pharmacogenomics is a study of genomics and pharmacology has paved the way for personalized medicine and treatment. Pharmacogenomic testing identifies genetic variants affecting the drug metabolism, efficacy, and toxicity to develop safe medications that can be prescribed based on an individual's requirement and need to get more effective treatment [38, 41]. Examples:

a. **Warfarin Sensitivity**: This condition has a low tolerance for the drug warfarin. Variants in CYP2C9 and VKORC1 genes influence the dosing results.
b. **Clopidogrel Resistance**: Clopidogrel resistance occurs when the antiplatelet drug called "clopidogrel" is less effective than expected; it can cause blood clots and other cardiovascular events. CYP2C19 polymorphisms impact antiplatelet therapy; it is particularly important or activation of clopidogrel.
c. **Cancer Therapies**: Tumor mutation profiling informs the use of PARP (Poly(ADP-ribose) polymerase) inhibitors and immune checkpoint inhibitors [31].

2. Epigenomics:

Epigenomics is a study of the whole set of epigenetic modifications on the genetic material of a cell that profiling examines modifications of DNA methylation and histone acetylation that regulates gene expression without altering the DNA sequence [19, 29]. Advances include:

a. **Biomarkers**: Hypermethylation of O^6-methylguanine-DNA-methyltransferase (MGMT) acts as a predictive marker for the glioblastoma therapy.

b. **Therapeutics**: Development of drugs that target epigenetic enzymes such as Histone deacetylase (HDAC) inhibitors for the T-cell lymphoma [36, 43].

3. Proteomics and Multi-Omics Integration:
The integration of genomics with proteomics, transcriptomics and metabolomics offers a holistic view of the biological systems. Multi-omics have studied:

a. Identified biomarkers for the diseases like Alzheimer's and Type-II Diabetes.
b. Uncovered novel pathways for therapeutic targeting in cancer.

a. Population Genomics:
Population genomics is a large-scale application of genomic technologies to study the populations of individuals, such as large-scale initiatives like the 100,000 Genomes Project and the All of Us Research Program, which have generated valuable genetic data for understanding the population-specific disease risks and tailoring interventions. It includes:

a. Identification of rare genetic variants in the under-represented populations.
b. Precision health initiatives focusing on preventive strategies for large scale.

Advancements in genomic the technologies continue to be redefine the landscape of medicine by enabling precision diagnostics, personalized therapies and novel drug development. By integrating cutting-edge tools such as NGS, Cas9-CRISPR, and AI-driven analytics, we are closer than ever to realizing the full potential of genomics in addressing the complex health challenges in the long term for the large-scale approach.

b. CRISPR-Cas Systems
The discovery of Clustered regularly interspaced short palindromic repeats – Cas9 (CRISPR-Cas9) as a gene-editing tool has transformed our ability to manipulate the genome with precision, allowing researchers to modify DNA in living organisms. Its genome editing system was adapted from bacteria. CRISPR technology allows researchers to add, delete, or modify specific DNA sequences with remarkable accuracy by offering promising therapeutic applications [18, 41].

(i). Applications:

1. **Gene Therapy**: Gene therapy is a treatment that modifies an individual's genes in order to treat or cure diseases. It corrects mutations in genes associated with monogenic disorders, like sickle cell anaemia and cystic fibrosis.
2. **Functional Genomics**: Functional Genomics is a field of molecular biology that studies how genes and their products (like RNA and proteins) function and interact within an organism. It aims to understand the complex relationship between genotype and phenotype by analyzing gene expression, regulation and protein activity on a genome-wide scale by using high-throughput CRISPR screens to identify essential genes and pathways in cancer and other diseases.
3. **Synthetic Biology**: Synthetic Biology is a scientific field that involves redesigning individuals to create new abilities, by engineering microbes for bioproduction of pharmaceuticals and biofuels.

(ii). Innovations:

a. **Base Editing**: Base Editing is a type of genome editing; CRISPR-based base editors enable the single nucleotide changes without creating double-strand breaks.
b. **Prime Editing**: This is a next-generation CRISPR tool that allows precise insertions, deletions and substitutions with fewer off-target effects [18].

10.3 Prenatal Genomic Screening

The integration of genomic technologies into prenatal care has revolutionized the early detection and management of genetic disorders, significantly improving both maternal and foetal results [35]. Prenatal genomic screening encompasses a range of techniques designed to assess the genetic makeup of the foetus by enabling the early diagnosis and intervention for various genetic conditions. Here, we will discuss the techniques, their applications, benefits and ethical considerations regarding prenatal genomic screening [49].

10.3.1 Used Techniques and Methods

Prenatal genomic screening takes into employment a variety of advanced techniques to analyse the foetal genetic material. These methods range from non-invasive engaging to more invasive diagnostic procedures that periodically depend on the clinical scenario and the information required for good outcomes as shown in Table 10.3.

1. Non-Invasive Prenatal Testing (NIPT):
NIPT includes the analysis of cell-free foetal DNA (cffDNA) which is circulating in the maternal/mother's bloodstream [23]. Key features include the following:

a. **Detection of Chromosomal Aneuploidies**: Commonly targeted aneuploidies include Trisomy 21 (Down syndrome), Trisomy 18 (Edwards syndrome), and Trisomy 13 (Patau syndrome).
b. **High Sensitivity and Specificity**: NIPT boasts over 99% sensitivity for detecting Trisomy 21, significantly reducing the need for invasive procedures.
c. **Expanded Panels**: Advanced NIPT panels now include microdeletion syndromes (e.g., DiGeorge syndrome) and other single-gene disorders.

2. Invasive Diagnostic Techniques:
When NIPT results indicate abnormalities or when detailed genetic information is required, invasive techniques [51] are employed as follows:

a. **Chorionic Villus Sampling (CVS)**: Performed between 10 to 12 weeks of gestation period, CVS involves obtaining placental tissue for chromosomal and molecular abnormalities analysis, majorly by using FISH (fluorescent *in situ* hybridization) or PCR (polymerase chain reaction). CVS is particularly performed in a transcervical or transabdominal manner. It takes place earlier than the amniocentesis or percutaneous umbilical cord blood sampling.
b. **Amniocentesis**: This is a medical practise where we primarily use the prenatal diagnosis of the genetic conditions of foetus. Conducted after 15 weeks of gestation period, this technique collects amniotic fluid containing foetal cells for featuring karyotyping; karyotyping is a chromosomal

Table 10.3 Techniques and methods used in prenatal genomic screening.

Technique	Description	Key features
Non-Invasive Prenatal Testing (NIPT)	Analysis of cell-free foetal DNA (cffDNA) circulating in the maternal bloodstream.	High sensitivity for chromosomal aneuploidies, reduced need for invasive procedures, expanded panels for microdeletion syndromes.
Chorionic Villus Sampling (CVS)	Obtaining placental tissue for chromosomal and molecular analysis (10-13 weeks).	Early detailed genetic information.
Amniocentesis	Collection of amniotic fluid containing foetal cells for analysis (after 15 weeks).	Detailed genetic data *via* karyotyping, microarray or WGS.
Foetal Blood Sampling (Cordocentesis)	Direct sampling of foetal blood, typically for diagnosing foetal anaemia or infections.	Used in rare cases.
Whole-Exome Sequencing (WES)	Sequencing the coding regions of the foetal genome to diagnose Mendelian disorders.	Useful for suspected congenital anomalies.
Whole-Genome Sequencing (WGS)	Comprehensive sequencing of the entire foetal genome, including non-coding regions.	Offers a broad view of genetic variations.
Targeted Panels	Analysis focused on specific gene sets associated with known conditions.	Useful for skeletal dysplasias and cardiac defects.

analyzing genetic test of cells as samples that identifies genetic problems that can cause disease or birth defects in foetus; it is a microarray analysis technique majorly used for interpreting data that is generated from experiments on DNA such as gene chip analysis, RNA and protein microarrays or Whole Genome Sequencing.

c. **Cordocentesis: Percutaneous umbilical blood sampling**: This is a test where diagnosis is done by examining a blood sample from the foetus to detect foetal abnormalities. Used in rare cases to diagnose conditions, because diagnostic procedures such as amniocentesis and Chorionic villus sampling are related to lower risk, Cordocentesis is used as diagnosis of foetal anaemia or infections.

3. Advanced Genomic Technologies:

a. **Whole-Exome Sequencing (WES)**: Increasingly utilized to diagnose Mendelian disorders when ultrasonographic findings suggest congenital anomalies, it can identify and analyse all genes in a single assay. Even after its findings, it is suggested to reanalyse the data of WES after 12 months. Still, 50-70% of patients are not able to receive a genetic diagnosis after WES.
b. **Whole-Genome Sequencing (WGS)**: WGS offers a brief view of the foetal genome that includes non-coding regions as foetus's DNA analyse entirely to detect potential genetic abnormalities in whole genome regions (coding regions and non-coding regions). The maximal potential of WGS is that it can identify genetic causes for complex foetal abnormalities.
c. **Targeted Panels**: Focused on specific gene sets associated with conditions like skeletal Dysplasias or cardiac defects. Panels can be predesigned or custom made as the target genes involved in specific pathways regulate the whole diagnosis.

10.3.2 Applications and Benefits of Clinical Usage

The clinical implementation of prenatal genomic screening has potential implications for maternal-foetal medicine by enabling personalized care as shown in Table 10.4 and improved pregnancy outcomes.

Table 10.4 Applications and benefits of clinical usage.

Application	Description	Examples
Early Detection of Aneuploidies	Identifying chromosomal abnormalities.	Down syndrome, Turner syndrome, Klinefelter syndrome.
Detection of Microdeletion Syndromes	Screening for sub-microscopic chromosomal deletions.	Prader-Willi, Angelman, Cri-du-chat syndromes.
Diagnosis of Monogenic Disorders	Identifying conditions caused by single gene mutations.	Cystic fibrosis, haemophilia, sickle cell anaemia.
Foetal Therapy Planning	Planning interventions for foetal conditions.	Spina bifida repair, twin-to-twin transfusion treatment.
Optimized Delivery Planning	Ensuring appropriate resources for high-risk deliveries.	Neonatal intensive care arrangements.
Psychosocial Benefits	Providing parents with actionable insights to reduce anxiety.	Improved decision-making and preparedness.
Public Health Impact	Reducing genetic disorder prevalence through informed reproductive choices.	Decreased healthcare costs and reduced need for postnatal interventions.

a. Early Detection of Genetic Disorders:

Prenatal screening facilitates the identification of a wide range of genetic conditions, including common to rare [49, 51]:

(i) **Aneuploidies**: It is a genetic disorder that occurs when an abnormal number of chromosomes are present in an organism; some may be missing or there may be duplication. Examples: Down syndrome, Turner syndrome and Klinefelter syndrome.

(ii) **Microdeletion and Duplication Syndromes**: Microdeletion and duplication syndromes are genetic disorders that occur when an organism's chromosome DNA got lost or duplicated. Examples: Prader-Willi, Angelman, and Cri-du-chat syndromes. This syndrome largely contributes to abnormalities regarding Development and Physical.

(iii) **Monogenic Disorders**: Monogenic disorders is a genetic disease which is caused by a mutation in a single gene and can affect multiple organs with a variety of symptoms at the same duration of time. Examples are Cystic fibrosis, Haemophilia and Sickle cell anaemia.

b. Personalized Pregnancy Management:

Early identification allows for the tailored/targeted prenatal care and its decision-making [35, 46]:

(i) **Foetal Therapy or Foetal Treatment**: Intra-uterine surgeries or medical treatments can be planned for conditions such as "spina bifida" or "twin-to-twin" transfusion syndrome.

(ii) **Optimized Delivery Planning**: Optimized delivery planning ensures appropriate resources for high-risk deliveries that include neonatal intensive care by planned delivery from 34 weeks' gestation duration onwards, which significantly reduced the risk of major maternal morbidity. To ensure care for both mother and foetus, a planned delivery or labour is necessary to align with medicine needs also.

c. Psychosocial and Reproductive Benefits:

(i) **Informed Decision-Making**: Parents gain crucial insights and understanding about their pregnancy by enabling themselves to prepare for the needs of an affected child or consider alternative options. It is a long process of choosing everything regarding their birthing plans by seeing all available options, risks and benefits with their healthcare provider.

(ii) **Reduced Anxiety**: High precision and accuracy of NIPT reduces false-positive rates by alleviating unnecessary parental stress. Open communication with patients, providing comprehensive information about the complete process of fostering a good and supportive environment, offers a relaxation period and reduces concerns and fears of being pregnant. An example is Cognitive behavioural therapy (CBT).

d. Public Health Impact:
Widespread adoption of prenatal genomic screening has the potential to:

(i) Decrease the incidence of genetic disorders through informed reproductive choices and having knowledge about genetic disorder analyses.
(ii) Reduce healthcare costs by minimizing the need for post-natal interventions and increasing the ability to get back to well-being and health.

10.3.3 Ethical Considerations

While prenatal genomic screening offers numerous benefits, it also raises significant ethical challenges that must be addressed to ensure equitable and responsible use of these techniques and analyses [18].

a. Informed Consent: Ensuring that parents fully understand the implications of genomic testing is important. Challenges include:

(i) Complexity of Genetic Information: Communicating results in an understandable manner and supporting easy accessible way to represent the results.
(ii) Implications for Reproductive Decisions: Balancing the provision of information with respect for parental autonomy; taking necessary steps as per need must align with respect of the decision made by the parents.
(iii) Equity and Accessibility: Having a good and unbiased environment for all pregnant women, regardless of their socio-economic status, race, ethnicity, etc.
(iv) Disparities in Access: High costs of advanced genomic technologies may limit access for underprivileged populations, so for equal access it is important to reduce the cost without disturbing the high quality and timing of maternity care services.
(v) Global Variability: Differences in cultural, geographical, ethical and even religious norms influence the acceptance and use of prenatal screening. The knowledge of being in a norm sometimes enables someone to make a large change but sometimes it creates a big hurdle [35].

b. Potential for Discrimination:

(i) **Selective Termination**: Ethical concerns arise over the potential misuse of genomic information to terminate pregnancies based on non-medical traits. Government regulations may influence the availability of the technique, which shows the necessity of having enough knowledge so that selective termination is not supported in the long term.

(ii) **Eugenics**: Fear of societal pressures to achieve "genetic perfection" or "social perfection" or having even a small chance of not being perfect, sometimes gives rise to a fear of being unfit for society—that is how achieving genetic perfection can also lead to social pressure regardless of whether a person has low or high literacy.

c. Data Privacy and Security:

(i) **Genomic Data Protection**: Safeguarding sensitive genetic information to prevent misuse is protected by laws and regulations along with the technical involvement organizations. These include Genetic Information and Nondiscrimination Act and Health Insurance Portability and Accountability Act.

(ii) **Long-Term Implications**: Due to considerations regarding the future use of foetal genomic data, including implications for insurability and employability with the involved organization, it is necessary to have data protection at every level, so that its security, like personal details, genetic detail, etc., is not breached.

d. Psychological Impact:

(i) **Parental Stress**: Uncertainty in interpreting Variants of Unknown Significance (VUS) can result in anxiety as having a genetic test for their child sometimes indicates unclear significance and makes parents worry about their child's future.

(ii) **Impacts on Family Dynamics**: Knowledge of a genetic condition may affect parental relationships and sibling interactions. Unclear and pro-negative results can create anxiety among close family members and sometimes results in strained relationships.

Prenatal genomic screening represents a transformative approach in maternal-foetal medicine that combines cutting-edge technologies, techniques with the clinical expertise to improve the pregnancy outcomes for mother and child both that enhances reproductive choices. While the benefits are substantial in addressing the associated ethical, social and psychological challenges, it is imperative to ensure its responsible integration into clinical practice for being available and secure for mother and foetus.

10.4 Paediatric Genomics: Early Detection of Genetic Disorders

The field of paediatric genomics holds the promise of early detection, precise diagnosis and tailored interventions for genetic disorders [34]. With the advancements in genetic technologies and an increasing understanding of leading hereditary diseases, paediatric care is evolving towards a more precise, accurate and preventive approach. This section delves into common paediatric genetic disorders and the role of genomics in preventive care in paediatrics, emphasizing that early intervention can dramatically improve results [30].

10.4.1 Common Paediatric Genetic Disorders

Paediatric genetic disorders show a wide range of conditions that manifest majorly during childhood and can significantly affect growth, development and quality of life. Understanding these disorders is crucial for early detection and management.

1. Chromosomal Disorders:

Chromosomal abnormalities arise from the numerical (deletion, duplication) or structural changes in chromosomes and are among the most common causes of genetic disorders in children [2]. Examples:

a. **Down Syndrome (Trisomy 21):** Characterized by speech difficulties, this is an intellectual disability that ranges from mild to severe. It includes distinct facial features such as slanted upward eyes and increased risk of congenital heart diseases along with chances of having cancer or autoimmune disorders. Early involvement programs focusing on speech, occupational and physical therapies can enhance developmental outcomes and increase the chances and possibilities to live healthy lives.

b. **Turner Syndrome:** This only occurs in females with a missing or incomplete X chromosome that results in short stature, webbed neck, infertility as ovarian issues and heart abnormalities. Hormone replacement therapy and growth hormone injections are common treatments but treatment completely depends on developmental characteristics such as absence or incomplete puberty, inability to conceive, problems with spatial perception along with non-verbal memory, etc., impaired attention and concentration, and more. Sometimes, having more physiological challenges make it more difficult for a few patients to be treated, like endocrine and autoimmune diseases, cardiac diseases, etc.

c. **Klinefelter Syndrome:** Males with an extra X chromosome (XXY) often have learning difficulties and reduced fertility, but sometimes some physical changes are more pronounced, such as delayed puberty, small testes with less production of testosterone, breast enlargement, weaker bones, feet are flat, curvy pinky fingers, along with a higher risk associated with cardiac diseases, autoimmune disorders, etc. Early educational support and testosterone replacement therapy can improve the chances of living a healthy life.

2. Single-Gene Disorders:

These disorders result from mutations in a single gene with effects varying based on the gene's function and expression as shown in Table 10.5.

a. **Cystic Fibrosis:** A life-threatening disorder causing thick mucus production that affects the lungs, pancreas and intestine, major deposition around the organs. Treatments include physiotherapy, enzyme supplements and new precision therapies like CFTR modulators such as ivacaftor. Importantly, it affects many organs including lungs, digestive system, etc., and also changes the electrolyte transport system. It can cause inflammation, exacerbations of bronchiectasis, etc. Early detection and care with diet, nutrient supplements and medications can improve the health in the long term.

b. **Sickle Cell Disease:** A haemoglobinopathy leading to anaemia, periodic pain and increased risk of having infections. Hydroxyurea therapy and bone marrow transplantation are key treatments. The symptoms and severity varies from person to person. Headaches, periodic pain, a rapid heartbeat, dizziness and fainting, etc., are the common symptoms that make it difficult on a daily basis.

c. **DMD:** A severe muscle-wasting disease. In the worst scenario, an individual can require a wheelchair by their teenage years and need it for the rest of their lives. A few individuals can show complications like cardiomyopathy and respiratory difficulties. Gene therapy and corticosteroids can slow disease progression. Early detection can make it durable with time, but cannot make it easy to live a healthy, independent life [12, 41].

Table 10.5 Common paediatric disorders.

Disorder	Mutation	Symptoms	Treatment
Cystic Fibrosis	CFTR gene	Thick mucus affecting lungs and digestion	Physiotherapy, enzyme supplements, CFTR modulators
Sickle Cell Disease	HBB gene (hemoglobin β gene)	Anaemia, pain crises, infection risk	Hydroxyurea therapy, bone marrow transplantation
Duchenne Muscular Dystrophy (DMD)	DMD gene	Progressive muscle wasting	Gene therapy, corticosteroids
Disorder	**Enzyme deficiency**	**Symptoms**	**Treatment**
Phenylketonuria (PKU)	Phenylalanine hydroxylase	Intellectual disability	Dietary management, sapropterin dihydrochloride
Maple Syrup Urine Disease (MSUD)	Branched-chain keto acid dehydrogenase	Neurological symptoms	Dietary restrictions, emergency treatment for crises

3. Metabolic Disorders:
Errors in metabolism due to enzyme deficiencies can lead to toxic accumulation or insufficient production of essential substances.

a. **Phenylketonuria (PKU):** PKU is a metabolic disorder that causes inability to break down phenylalanine, leading to intellectual disability if left untreated over time. PKU is a result of a mutation in phenylalanine hydroxylase gene resulting in an increase in phenylalanine in the blood. Dietary management is the cornerstone of treatment, supplemented by sapropterin dihydrochloride in responsive cases.
b. **Maple Syrup Urine Disease (MSUD):** An inherited disorder of the branched-chain amino acid metabolism that results in severe neurological symptoms. Early symptoms may be a sweet odour, poor feeding, lack of energy and nausea for a long time. Management includes dietary restrictions and prompt treatment during metabolic crises. A long-term untreated disorder can result in seizures, coma and death.

4. Neurodevelopmental Disorders:
Genetic mutations often contribute to neurodevelopmental challenges.

a. **Fragile X Syndrome:** Fragile X Syndrome is a leading inherited genetic condition that often leads to intellectual disabilities and is linked to autism spectrum disorders. It results from a mutation in the FMR1 gene on the X chromosome, causing a deficiency of the fragile X mental retardation protein (FMRP), an important factor for brain development. Symptoms may include learning difficulties, behavioural issues and social anxiety. Distinctive physical features, such as an elongated face, large ears, and hyper-flexible joints, aid in diagnosis. Understanding this syndrome is vital for understanding affected individuals and their family histories. Early educational involvements and medications targeting behavioural symptoms are essential.
b. **Rett Syndrome:** Rett syndrome is a neurodevelopmental disorder, a condition almost exclusively affecting females that is marked by severe cognitive and motor impairments. Supportive therapies, including physical and communication

aids, play a vital role in care. Early signs of Rett syndrome is Hypotonia, i.e., loss of muscle tone, difficulty during feeding, jerkiness in limb movement, etc.

5. Cancer Predisposition Syndromes:
Genetic mutations can increase susceptibility to certain childhood cancers.

a. **Li-Fraumeni Syndrome:** Li-Fraumeni syndrome is one of the most rare, an inherited condition that is co-related with the high risk of the patient developing cancer. It is associated with mutations in the TP53 gene which is also known as p53 gene. This is a tumour suppressor gene, leading to early-onset cancers. Regular whole-body imaging and clinical surveillance are essential as they give initial and early information about the gene expression.
b. **Retinoblastoma:** Caused by mutations in the RB1 gene that result in a rare eye cancer; shows a white colour in the center circle of the eye. There are majorly two types of growth patterns: Endophytic and Exophytic. Treatment options include laser therapy, chemotherapy or combination therapy as it comes as family history and enucleation depending on disease severity [31, 33].

6. Advances in Detection:
Early diagnosis is crucial for effective management. Advance techniques such as next-generation sequencing (NGS), chromosomal microarray analysis and gene panels have revolutionized the identification of paediatric genetic disorders [34]. For instance:

a. **NGS Applications:** Enables rapid detection of mutations in multiple genes simultaneously and aids in the diagnosis of conditions like epilepsy syndromes. It offers advances in identifying genomic mutations, detecting and quantifying RNA and researching genetics [16].
b. **Chromosomal Microarray Analysis:** Particularly useful for detecting the microdeletions and microduplications in disorders like DiGeorge syndrome. It can detect Foetal chromosomal imbalances, genetic causes of illness and physical developmental problems and other copy number variants associated conditions [23].

c. **Gene Panels:** Targeted panels for disorders such as muscular dystrophies provide high diagnostic yield, streamlining therapeutic decisions. Often, it is used to diagnose rare diseases. It helps to identify point mutations, insertions, deletions and copy number variants, etc. Compared to other techniques it is more economical, flexible and efficient to use than other traditional techniques.

10.4.2 Role of Genomics in Preventive Paediatric Care

The integration of genomics with preventive paediatric care has transformed our ability to predict, diagnose and manage genetic disorders Example: PKU, Huntington's disease. By leveraging the genomic insights, healthcare providers can tailor interventions to people's needs for improving results and reducing the consequences of the disease.

a. Genomic Screening and Risk Assessment:
Preventive care begins with identifying the risk of an individual's health before symptoms arise.

(i) **Newborn Screening/Expanded Carrier screening:** Universal genomic screening identifies metabolic and endocrine disorders early, and that allows immediate treatment to prevent complications. Examples: Whole genome sequencing that is used as a first tool to identify genetic variants. Syndromes such as PKU and Congenital hypothyroidism are periodically screened in many countries.

(ii) **Carrier Screening:** Carrier Screening is a genetic test that determines if a person is a carrier for a specific genetic disease. In other words, they carry one copy of a mutated gene that could potentially lead to a genetic disorder in the child if their partner also works as carrier for the same condition. Preconception and prenatal screening identify carriers of recessive conditions by enabling informed reproductive decisions. Both parents need to be carriers to have an affected child [6, 8].

(iii) **Predictive Testing:** Predictive testing is a test that identifies a person who is asymptomatic but has more risk of developing a genetic disease in the future, i.e., after analysing DNA, this is usually because of a history of the condition in the generational family. In other words, for families with a history of genetic disorders, testing can identify asymptomatic children at risk of developing conditions such as Huntington's disease or familial hypercholesterolemia.

b. Personalized Medical Management:
Genomic data enabling tailored care plans that address specific needs based on the genetic findings.

(i) **Targeted Interventions:** In simple words, tailoring medical treatments specifically to an individual patient's unique characteristics, such as an enzyme replacement therapy for lysosomal storage disorders such as Gaucher disease.
(ii) **Pharmacogenomics:** In other words, how genetics impact an individual's drug response with time. Having understanding of a child's genetic makeup can guide medication choices and dosages for reducing adverse reactions and improving efficacy. It is based on targeted and needed treatment plans to a patient's genetic analysing profile.
(iii) **Lifestyle Modifications:** Early detection of conditions such as Marfan syndrome allows for lifestyle changes that eventually minimize risk for the long term, such as avoiding contact sports, having a good diet, including exercise, changes in sleep habits, etc [14].

c. Enhancing Surveillance and Monitoring:
Genomic insights guide surveillance strategies for early detection of complications.

(i) **Cancer Risk Management:** Children with Li-Fraumeni Syndrome gets benefit from periodically imaging and physical clinical evaluations to detect malignancies and developmental stages early, such as Whole body MRIs.
(ii) **Cardiac Monitoring:** Conditions such as Long QT Syndrome requires continuous monitoring and preventive measures to avoid sudden cardiac diseases. In suspected LQTS cases, genetic testing can help identify the specific genetic mutation that causes this situation.

d. Improving Outcomes Through Early Intervention:
Early genomic diagnosis has been shown to improve the outcomes in numerous conditions.

(i) **Spinal Muscular Atrophy (SMA):** It is a genetic disorder that causes progressive muscle weakness and wasting. Gene therapy, such as "onasemnogene abeparvovec", has revolutionized treatment especially when the expression is administered before symptom onset.

(ii) **Severe Combined Immunodeficiency (SCID)/Bubble boy disease:** It is a rare inherited disorder that results in an impairment of the immune system. SCID's well-known form is caused by Adenosine Deaminase (ADA) deficiency. Early bone marrow transplantation can be curative, if it is performed before the infections occur.

e. Ethical and Social Considerations:

While genomics offers immense benefits it also raises ethical challenges in paediatrics.

(i) **Informed Consent:** A potential implication to understand is that parents must fully analyse genomic testing for their child and family.

(ii) **Equity in Access:** Efforts must ensure that genomic analysing advancements benefit all children regardless of socio-economic status, i.e., everyone should have fair participation without any bias.

(iii) **Psychological Impact:** Every family needs support in coping with genetic diagnoses and the uncertainties associated with variants of unknown significance. The test includes a wide range of emotions such as anxiety, depression, guilt and stress.

Paediatric genomics is revolutionizing initial detection and preventive care for genetic disorders. Leveraging advanced techniques and integrating genomics into regular care with healthcare providers can improve health outcomes and enhance the quality of life for affected children. Recent researches have shown results coupled with ethical considerations and equitable access that will be critical in maximizing the potential of genomics in paediatrics.

10.5 Personalized Medicine and Targeted Interventions

10.5.1 Needed Therapeutic Approaches for Desired Results

Personalized medicine has emerged as a revolutionary approach in modern healthcare, targeting therapeutic strategies to individual characteristics of patients. This field has gained momentum due to advancements in genomics, proteomics, and bioinformatics, which enable a deeper understanding of disease mechanisms at the molecular level. The goal of personalized medicine is to optimize treatment efficacy while minimizing the adverse effects, particularly in complex diseases such as cancer, diabetes and autoimmune disorders [31, 41].

Precision in Cancer Therapies:
In oncology, the advent of personalized or targeted medicine has been significantly transformed therapeutic paradigms. Traditional "one-size-fits-all" approaches have often yielded suboptimal results due to the heterogeneity of tumor or cancer biology among patients. Example: Glioblastoma. Personalized or targeted medicine always addresses this issue by stratifying the patients based on their genetic, epigenetic and proteomic profiles. For instance, the therapies targeting specific mutations in the EGFR, HER2 or BRAF genes have shown remarkable efficacy in their subsets of patients with lung, breast and melanoma cancers, respectively [7]. Additionally, the immune checkpoint inhibitors such as PD-1/PD-L1 and CTLA-4 blockers simply explains how understanding the tumor microenvironment or supporting environment of cancer can lead to more effective interventions.

- **Role of Pharmacogenomics:**

Pharmacogenomics, the study of how genetic variations influence the drug response, plays an important role in personalizing treatment. Variants in genes encoding drug-metabolizing enzymes such as CYP450 family, transporters and targets/receptors can lead to variations in drug efficacy and toxicity with response time. For instance, individuals with certain CYP2D6 polymorphisms may metabolize drugs, such as tamoxifen that is used in breast cancer, or codeine, differently, necessitating dosage adjustments or alternative therapies. Incorporating pharmacogenomic data with clinical/medical practice ensures that patients receive the right drug at the right dose, thereby improving outcomes [38].

- **Biomarker-Driven Strategies:**
The identification and validation of biomarkers have been instrumental in guiding personalized interventions. Predictive biomarkers such as BRCA1/2 mutations for PARP inhibitors or ALK rearrangements for ALK inhibitors help to select patients who are most likely to benefit from these specific therapies. An example is breast cancer [31]. Moreover, monitoring biomarkers such as circulating tumor DNA (ctDNA) and minimal residual disease (MRD) enables dynamic assessment of treatment response and disease progression [41].

- **Combining Therapeutic Modalities:**
Combining different therapeutic modalities such as targeted therapies, immunotherapies and conventional chemotherapy or radiotherapy offers synergistic effects in needed/personalized medicine. For instance, combining BRAF inhibitors with MEK inhibitors in melanoma patients eventually reduces the risk of resistance by enhancing the long-term health results. Similarly, integrating the genomic and proteomic data to design personalized vaccine based therapies is an emerging area of interest to get the results based on need for future patients. Recent studies have mentioned that combination therapies are the future, having good genomic analyses and family history. An example is prostate cancer.

- **Overcoming Challenges:**
Despite its overall potential, implementing personalized medicine faces several challenges. High costs of genetic testing, targeted therapies that lack universal access to advanced diagnostic tools, and variability in healthcare infrastructure limit its widespread adoption to meet future needs.

10.5.2 Impact of Early Genetic Detection on Treatment

Early genetic detection has revolutionized the approach to diagnosing and managing various diseases by enabling the interventions before the onset of clinical or medical symptoms. This paradigm shift from a reactive to proactive healthcare leverages genetic insights to prevent or mitigate disease progression, particularly in hereditary and complex multifactorial conditions [25, 41].

- **Genetic Screening in Oncology**
In cancer, early genetic detection plays a critical role in risk assessment, prevention and treatment. For instance, identifying germline mutations

in BRCA1 and BRCA2 genes enables implementation of risk-reducing strategies such as prophylactic mastectomy or salpingo-oophorectomy in high-risk individuals. Similarly, by detecting Lynch syndrome associated mutations allows for enhanced surveillance and an early intervention in colorectal and endometrial cancers. Beyond the risk assessment with time, early detection of the somatic mutations in tumors aids in selecting the targeted therapies, thereby improving survival rates [31].

- **Prenatal and Neonatal Genetic Screening**

Advancements in genetic testing techniques such as NIPT and WGS have facilitated the early detection of genetic anomalies in foetuses and newborns. Conditions such as Down syndrome, cystic fibrosis and congenital heart diseases can now be identified with high accuracy [51]. Early diagnosis enables parents and clinicians to plan appropriate medical or surgical interventions that improve the quality of life for affected individuals in the future [49].

- **Cardiovascular Disease Risk Assessment**

Early genetic detection has also impacted the management of cardiovascular diseases (CVDs). Detecting mutations associated with familial hypercholesterolemia (FH), hypertrophic cardiomyopathy (HCM) or arrhythmogenic right ventricular cardiomyopathy (ARVC) enables periodic lifestyle modifications and pharmacological interventions for having clinical benefits for good living. Initiating a statin therapy in an individual's lifestyle with FH, for example, reduces the risk of premature cardiovascular diseases.

- **Genetic Testing in Rare Diseases**

For rare genetic disorders, early detection through newborn screening programs significantly improve outcomes for the health of both mother and foetus. Disorders like phenylketonuria (PKU) and spinal muscular atrophy (SMA) can be managed effectively, if diagnosed early. For example, dietary and routine exercise modifications in PKU patients or gene therapy in SMA patients can prevent severe complications and improve a person's life.

- **Ethical and Practical Considerations**

While early genetic detection offers significant benefits it also raises ethical and practical or clinicians' concerns. The interpretation of genetic variants of uncertain significance (VUS) leads to challenges in decision-making for the long term. Moreover, the psychological impact of detecting genetic risks, particularly for conditions with no known cure, requires careful counselling and support for complete understanding. Ensuring the precision, accuracy, affordability, and accessibility of genetic tests is critical for their effective integration in healthcare systems [18].

- **Future Directions**

Advances in a CRISPR-based gene editing technologies and artificial intelligence–driven analytics hold promise for improving the accuracy and scope of genetic detection and analyses. Integrating multi-omics data that includes genomics, metabolomics and transcriptomics will provide a more comprehensive understanding of disease mechanisms by enabling even earlier and more precise interventions. Policymakers and healthcare providers must collaborate to establish the guidelines and frameworks that will ensure the ethical and equitable implementation of early genetic detection in clinical practice.

10.6 Challenges in Genomic Implementation

10.6.1 Ethical and Psychological Implications of Genomic Implementation

The implementation of genomic medicine brings a transformative potential to the healthcare system but also introduces significant ethical and psychological challenges ahead of disease or disorder [44, 48]. These challenges come from the deeply personal nature of genetic analysis and information and its far-reaching implications for individuals and society as shown in Figure 10.1.

- **Privacy and Data Security**

One of the foremost ethical concerns is the potential misuse of genetic data, majorly related to personal details. Genomic information is highly sensitive, as it can result in privacy violations if it is not securely handled. The risk of an unauthorized access or a breaching of data could result in genetic discrimination and eventually can affect employment, insurance or social relationships over a period of time [45]. The establishment of robust data encryption methods and firm regulations like GINA (Genetic Information Nondiscrimination) in the U.S.A, is really important in safeguarding patient privacy [17, 46].

- **Informed Consent and Autonomy**

The principle of informed consent is essential to ethical genomic implementation. Patients must understand the potential benefits for upcoming

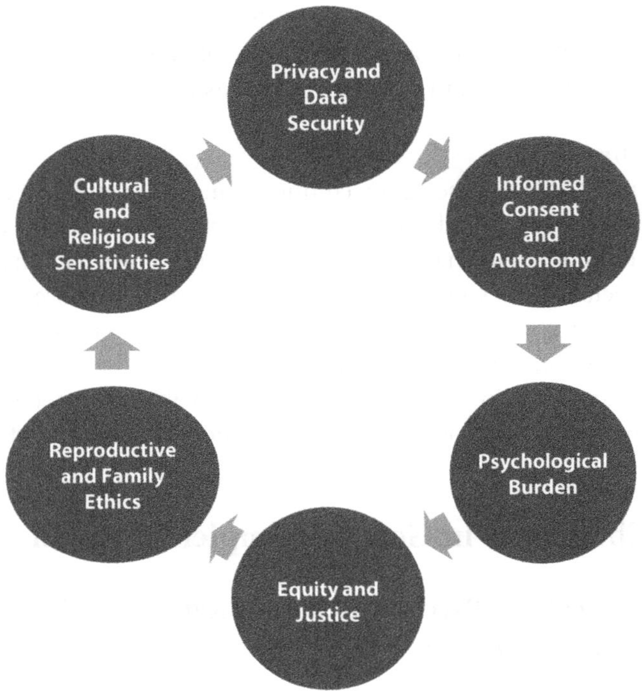

Figure 10.1 Challenges in ethical and psychological implications of genomic implementation.

risks and limitations of genetic testing. However, the complexity of its genomic information often makes this whole process challenging, especially for individuals who are closest to the patient [27]. By ensuring that individuals have complete autonomy to decide whether they will undergo testing and how to use their genetic information, there will be no force or stress regarding whether to choose these kinds of techniques [17]. The role of genetic counsellors becomes indispensable in this context, where even having a basic level of knowledge is completely dependent on the socioeconomic background of individual.

- **Psychological Burden of Genetic Knowledge**

Receiving genetic information must have profound psychological effects, particularly if it reveals a predisposition to a genetic disease or severe diseases. For instance, having anxiety, depression and feelings of fatalism are common responses among individuals who scientifically learn about their genetic risks. This psychological burden is further multiplied when the implications extend risks to their family members, those who may carry similar risks, and also for future generations. Providing adequate mental health support at the needed time, including counselling services, is essential to address these issues [18, 41].

- **Reproductive and Family Ethics**

Genomic implementation raises an ethical dilemma in the context of the reproductive choices for being in a continuous family. Preimplantation of genetic testing and prenatal screening allow parents to select the embryos free of specific genetic conditions, which sparks debates over "designer babies" and eugenics [40]. Additionally, the responsibility to inform family members of genetic risks faces a moral challenge, as it conflicts with the patient's right to privacy and personal details. Establishing guidelines for handling such familial obligations is necessary to navigate these complex scenarios; ethics is needed in science [18, 35].

- **Equity and Justice**

The equitable distribution of genomic advancements is another ethical concern. Socioeconomic disparities often limit access to genetic testing and personalized or targeted therapies, thus exacerbating health inequities. Ensuring that genomic medicine benefits all populations regardless of socioeconomic status, race or geographic location is a pressing ethical imperative [27, 46].

- **Cultural and Religious Sensitivities**

Cultural and religious beliefs exert an influence on perceptions of genomic medicine. Some communities may have a view of genetic testing and interventions as interfering with natural processes or divine will. Engaging with diverse cultural and religious groups to promote an understanding and acceptance of genomic technologies is crucial for their successful implementation.

10.6.2 Accessibility and Equity in Genomic Medicine

The promise of genomic medicine is to revolutionize the healthcare infrastructure by offering tailored interventions for every patient's potential. However, disparities in access threaten to widen the gap between those who can benefit from these advancements and those who cannot. Addressing accessibility and equity challenges is essential to ensure that genomic medicine fulfils its potential as a tool for global health improvement in consideration for future perspectives as shown in Table 10.6.

Table 10.6 Challenges to accessing genomic medicine.

Challenges	Description	Possible solutions
Socioeconomic Barriers	High costs of testing and limited insurance coverage.	Lower sequencing costs, government funding, and insurance policy reforms.
Geographic Disparities	Limited genomic infrastructure in LMICs.	International collaborations and training programs.
Underrepresentation in Research	Bias towards populations of European ancestry.	Increase diversity in genomic studies.
Education and Awareness	Lack of training for healthcare professionals.	Implement genomic literacy programs for clinicians.
Infrastructure Constraints	Inadequate sequencing and bioinformatics facilities.	Public-private partnerships for infrastructure development.
Policy and Regulation Issues	Lack of standardized genomic policies.	Develop clear global regulatory frameworks.

- **Socioeconomic Barriers**

High costs associated with genomic testing and targeted therapies pose significant barriers to accessibility. Whole-genome sequencing and precision treatments like Chimeric Antigen Receptor-T (CAR-T) cell therapy remain prohibitively expensive for many patients. Insurance coverage often excludes or inadequately reimburses such interventions, placing them out of reach for economically disadvantaged populations. Creating cost-effective diagnostic tools and advocating for wider insurance coverage are necessary steps to address this problem for a better future approach, besides giving detailed knowledge of necessary treatment to patients [41, 45].

- **Geographic Disparities**

Access to genomic medicine is a non-uniform distribution across the global regions. High-income countries typically have a better health infrastructure, with skilled personnel and advanced technologies, while low- and middle-income countries (LMICs) have been facing significant limitations for long-term outcomes in establishing the same planned health infrastructure. The lack of well-planned genomic research centres, trained genetic counsellors and clinical laboratories in the LMICs hampers the implementation of genomic medicine uniformly. International approaches, collaborations and the capacity for building initiatives can help to bridge this gap, as soon as possible [25, 41].

- **Underrepresentation in Genomic Research**

The underrepresentation of diverse populations in genomic research contributes to health disparities. Most genomic studies have predominantly included individuals of European ancestry, thus limiting the applicability of findings to other ethnic groups, located globally. This bias affects the development of the related genetic risk scores and the identification of a population with specific biomarkers. Increasing diversity in genomic research is a primary basis to ensure that a personalized or targeted medicine benefits all populations without bias [24].

- **Education and Awareness**

Limited knowledge, awareness and understanding of genomic medicine among healthcare providers and patients contribute to accessibility challenges. Many clinicians lack the training to interpret the genomic data and incorporate it into clinical practice or medical implementation, while patients may not understand the benefits or implications of genetic analysing and genetic testing. For long-term beneficial results, investing in education and training programs for healthcare professionals and public awareness campaigns can address these gaps and make a suitable bridge to address the large gaps.

- **Infrastructure and Resource Constraints**

Implementing genomic medicine requires a robust healthcare infrastructure including sequencing facilities, bioinformatics platforms and more focus on data storage systems. Many healthcare systems are in resource-limited settings that lack these essential components, as infrastructure is limited [25]. Governments and private sectors must invest in developing the necessary infrastructure to support genomic medicine for better treatments with lower costs and without bias.

- **Policy and Regulatory Challenges**

The absence of standardized policies and rigid regulations for genomic medicine further exacerbates the challenges. Establishing clear guidelines for genetic analysing and testing data sharing and reimbursement policies is critical to ensure equitable access. Collaborative efforts among policy-makers, healthcare providers and patient advocacy groups are essential to create a regulatory framework that promotes fairness and inclusivity in society [41, 45].

10.6.3 Cost-Effectiveness of Genomic Screening Programs

Genomic screening programs have the potential to improve the healthcare system and give results by enabling early identification and prevention of diseases. However, their cost-effectiveness is a critical factor in determining their feasibility and scalability, particularly in resource-mannered healthcare systems as shown in Figure 10.2.

- **Direct Costs of Genomic Screening**

The costs of genomic screening include expenses related to the sequencing techniques, laboratory infrastructure and personnel. While the cost of sequencing technique has been significantly reduced over the last decade, it remains considerable for large-scale programs [50]. Additionally, the costs of the confirmatory testing, counselling and follow-up interventions add to the financial burden. Evaluating the long-term benefits of the genomic screening against these initial expenses is essential to justify its implementation and approach for future [46].

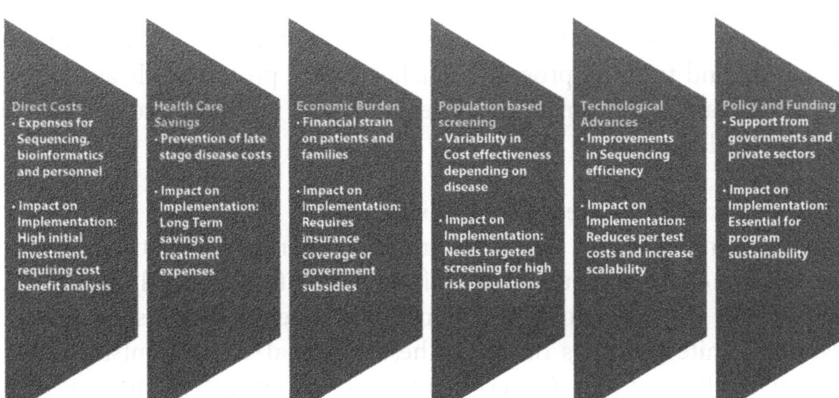

Figure 10.2 Cost analysis of genomic screening.

- **Healthcare Savings from Early Detection**

Genomic screening can lead to significant healthcare savings by detecting all individuals at risk for diseases before symptoms arise [25]. For example, an early detection of hereditary cancer syndromes such as BRCA-associated breast cancers, i.e., BRCA1, BRCA2 and ovarian cancers enables preventive measures by reducing treatment costs and improving survival rates in overall population [32]. Also, identifying carriers of these genetic conditions like familial hypercholesterolemia (FH) allows for routine interventions to prevent overall cost complications [1].

- **Economic Impact on Patients and Families**

Besides the healthcare savings, a genomic screening can reduce the economic/financial burden on patients and their families by preventing disease-treatment-related expenses. For instance, an early diagnosis of rare genetic disorders enables timely treatment by avoiding extended hospital stays and its associated costs [39]. Moreover, the affordability of the genomic tests remains an important concern for many patients and families that particularly applies to low-income settings without bias in society. Consequently, subsidizing genomic testing or incorporating it into universal healthcare schemes can remove this burden [46].

- **Cost-Effectiveness of Population-Based Programs**

The cost-effectiveness of the genomic screening programs varies, depending on the target population and the conditions being screened. For instance, newborn screening programs for metabolic disorders have been proven to be highly cost-effective due to the significant benefits of early treatment. In contrast, a wide population genomic screening for a multifactorial or complex diseases requires careful evaluation and screening to determine its value relative to other public health interventions. This helps us to better understand the dynamics of large populations [39].

- **Technological Advances and Cost Reduction**

Precision developments in the sequencing techniques and bioinformatics have the potential to reduce the overall costs of genomic screening programs. For instance, the development of a high-throughput sequencing platforms and machine learning algorithms the data analysis can improve its efficiency and scalability, from average to higher levels. Emerging innovation and competition in the genomics industry is necessary to drive cost reductions [50].

- **Policy and Funding Considerations**

Implementing cost-effective genomic screening programs requires more supportive policies and sustainable funding mechanisms. Governments should allot resources to prioritize genomic medicine as part of its national healthcare strategies. Public-led private partnerships along with international collaborations can provide us with additional funding and expertise to support these initiatives at every level. Establishing clear criteria for cost-effective evaluations can help policymakers to make uniform decisions [41].

- **Ethical and Social Implications**

The cost-effectiveness of the genomic screening programs must be balanced and uniform with ethical and social considerations. Ensuring that the cost-saving measures do not compromise the quality of care or exclude its vulnerable populations is an essential element. Transparent decision-making processes that includes stakeholders, patients, healthcare providers and ethicists can promote fairness and public trust along with high-quality treatment [18].

Future Directions: The emerging future of cost-effective genomic screening is completely integrating with multi-omics approaches and personalized healthcare strategies. Combining genomic data with proteomics, metabolomics and environmental factors by screening the programs can provide more accurate and precise risk assessments and targeted or unique interventions. Policymakers and researchers must work together to develop frameworks that can maximize the benefits of genomic screening while minimizing the costs of overall treatment and every kind of disparity.

10.7 Future Directions and Innovations

10.7.1 Emerging Genomic Technologies

The field of genomics continues to advance at an extraordinary pace that is driven by technological innovations that uphold the potential to transform the healthcare infrastructure. Emerging genomic technologies are composed to overcome existing challenges and smooth the way for groundbreaking applications in medicine and biology.

a. Single-Cell Genomics

Single-cell genomics enables the study of an organism's cells by providing the insights into a cellular heterogeneity within the tissues. This technology is revolutionizing our understanding of cancer/tumour, immune responses and developmental biology. By identifying a unique cellular profiles, single-cell genomics can improve the precision or accuracy of diagnostics and its therapies. For instance, mapping the tumour microenvironment or cancer microenvironment at a single-cell level aids in developing targeted/personalized immunotherapies.

b. Long-Read Sequencing

In contrast to traditional short-read sequencing, long-read sequencing offers the ability to resolve the complex genomic regions, the structural variations and their epigenetic modifications. Technologies like nanopore and PacBio sequencing are becoming increasingly accessible and affordable. These methods enhance the detection of rare genetic mutations and provide a more comprehensive understanding of the genome by enabling breakthroughs in genetic disease research.

c. CRISPR-Based Technologies

CRISPR and its associated tools have revolutionized genome editing. Advances such as nucleotide base editing and prime editing allow for precision genetic modifications with minimal off-target effects. Beyond gene therapy, CRISPR-based diagnostics, like SHERLOCK and DETECTR are emerging as rapid and cost-effective tools for detecting genetic mutations and infectious diseases.

d. Multi-Omics Integration

Integrating genomics with other "omics" categories, such as proteomics, metabolomics, and transcriptomics, provides a holistic view of the biological systems. Multi-omics approaches enable the identification of biomarkers, unravelling disease mechanisms and predicting therapeutic responses. For instance, combining genomics and proteomics data can enhance the accuracy of cancer prognostics and drug development.

e. Synthetic Biology

Synthetic biology leverages genomic engineering to design and construct biological systems with new functions. Applications include the development of gene circuits for therapeutic purposes, engineering microbes for bioproduction and creating synthetic genomes. Advances in synthetic biology are expanding possibilities in regenerative medicine, vaccine development and sustainable bioengineering.

f. Quantum Computing in Genomics

Quantum computing contains the promise of accelerating complex genomic analyses. Its unparalleled computational power can enhances the genome assembly, drug discovery and personalized medicine. Although still in its early stages, the integration of quantum computing into genomics is expected to address the challenges of big data processing, analysis and predictive modelling.

g. Challenges and Future Directions

While emerging genomic technologies offer immense potential for facing challenges, scalability, cost and ethical considerations remain constant. Continued investment in research and development coupled with multidisciplinary collaborations is essential to translate these innovations into clinical/medical practices for better goal achievement. Policymakers must also ensure the advancements are uniformly distributed to prevent widening healthcare bias.

10.7.2 The Role of Artificial Intelligence and Big Data

Artificial Intelligence (AI) and big data are transforming genomics by enabling the analysis and evaluation of immense amounts of genetic information. These technologies play an essential role in advancing precision medicine, accelerating research and optimizing healthcare delivery.

a. AI in Genomic Data Analysis

AI algorithms are exclusively for ML and deep learning that are instrumental in analysing complex genomic datasets. They facilitate specific tasks like variant calling, gene expression analysis and structural variant identification. AI-powered tools such as Google DeepVariant and IBM Watson Genomics are improving precision, accuracy and efficiency of the genomic interpretations, resulting in better diagnostic outcomes.

b. Predictive Genomics

Artificial Intelligence enables a predictive modelling of the disease-related risks based on the genomic data. Polygenic risk scores that arise from Artificial Intelligence algorithms are being used to assess an individual's bias to conditions such as diabetes, cancer and CVDs. This divining capability allows for personalized deterring strategies and their early interventions.

c. Drug Discovery and Development

Big data and Artificial Intelligence are revolutionizing drug discovery by identifying its potential therapeutic targets and predicting the drug responses for a better approach. Artificial Intelligence is driven by platforms genomic analysis and clinical data to uncover the biomarkers for the drug efficacy and its safety. These techniques are accelerating the development of precision/targeted medicines and decrease the amount of time needed and cost of bringing new drugs to market, as billions of dollars have been involved in making a single drug with high precision.

d. Integration with Electronic Health Records (EHRs)

The assimilation of genomic data with Electronic Health Records through AI enables a comprehensive understanding of patient health at every analytical stage. AI algorithms can identify the patterns and make a correlation between the genetic factors and clinical/medical outcomes by enhancing personalized treatment schedules.

e. Ethical and Privacy Considerations

The utilization of Artificial Intelligence and big data in genomics raises ethical concerns regarding data privacy, bias and transparency in healthcare systems. Ensuring that Artificial Intelligence algorithms are interpretable, unbiased and secure is critical to building trust among stakeholders and trustees. Establishing a stringent data governance with frameworks and diverse populations in genomic studies is an essential component to address these challenges with recent problems arising regarding data storage.

f. Future Opportunities

The harmony between AI, big data and genomics is expected to advance fields like gene therapy, synthetic biology and population health genomics. Continued investment in computational infrastructure, interdisciplinary training and international organization collaborations drives innovation and improvement of global health infrastructure.

10.7.3 Genomic Counselling and Education

As genomic medicine becomes more prominent in the healthcare system, the roles of genomic counselling and education are becoming a necessary part of healthcare on a daily basis. These components are essential for empowering patients, training healthcare professionals and implementing a good public understanding of genomics.

a. The Role of Genomic Counsellors

Genomic counsellors act as intermediaries between patients and complex genetic information for better understanding and also for reaching patients of all ages and genders. They provide personalized guidance on its benefits, related risks and the limitations of genetic testing, by ensuring an informed decision-making for treatment. As the demand for genomic testing develops rigorously, the demand for trained counsellors is also increasing. Expanding genomic counselling services and opportunities for integrating them into healthcare systems is crucial for supporting patients and their families.

b. Patient Education and Empowerment

Educating patients about genomics and its implementation can enhance their ability to make informed choices and engage them in their healthcare decisions. Accessible resources like brochures, videos and digital tools can help patients in understanding their genetic risks and implications of their treatment for the long term. Culturally sensitive approaches are necessary to address and implement the diverse needs and preferences of the patients from different backgrounds and medical histories.

c. Training Healthcare Professionals

Many healthcare professionals lack the knowledge and skills to interpret, analyse and apply genomic data in clinical practices. Comprehensive training programs including workshops, online courses and certifications are essential to bridge this gap of knowledge. Integrating genomics education into medical and nursing curricula is really important to ensure that future practitioners are equipped to meet the demands and accuracy of genomic medicine for needed treatment.

d. Public Awareness and Literacy

Encouraging public awareness of genomics fosters acceptance and participation in genetic testing and research specially for rare genetic disorders. Awareness programs, community engagement initiatives and media campaigns can illustrate genomics and highlight its benefits and implementation in treatment. Addressing misconceptions and ethical considerations is vital for building public trust and encouraging informed participation for a better outreach to all ages in a population.

e. Ethical and Cultural Considerations

Genomic counselling and education must consider ethical and cultural factors, as they are important components of society. Respecting every patient's autonomy by ensuring their privacy and addressing stigmatization are fundamental principles during considerations in the time period of treatment. Collaborating with community leaders, impactful people and organizations that can enhance the cultural competence of counselling and education efforts is essential in order to reach people of all categories in a short time.

f. Innovations in Education

Developing technology or techniques always plays a significant role in advancing genomic education and literacy. Virtual reality simulations, interactive platforms and AI-powered tools are providing a riveting and personalized learning experiences for everyone. These innovations make the genomic concepts more accessible and engaging for both patients and professionals, developing its reach and making it more accessible.

g. Future Directions

The future of genomic counselling and education lies in their integration into holistic healthcare models. Interdisciplinary approaches that combine counselling with psychological, social and ethical support will enhance patient care. Establishing global standards for counselling practices and leveraging digital technologies will further strengthen the impact of genomic education on society.

10.8 Conclusion

10.8.1 The Future of Genomics in Prenatal and Paediatric Medicine

Genomics has revolutionized the whole healthcare system. It offers unparalleled insights into human biology and the potential to personalize medical interventions. In prenatal and paediatric medicine, genomics is a poised to redefine preventive care, diagnostics and treatment, promising a healthier future for generations to come with better genomic analyses and data prevention, leading to better understanding in the long term.

a. Transforming Prenatal Care

Genomics in a prenatal care is enabling the early detection of genetic disorders that allow families to make informed decisions with better understanding and prepare for potential challenges. NIPT, which analyzes foetal DNA circulating in the mother's bloodstream, has become an important standard tool for screening chromosomal abnormalities. Strong future advancements may extend the NIPT to include single-gene disorders such as Cystic fibrosis and multifactorial conditions, further enhancing its scope in incoming issues.

Recent, emerging technologies like preimplantation genetic testing (PGT) during *in vitro* fertilization (IVF) are helping the prospective parents to select embryos free of specific genetic conditions. Such interventions offers hope for reducing the burden or misinformation of hereditary diseases while raising ethical and social considerations that require careful navigation and prevention that will result in easy and better living.

b. Enhancing Paediatric Diagnostics

In paediatric medicine, genomics facilitates the diagnosis of a rare and complex conditions that might otherwise remain undetected and lead to more complications. WGS and WES are being increasingly used to detect genetic causes of developmental delays, metabolic disorders and congenital abnormalities. Rapid sequencing technologies now allow critically ill newborns to be diagnosed within days or weeks by enabling timely and targeted interventions with better perspectives of technology usage.

As recent researches provide a growing understanding of gene environment interactions, genomics may help to predict how children respond to environmental factors such as toxins or allergens, paving the way for personalized preventive strategies. Early genomic insights can guide lifestyle adjustments and medical interventions to mitigate long-term health risks.

c. Personalized Therapies for Paediatric Patients

The integration of genomics into treatment planning is transforming paediatric care. Pharmacogenomics, which studies how genetic variations influence drug responses, is guiding the selection of medications and dosages targeted to individual patients for leading a better lifestyle. Even though there are some difficulties, it still gives a higher chance of survival. This approach may minimize the adverse effects and maximize therapeutic efficacy and accuracy, especially in conditions such as epilepsy, cancer and asthma. Gene therapy is arising as a transformative for treating genetic disorders in children of all ages. Conditions like spinal muscular atrophy (SMA) and severe combined immunodeficiency (SCID) can now be treated with it, with gene-editing techniques that can correct the fundamental genetic defects. Leading advancements in delivery systems such as adeno-associated viruses (AAVs) are enhancing the safety, accuracy and efficiency of these therapies with the precision of a complete process.

d. Ethical Considerations and Equity

Genomics has become an important conventional component in prenatal and paediatric medicine, and ethical considerations must be at the forefront of future treatment of genetic disorders. Challenges such as informed consent, privacy, data storage and the potential for genetic discrimination require rigid regulatory frameworks to make sure of efficacy at every stage. Families must have access to genetic counselling to understand the implications of genomic information and make decisions aligned with their values and beliefs regarding treatment. Ensuring equitable access to genomic technologies is another most prevalent challenge in recent times. Disparities in healthcare infrastructure and socioeconomic status can limit the availability of genomic testing and interventions in the underserved populations in the long term. Policymakers must therefore prioritize the initiatives that bridge these gaps for fostering inclusivity in genomic medicine.

e. Innovations on the Horizon

The future of genomics in prenatal and paediatric care is bright, and has been driven by rapid technological advancements in recent decades, such as, AI and ML. These are an essential component in analysing large genomic datasets for uncovering patterns that enhance our understanding of disease mechanisms. Multi-omics approaches which integrate genomics with proteomics, metabolomics and transcriptomics are providing comprehensive insights into health and disease both. Synthetic biology and gene editing hold the maximum potential to not only treat the disease but also prevent genetic disorders for later. For instance, the development of synthetic gene circuits may enable precise control over a gene expression, reducing the risk of off-target effects. Collaborative efforts between academia, industry and governments are essential to bring these innovations to fruition along with promising results.

f. Preparing for the Genomic Era

To fully implement the potential of genomics in prenatal and paediatric medicine, a multidisciplinary approach is required for the time being. This includes training and more analytics for healthcare professionals to interpret and apply the genomic data in analysing and for facilitating public literacy about genomics and establishing ethical guidelines that can balance an innovation with responsibility and privacy, with protection of genomic data. Investment in infrastructure such as genomic sequencing installation and bioinformatics platforms are also an important component to support the universal adoption of a genomic medicine in a daily routine of life.

A Vision for the Future

The assimilation of genomics with prenatal and paediatric medicine represents a constant shift in a healthcare system. Enabling early identification helps for personalized or targeted treatment and preventive care as genomics has the potential to significantly improve the outcomes and quality of life for children and families and for future generations also. For upcoming decades, collaboration, innovation and ethical supervision will be an important key for ensuring the benefits of genomics that are implemented, for everyone, without bias [25, 45].

AI disclosure statement: The author acknowledges the use of AI-based tools for minor language editing or grammatical correction. However, all intellectual contributions, analyses, and conclusions presented in this chapter are entirely the author's own.

References

1. Drozdz, D., Alvarez-Pitti, J., Wójcik, M., *et al.*, Obesity and Cardiometabolic Risk Factors: From Childhood to Adulthood. *Nutrients*, 13, 11, 4176, Published 2021 Nov 22, doi: 10.3390/nu13114176.
2. Rosenfeld, J.A., Coe, B.P., Eichler, E.E., Cuckle, H., Sha!er, L.G., Estimates of penetrance for recurrent pathogenic copy-number variations. *Genet. Med.*, 15, 6, 478–481, 2013, doi: 10.1038/gim.2012.164.
3. Suo, C., Dann, E., Goh, I., *et al.*, Mapping the developing human immune system across organs. *Science*, 376, 6597, eabo0510, 2022, doi: 10.1126/science.abo0510.
4. Monies, D., Abouelhoda, M., Assoum, M., *et al.*, Lessons Learned from Large-Scale, First-Tier Clinical Exome Sequencing in a Highly Consanguineous Population [published correction appears in Am J Hum Genet. 2019 Oct 3;105(4):879. doi: 10.1016/j.ajhg.2019.09.019.]. *Am. J. Hum. Genet.*, 104, 6, 1182–1201, 2019, doi: 10.1016/j.ajhg.2019.04.011.
5. Hoyme, H.E., Kalberg, W.O., Elliott, A.J., *et al.*, Updated Clinical Guidelines for Diagnosing Fetal Alcohol Spectrum Disorders. *Pediatrics*, 138, 2, e20154256, 2016, doi: 10.1542/peds.2015-4256.
6. Bell, C.J., Dinwiddie, D.L., Miller, N.A., *et al.*, Carrier testing for severe childhood recessive diseases by next-generation sequencing. *Sci. Transl. Med.*, 3, 65, 65ra4, 2011, doi: 10.1126/scitranslmed.3001756.
7. Carrera, P., Calzavara, S., Magistroni, R., *et al.*, Deciphering Variability of PKD1 and PKD2 in an Italian Cohort of 643 Patients with Autosomal Dominant Polycystic Kidney Disease (ADPKD). *Sci. Rep.*, 6, 30850, Published 2016 Aug 8, doi: 10.1038/srep30850.

8. Zhao, S., Xiang, J., Fan, C., *et al.*, Pilot study of expanded carrier screening for 11 recessive diseases in China: results from 10,476 ethnically diverse couples. *Eur. J. Hum. Genet.*, 27, 2, 254–262, 2019, doi: 10.1038/s41431-018-0253-9.
9. Lapehn, S. and Paquette, A.G., #e Placental Epigenome as a Molecular Link Between Prenatal Exposures and Fetal Health Outcomes through the DOHaD Hypothesis. *Curr. Environ. Health Rep.*, 9, 3, 490–501, 2022, doi: 10.1007/s40572-022-00354-8.
10. Huang, S., Liu, S., Huang, M., *et al.*, #e Born in Guangzhou Cohort Study enables generational genetic discoveries. *Nature*, 626, 7999, 565–573, 2024, doi: 10.1038/s41586-023-06988-4.
11. Korver, A.M., Smith, R.J., Van Camp, G., *et al.*, Congenital hearing loss. *Nat. Rev. Dis. Primers*, 3, 16094, Published 2017 Jan 12, doi: 10.1038/nrdp.2016.94.
12. Oates, E.C., Jones, K.J., Donkervoort, S., *et al.*, Congenital Titinopathy: Comprehensive characterization and pathogenic insights. *Ann. Neurol.*, 83, 6, 1105–1124, 2018, doi: 10.1002/ana.25241.
13. Muhle, R., Trentacoste, S.V., Rapin, I., The genetics of autism. *Pediatrics*, 113, 5, e472–e486, 2004, doi: 10.1542/peds.113.5.e472.
14. Rauen, K.A., Cardiofaciocutaneous Syndrome, in: *GeneReviews*®, M.P. Adam, J. Feldman, G.M. Mirzaa, R.A. Pagon, S.E. Wallace, A. Amemiya (Eds.), University of Washington, Seattle, Seattle (WA), January 18, 2007.
15. Byrne, A.B., Arts, P., Ha, T.T., *et al.*, Genomic autopsy to identify underlying causes of pregnancy loss and perinatal death [published correction appears in Nat Med. 2024 Jan;30(1):302. doi: 10.1038/s41591-023-02487-1.]. *Nat. Med.*, 29, 1, 180–189, 2023, doi: 10.1038/s41591-022-02142-1.
16. Bertoli-Avella, A.M., Beetz, C., Ameziane, N., *et al.*, Successful application of genome sequencing in a diagnostic setting: 1007 index cases from a clinically heterogeneous cohort. *Eur. J. Hum. Genet.*, 29, 1, 141–153, 2021, doi: 10.1038/s41431-020-00713-9.
17. Diemer, E.W., Tuhkanen, J., Sammallahti, S., *et al.*, Epigenome-wide meta-anal-ysis of prenatal vitamin D insufficiency and cord blood DNA methylation. *Epigenetics*, 19, 1, 2413815, 2024, doi: 10.1080/15592294.2024.2413815.
18. Mattar, C.N.Z., Labude, M.K., Lee, T.N., Lai, P.S., Ethical considerations of preconception and prenatal gene modification in the embryo and fetus. *Hum. Reprod.*, 36, 12, 3018–3027, 2021, doi: 10.1093/humrep/deab222.
19. Lovinsky-Desir, S. and Miller, R.L., Epigenetics, asthma, and allergic diseases: A review of the latest advancements. *Curr. Allergy Asthma Rep.*, 12, 3, 211–220, 2012, doi: 10.1007/s11882-012-0257-4.
20. Akiyama, N., Shimura, M., Yamazaki, T., *et al.*, Prenatal diagnosis of severe mitochondrial diseases caused by nuclear gene defects: A study in Japan [published correction appears in Sci Rep. 2021 Nov 16;11(1):22682. doi: 10.1038/s41598-021-02108-2.]. *Sci. Rep.*, 11, 1, 3531, Published 2021 Feb 11, doi: 10.1038/s41598-021-81015-y.
21. Wang, J., Xu, Y.F., Fu, Q.H., *Zhonghua Yu Fang Yi Xue Za Zhi*, 55, 9, 1023–1027, 2021, doi: 10.3760/cma.j.cn112150-20210427-00420.

22. Reilly, K., Sonner, S., McCay, N., *et al.*, The incremental yield of prenatal exome sequencing over chromosome microarray for congenital heart abnormalities: A systematic review and meta-analysis. *Prenat. Diagn.*, 44, 6-7, 821–831, 2024, doi: 10.1002/pd.6581.
23. Sarac Sivrikoz, T., Basaran, S., Has, R., *et al.*, Prenatal sonographic and cytogenetic/molecular findings of 22q11.2 microdeletion syndrome in 48 confirmed cases in a single tertiary center. *Arch. Gynecol. Obstet.*, 305, 2, 323–342, 2022, doi: 10.1007/s00404-021-06125-4.
24. Chong, S.C., Cao, Y., Fung, E.L.W., *et al.*, Expansion of the clinical and molecular spectrum of WWOX-related epileptic encephalopathy. *Am. J. Med. Genet. A*, 191, 3, 776–785, 2023, doi: 10.1002/ajmg.a.63074.
25. Wise, A.L., Manolio, T.A., Mensah, G.A., *et al.*, Genomic medicine for undiagnosed diseases. *Lancet*, 394, 10197, 533–540, 2019, doi: 10.1016/S0140-6736(19)31274-7.
26. Maury, E.A., Jones, A., Seplyarskiy, V., *et al.*, Somatic mosaicism in schizophrenia brains reveals prenatal mutational processes. *Science*, 386, 6718, 217–224, 2024, doi: 10.1126/science.adq1456.
27. Timberlake, A.T., Griffin, C., Heike, C.L., *et al.*, Haploinsufficiency of SF3B2 causes craniofacial microsomia. *Nat. Commun.*, 12, 1, 4680, Published 2021 Aug 3, doi: 10.1038/s41467-021-24852-9.
28. Verbitsky, M., Krishnamurthy, S., Krithivasan, P., *et al.*, Genomic Disorders in CKD across the Lifespan. *J. Am. Soc. Nephrol.*, 34, 4, 607–618, 2023, doi: 10.1681/ASN.2022060725.
29. Martino, D. and Prescott, S., Epigenetics and prenatal influences on asthma and allergic airways disease. *Chest*, 139, 3, 640–647, 2011, doi: 10.1378/chest.10-1800.
30. Heindel, J.J. and Vandenberg, L.N., Developmental origins of health and disease: A paradigm for understanding disease cause and prevention. *Curr. Opin. Pediatr.*, 27, 2, 248–253, 2015, doi: 10.1097/MOP.0000000000000191.
31. Gao, J., Wu, Y., Yu, J., *et al.*, Impact of genomic and epigenomic alterations of multigene on a multicancer pedigree. *Cancer Med.*, 13, 13, e7394, 2024, doi: 10.1002/cam4.7394.
32. Rumballe, B., Georgas, K., Wilkinson, L., Little, M., Molecular anatomy of the kidney: What have we learned from gene expression and functional genomics? *Pediatr. Nephrol.*, 25, 6, 1005–1016, 2010, doi: 10.1007/s00467-009-1392-6.
33. Acharya, S., Jayabose, S., Kogan, S.J., *et al.*, Prenatally diagnosed neuroblastoma. *Cancer*, 80, 2, 304–310, 1997, doi: 10.1002/(sici)1097-0142(19970715)80:2<304::aid-cncr19>3.0.co;2-y.
34. Abou Tayoun, A.N., Krock, B., Spinner, N.B., Sequencing-based diagnostics for pediatric genetic diseases: Progress and potential. *Expert Rev. Mol. Diagn.*, 16, 9, 987–999, 2016, doi: 10.1080/14737159.2016.1209411.

35. Garcia-Herrero, S., Simon, B., Garcia-Planells, J., The Reproductive Journey in the Genomic Era: From Preconception to Childhood. *Genes (Basel)*, 11, 12, 1521, 2020Published 2020 Dec 19, doi: 10.3390/genes11121521.
36. Nguyen, K.V., Potential epigenomic co-management in rare diseases and epigenetic therapy. *Nucleosides Nucleotides Nucleic Acids*, 38, 10, 752–780, 2019, doi: 10.1080/15257770.2019.1594893.
37. van den Veyver, I.B. and Eng, C.M., Genome-Wide Sequencing for Prenatal Detection of Fetal Single-Gene Disorders. *Cold Spring Harb. Perspect. Med.*, 5, 10, a023077, Published 2015 Aug 7, doi: 10.1101/cshperspect.a023077.
38. Leeder, J.S., Developmental and pediatric pharmacogenomics. *Pharmacogenomics*, 4, 3, 331–341, 2003, doi: 10.1517/phgs.4.3.331.22693.
39. Antonson, A.M., Evans, M.V., Galley, J.D., *et al.*, Unique maternal immune and functional microbial pro"les during prenatal stress. *Sci. Rep.*, 10, 1, 20288, Published 2020 Nov 20, doi: 10.1038/s41598-020-77265-x.
40. Liao, C.H., Chang, M.Y., Ma, G.C., *et al.*, Preimplantation Genetic Diagnosis of Neurodegenerative Diseases: Review of Methodologies and Report of Our Experience as a Regional Reference Laboratory. *Diagnostics (Basel)*, 9, 2, 44, Published 2019 Apr 23, doi: 10.3390/diagnostics9020044.
41. Hayeems, R.Z., Dimmock, D., Bick, D., *et al.*, Clinical utility of genomic sequencing: A measurement toolkit. *NPJ Genom. Med.*, 5, 1, 56, Published 2020 Dec 15, doi: 10.1038/s41525-020-00164-7.
42. Dhombres, F., Morgan, P., Chaudhari, B.P., *et al.*, Prenatal phenotyping: A community e!ort to enhance the Human Phenotype Ontology. *Am. J. Med. Genet. C Semin. Med. Genet.*, 190, 2, 231–242, 2022, doi: 10.1002/ajmg.c.31989.
43. Lim, T.B., Foo, S.Y.R., Chen, C.K., The Role of Epigenetics in Congenital Heart Disease. *Genes (Basel)*, 12, 3, 390, Published 2021 Mar 9, doi: 10. 3390/genes12030390.
44. Agranier, M., Demurger, F., Dubourg, C., et al., Prenatal diagnosis of mucopolysaccharidosis type I on hepatosplenomegaly and coarse features: A case-report. *BMC Pregnancy Childbirth*, 25, 1, 3, Published 2025 Jan 3, doi: 10.1186/s12884-024-07115-5.
45. Trosman, J.R., Weldon, C.B., Slavotinek, A., Norton, M.E., Douglas, M.P., Phillips, K.A., Perspectives of US private payers on insurance coverage for pediatric and prenatal exome sequencing: Results of a study from the Program in Prenatal and Pediatric Genomic Sequencing (P3EGS). *Genet. Med.*, 22, 2, 283–291, 2020, doi: 10.1038/s41436-019-0650-7.
46. Shear, M.A., Robinson, P.N., Sparks, T.N., Fetal imaging, phenotyping, and genomic testing in modern prenatal diagnosis. *Best Pract. Res. Clin. Obstet. Gynaecol.*, 98, 102575, 2025, doi: 10.1016/j.bpobgyn.2024.102575.
47. Butchbach, M.E., Applicability of digital PCR to the investigation of pediatric-onset genetic disorders. *Biomol. Detect. Quantif.*, 10, 9–14, Published 2016 Aug 8, doi: 10.1016/j.bdq.2016.06.002.

48. Sikdar, S., Joehanes, R., Joubert, B.R., *et al.*, Comparison of smoking related DNA methylation between newborns adults from personal smoking. *Epigenomics,* 11, 13, 1487, Äì1500, 2019, doi: 10.2217/epi-2019-0066.
49. Bianchi, D.W., *From prenatal genomic diagnosis to fetal personalized medicine: Progress and challenges.* 2012 Jul 6, doi: 10.1038/nm.2829.
50. Datkhaeva, I., Arboleda, V.A., Senaratne, T.N., *et al.*, Identification of novel PIEZO1 variants using prenatal exome sequencing and correlation to ultrasound and autopsy findings of recurrent hydrops fetalis. *Am. J. Med. Genet. A*, 176, 12, 2829–2834, 2018, doi: 10.1002/ajmg.a.40533.
51. Cunniff, C., American Academy of Pediatrics Committee on Genetics. Prenatal screening and diagnosis for pediatricians. *Pediatrics,* 114, 3, 889–894, 2004, doi: 10.1542/peds.2004-1368.

11

Integrating Genomic Data into Electronic Health Records (EHRs)

Raju C.[1*], Thrisha Lakkireddy[1], Reddi Mohan Naidu K.[2], Sivasubramanyam Medasani[3] and Bukke Pasunath Naik[4]

[1]*Department of Electronics and Communication Engineering, Sri Venkateswara College of Engineering, Tirupati, Andhra Pradesh, India*
[2]*Department of Fiber System Engineering, Yeungnam University, Gyeongsan, Republic of Korea*
[3]*Department of Computer Science and Engineering, K. S. School of Engineering and Management, Bengaluru, India*
[4]*Department of ECE, Indian Institute of Information Technology Design and Manufacturing (IIITDM), Kancheepuram, Nellikuppam, Tamil Nadu, India*

Abstract

Clinical genomics stands to gain a great deal from the broad use of electronic health records. The results are sent back to the consultant and the genetic detections can be used for clinical decisions and the data can be collected in the electronic health record (EHR) to offer the phenotypic information for study. The technologies in Genomic medicine have rapidly advanced, involving the delivery of precision medicine in all areas of healthcare. The EHR has turned out to be a standard practice in tandem with the advancement of genetic medicine. The integration of genetic data into EHRs is changing healthcare by utilizing personalized medication and clinical outcomes. The American College of Medical Genetics and Genomics (ACMG) issued a framework for the integration of genetic data into the EHRs. The principal difficulties in the integration of data include the privacy issues related to genetic testing and uncertainty in interpreting the regulatory requirement for the return result. The present EHR systems need to be meaningfully reformed in order to effectively include the genetic data. The nominal functional requirements that an EHR must achieve in order to ease the successful integration of the genetic results are conferred along with the other unanswered challenges. This review chapter gives a systematic

**Corresponding author*: c.raju1112@gmail.com

Swetha Kumari Koduru, Sarad Pawar Naik Bukke, Hussaini Bello, Muhamad Mustafa Mundu and Ayush Madan (eds.) Genomics and Precision Healthcare: Innovations, Applications and Challenges, (351–372) © 2026 Scrivener Publishing LLC

summary of the state of genomic data integration in EHRs today, critically analyzing the various strategies, difficulties, and possibilities related to this quickly developing theme. The focus on improving data security, interoperability, emerging clinical capability and encouraging patient involvement can be done in future for potential utilization of genomic-informed and customized medicine.

Keywords: Electronic health record (EHR), genomics data processing, genetic testing, genomic variants, privacy and security, personalized medicine

11.1 Introduction

Electronic health records, or EHRs, have transformed healthcare across the globe due to digitization for patient care. The paper records of the patient can be removed by integrating the data, which has encouraged smooth communication among healthcare experts. This will increase patient safety, minimize medical errors and significantly improve care coordination. EHRs make it possible to acquire vital information more rapidly, which speeds up the diagnosis and treatment strategies.

EHRs are not only essential for public health but also for the individual. EHRs are used in population health management by contributing information for various disease epidemics, keeping an eye on the health trends and evaluating the success of public health campaigns. The significance of EHR is depicted in Figure 11.1. Medical progress is accelerated by researchers using EHR data for clinical studies. Additionally, by giving patients access to their medical records and facilitating active engagement in their care, EHRs empower people *via* portals. Clinical decision support systems are integrated into EHRs to help doctors practice evidence-based medicine. In the end, EHRs are essential to creating a global healthcare system that is more effective, accessible, and patient-centered.

The market for electronic health records is expanding significantly on a global scale. About 96% of non-federal acute care hospitals in the United States used certified Electronic Health Record (EHR) technology in 2021, making it nearly universal. Although slightly lower, office-based physicians also demonstrated strong uptake at 78%. The acknowledged advantages of digital health records and government incentives are responsible for the notable surge in EHR use over the last ten years. The global market for electronic health records is expected to reach USD 43.66 billion by 2034, growing at a compound annual growth rate (CAGR) of 4.32% from 2024 to 2034. The market is currently valued at USD 28.60 billion. Cloud-based systems are controlling the market. The Global EHR Adoption over time is shown in Figure 11.2, and Table 11.1 shows the statistics globally.

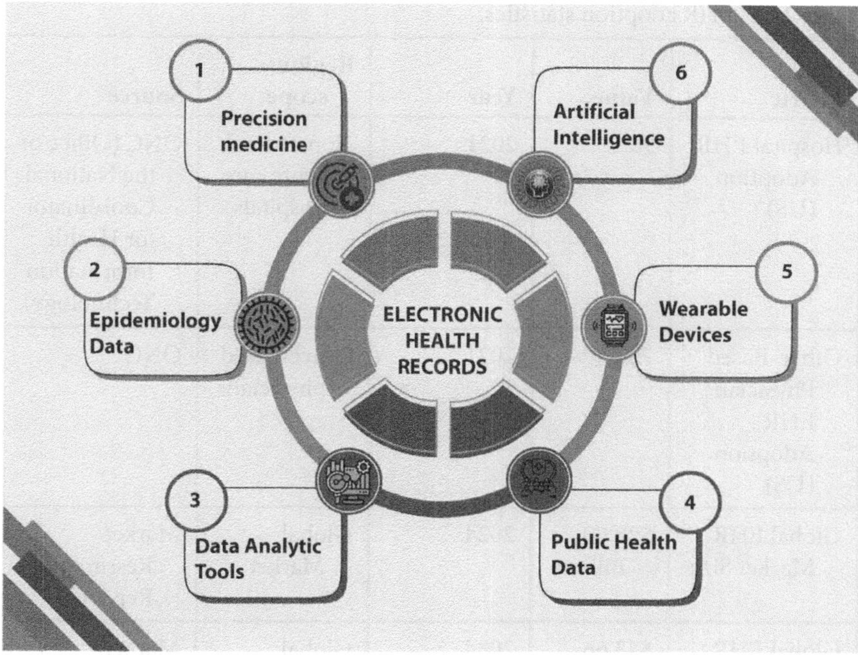

Figure 11.1 Significance of EHRs.

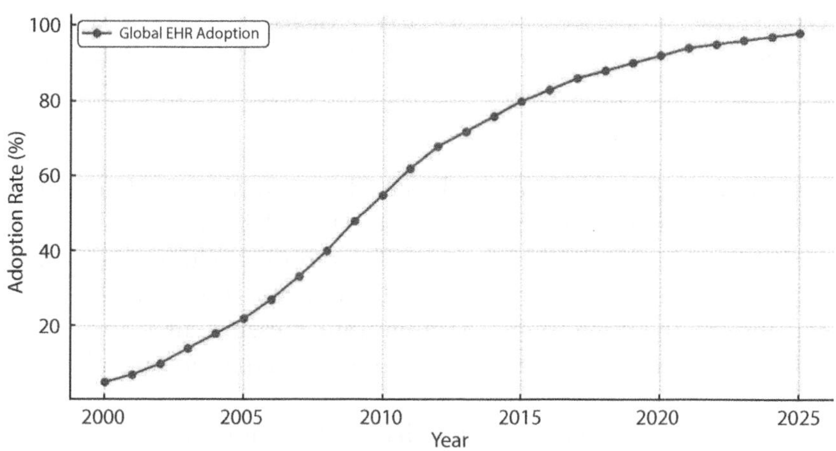

Figure 11.2 Global EHR adoption over time.

Table 11.1 EHR adoption statistics.

Metric	Value	Year	Region/ scope	Source
Hospital EHR Adoption (US)	96%	2021	Non-federal acute care hospitals	ONC (Office of the National Coordinator for Health Information Technology)
Office-Based Physician EHR Adoption (US)	78%	2021	Office-based physicians	ONC
Global EHR Market Size	$28.60 Billion	2024	Global Market	Market Research Reports
Global EHR Market Projected Size	$43.66 Billion	2034	Global Market	Market Research Reports
Global EHR Market CAGR	4.32%	2024 - 2034	Global Market	Market Research Reports

The shift to digital records in the health care sector is in line with the trend of the majority of other businesses [1–4]. With little chance of the return to paper records, the adoption of electronic health records (EHRs) has doubled nationally during the last few years. The potential of "big data" and creative applications to significantly measure, alter, and enhance healthcare delivery and measurement comes with the widespread usage of EHRs.

The availability of genomic data has significantly expanded, particularly with the introduction of genetic sequencing technology, which is less expensive [5]. Projects like the Electronic Medical Records and Genomics (eMERGE) collaboration have investigated how genetic data might be included into EHRs to influence patient care, as well as determining when results might be returned [6, 7]. Figure 11.3 shows the integration of genomic data in EHR.

To completely incorporate genomic medicine into patient care, an infrastructure must support each step of genetic delivery, including patient identification, test ordering, result delivery, clinical decision support (CDS), and billing/reimbursement [8]. Building this infrastructure is known to need informatics-based techniques, and because of the widespread use of EHRs in clinical care, a lot of attention has been focused on it [9–15]. The current status of data collection and storage in EHRs is discussed in this study, along with real-world obstacles to integrating genetic and genomic data into EHRs.

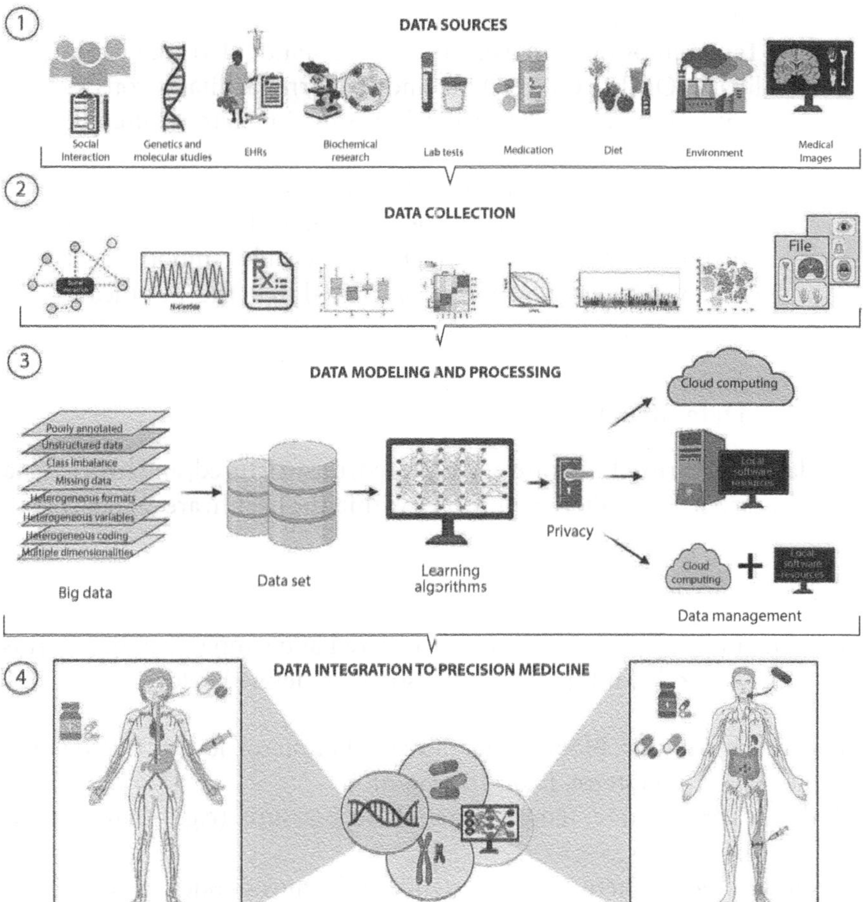

Figure 11.3 Integration of genomic data into EHRs (Source: www.google.com).

11.2 EHRs: Structure and Standards

EHRs are made to organize the storage, management, and exchange of patient health data. To assure the efficiency and interoperability, EHRs adhere to specific standards and frameworks.

- **Data Structure:** EHRs normally include structured data like prescriptions, lab results and unstructured data like clinical notes, imaging reports.
- **Standardization Protocols:** There are some standard protocols for EHRs. Some of the standards, like HL7 Fast Healthcare Interoperability Resources (FHIR), DICOM (for imaging data), and LOINC (for lab results) ensure seamless integration and exchange of health data.
- **Terminology Standards:** The use of standard vocabularies like SNOMED CT, ICD-10, and RxNorm facilitates consistent representation of medical information across different systems.
- **Interoperability Solutions:** EHRs must have the provision for interoperability frameworks, like HL7 and Clinical Document Architecture (CDA) to enable smooth data exchange between genomic databased and the healthcare organizations.

11.2.1 Data Structure in EHRs

The data structure in EHRs are of structured, unstructured, and hybrid data models, each having different purposes within the healthcare management.

11.2.1.1 Structured Data

The structured data type is extremely ordered and simply searchable. This has numbers, coded information, and some predefined fields such as:

Laboratory Results: Blood sugar levels, cholesterol levels, hemoglobin counts (e.g., "Cholesterol, LDL: 100 mg/dL").
Vital Signs: Heart rate, temperature, blood pressure, (e.g., "temperature: 101^0 F").
Diagnosis Codes: There are some standard diagnosis codes, such as ICD-10 for diseases.

(e.g., "E11.9" for Type 2 Diabetes).
Medications and Allergies: Lists of prescribed drugs using Rx Norm codes (e.g., "Metformin 500mg daily").
Example: The structured entry for the diabetic patient's EHR
"E11.9 (Type 2 Diabetes) | HbA1c: 7.2% | Rx: Metformin 500 mg BID."

11.2.1.2 Unstructured Data

The unstructured data does not follow any predefined format and consists of free-text clinical notes, imaging reports.

Physician Notes: These are the exhaustive descriptions of patient symptoms and history.
Radiology Reports: These include the textual interpretations of images (e.g., "CT scan displays the mild atherosclerosis in coronary arteries").
Surgical and Discharge Summaries: These are the normal text reports on the procedures and discharge instructions of the patient.
Example: A doctor may write, "Patient reports persistent headaches worsening over the past month, unresponsive to over-the-counter (OTC) pain relievers."

11.2.1.3 Hybrid Data Models

This hybrid data combines both structured and unstructured data for improved functionality.

Example: A pathology report of a patient may include structured biomarker results (e.g., "HER2: Positive") along with a physician's interpretation in free-text format.
Example: An oncology patient's EHR may contain structured genomic test results indicating the presence of a BRCA1 mutation, along with a physician's recommendation in an unstructured note advising preventive measures.

This hybrid approach is attractive and more common, particularly with advancement in the field of artificial intelligence (AI) and natural language processing (NLP), which enable automated extraction of meaningful information.

11.2.2 Standardization Protocols

Standard protocols are required for ensuring smooth data interchange across various healthcare systems. To encourage platform compatibility, data integrity, and interoperability, there are various standardization protocols available.

11.2.2.1 Health Level Seven (HL7) Standards

The international standard for electronic health information exchange is Health Level Seven (HL7).

HL7 v2: This is used for clinical messaging, allowing communication among various hospital systems.
HL7 FHIR (Fast Healthcare Interoperability Resources): This standard is mainly designed for web-based interoperability which uses the RESTful APIs for data exchange.
Example: A company using HL7 FHIR APIs can directly send the patient-specific pharmacogenomic results to an EHR, for ensuring clinicians can access the relevant genomic insights in real time.

11.2.2.2 Observational Medical Outcomes Partnership (OMOP) Common Data Model

For large-scale observational research, the OMOP standard is used. This enables investigators to analyze patient records across various organizations while guaranteeing uniformity.
Example: The study examining the impact of the genetic variants on drug response may use OMOP-formatted EHR data.

11.2.2.3 Digital Imaging and Communications in Medicine (DICOM)

For managing, archiving, and transferring the medical imaging data, including CT, MRI, and X-ray scans, DICOM Standard is used. It confirms that the imaging data from different equipment manufacturers remains interoperable.
Example: A radiologist reviewing an MRI scan from a different hospital can access it in their EHR system due to DICOM compatibility.

11.2.2.4 Logical Observation Identifiers Names and Codes (LOINC)

LOINC regulates lab test results and clinical notes to enable uniformity in reporting.
Example: A patient's cholesterol level measured at one hospital and later at another hospital will be documented uniformly as "LOINC 2093-3".

11.2.2.5 RxNorm

This provides standardized names and unique identifiers for medicines, ensuring uniformity across various systems.
Example: A patient's prescription for atorvastatin 20 mg at Hospital A will be documented as the same medication at Hospital B because of its RxNorm identifier.

HL7, OMOP, DICOM, LOINC, and RxNorm are some standard protocols that show a vital role in ensuring genomic data integration into EHRs and are efficient and usable across different healthcare settings. These enable enhanced data accuracy, streamlined workflows, and better patient outcomes.

11.2.3 Terminology Standards

EHRs contains standardized vocabularies which offer a public language for medical data in order to assure uniformity in clinical terminology. These terminologies enable accurate communication amongst regulatory agencies, research organizations, and healthcare providers.

11.2.3.1 SNOMED CT (Systematized Nomenclature of Medicine – Clinical Terms)

A comprehensive and multilingual healthcare terminology which provides standardized terms for diseases, symptoms, procedures, and clinical findings. This is used universally to improve the interoperability between healthcare organizations.
Example: A doctor identifying a patient with chronic obstructive pulmonary disease (COPD) will use the SNOMED CT code "13645005," confirming uniformity in documents.

11.2.3.2 ICD-10 (International Classification of Diseases, 10th Revision)

The World Health Organization (WHO) developed ICD-10, which is used for the classification of disease and medical billing. This gives standardized codes for diagnosis which helps in statistical analysis and epidemiological studies.
Example: A patient who is diagnosed with Type 2 Diabetes Mellitus will have the ICD-10 code "E11.9" documented in the EHR.

11.2.3.3 Unified Medical Language System (UMLS)

This integrates multiple health terminology standards to facilitate interoperability between different coding systems which is used in applications such as AI-driven clinical decision support and research data analysis.
Example: A system using UMLS can link SNOMED CT disease codes with ICD-10 billing codes, improving data integration.

By exploiting these standardized terminologies, EHRs guarantee that the genomic, clinical and pharmacological data is precisely characterized, improving patient care and thus enabling large-scale health data analysis. These are important for confirming that genomic data integration into EHRs is accurate, interpretable, and usable across all the healthcare systems.

11.2.4 Interoperability Solutions

In order to ease smooth data transfer among different EHR systems and assure that medical specialists can successfully access and analyze patient data, interoperability solutions are critical. Interoperability in EHRs is enabled by a wide variety of technology and strategies.

11.2.4.1 Semantic Interoperability

This allows various EHR systems to understand and use the exchanged data implicitly.
Example: A hospital using SNOMED CT can share a patient's diagnosis with another hospital using ICD-10, where an automated system maps the SNOMED CT code to a corresponding ICD-10 code for reliable interpretation.

11.2.4.2 Syntactic Interoperability

This allows that the data is formatted and structured in a way that various systems can process.
Example: HL7 FHIR permits an EHR in one hospital to exchange the patient information with another hospital *via* a standardized XML or JSON format.

11.2.4.3 Interoperability Frameworks and Networks

CommonWell Health Alliance: The network which allows EHR vendors to share the patient data across various platforms.
Carequality: This provides a national interoperability framework that allows healthcare institutes to share records.
Example: Information about a patient who is being treated at Hospital A can be retrieved by Hospital B using Carequality's framework.

11.2.4.4 Application Programming Interfaces (APIs)

APIs allow various EHR systems to exchange the patients' data securely in real time.
Example: Apple's HealthKit API permits the patients to access their EHR data from various providers on their smartphones.

11.2.4.5 Health Information Exchanges (HIEs)

HIEs enable the sharing of electronic health information among providers, payers and the health institutes.
Example: A state-level HIE permits various hospitals, labs, and clinics to access the shared patient database for improving coordination of care.

11.2.4.6 Cloud-Based Interoperability Solutions

Cloud platforms permit safe data storage and exchange amongst various healthcare systems, reducing dependence on local servers.
Example: Google Cloud Healthcare API allows EHRs to store and recover patients' genomic data and imaging effectively.

Healthcare institutes may encourage the integration of genetic data into EHRs, improve patient care and data exchange by using these interoperability solutions. Irrespective of the specific EHR system being used, these solutions assure that vital health information is accessible when and where it is required.

11.3 Genomics and Impact in Medicine

Modern medicine has been renovated by genomics—the study of an organism's entire DNA which includes all of its genes. Healthcare systems can provide more precise, competent, and individualized therapies by having a better grasp of the structure, function, and variability in genomes. Over the past 20 years, this has significantly transformed, moving from fundamental genetic research to a vital role in clinical treatment.

When the Human Genome Project finished in 2003, it was an important genomics accomplishment. The complete human genome, which consists of between 20,000–25,000 genes and 3 billion base pairs, was mapped effectively. The basis for comprehending how genetic variants affect health and disease was recognized by this huge endeavor [16]. Nowadays, genomic data helps as the foundation for many medical developments, directing the creation of risk assessments, personalized treatments, and diagnostic instruments.

Perhaps the most significant accomplishment of genomics is precision medicine which coordinates care and prevention based on the genomic profile of the individual. The field contributes to the way in which it separates the genetic variations contributing to the regulation of pharmacodynamics and predisposition to disease, thus abandoning the ahistorical and universalist one-size-fits-all approach to the practice of medicine. In oncology one example is that clinicians use a tumor-specific genetic profile to choose agents that treat local mutations. The clinical effectiveness of therapeutic agents is essential; trastuzumab (Herceptin), against HER2-positive breast tumor and imatinib (Gleevec) against chronic myeloid cancer have significantly changed patient outcomes [17].

In genomics, the subfield studying the effect of genetic variation on drug response is pharmacogenomics. Pharmacogenomic studies allow prescription and prescribing to be safer and effective due to the clarification of genetic factors leading to the processing of drugs. Genetic variations in the CYP2C19 gene, such as those involved in the metabolism of the antiplatelet clopidogrel; the low-functioning variants can inhibit formation of active drug and increase the rate of cardiovascular events [18]. Incorporating pharmacogenomic information in the Electronic Health Records (EHRs) enables providers to make small changes in treatment regimen and minimize negative responses and maximize therapeutic effect.

There has also been a transformation in early diagnosis and prevention of the illnesses through genomic methods. Genetic testing can be used to identify patients who have germline mutations that predispose them to cancer. Certain mutations such as in BRCA1 and BRCA2 genes significantly enhance the risk of malignancies of the breast and the ovaries [19]. Preventive screening guidelines and risk reducing surgeries are thus made possible. At the same time, newborn screening programs identify hereditary diseases, such as cystic fibrosis and phenylketonuria (PKU) in the newborn period, so in time effective intervention is done and the worst sequelae are considerably improved [20].

The same has happened to the terrain of the inherited disease diagnosis. Whole genome sequencing (WGS) and whole exome sequencing (WES) are currently used to treat patients presenting with rare phenotypes avoiding classical diagnostic fashion. These sophisticated technologies are being used in collaborative programs like the Undiagnosed Diseases Network (UDN) in order to establish etiological mutations. An example of such cases was the characterization of Glycogen Storage Disease Type IX through WES in undiagnosed persons using PHKA2 as the causative gene, and thus the necessary clinical follow-up [21].

At its core, cancer is a genetic disease caused by mutations that cause cells to grow out of control. Targeted therapy is guided by actionable mutations found through tumor genomic analysis. Tyrosine kinase inhibitors such as erlotinib and gefitinib, for example, are effective in treating non-small cell lung cancer (NSCLC) patients with EGFR mutations [22]. Furthermore, patients with malignancies with high microsatellite instability (MSI-H), high tumor mutational burden (TMB) respond better to immune checkpoint inhibitors like pembrolizumab [23].

Public health and the management of infectious diseases have been revolutionized by genomics. Rapid outbreak tracking and identification are made possible by pathogen genome sequencing. Genomic surveillance was essential in tracking the development of the SARS-CoV-2 virus during the COVID-19 pandemic, detecting variations such as Delta and Omicron [24]. The expansion of vaccines and public health initiatives were inspired by this data. Genomic characterisation of Mycobacterium tuberculosis has enabled proper drug resistance prediction, hence providing the basis to come up with effective treatment plans [25].

Despite its great clinical potential, a wide variety of moral, legal and societal implications are associated with genomic medicine. The main issues are informed consent, data security, and genetic privacy. The Genetic Information Nondiscrimination Act (GINA) is a law that aims to save citizens from genetic discrimination in the United States. In addition, there is a need to tackle the issue of equal access to the just delivery of genomic testing and related treatment by embarking on concerted regulatory reforms to ensure that both the poor and rich can access genomic testing and related treatment [26].

11.4 AI and ML in Integration of Genomic Data in EHRs

Inclusion of genetic data in Electronic Health Records (EHRs) is a relevant breakthrough of modern healthcare, allowing personalized pharmacotherapy and patient outcomes enhancement. The combination of genomics with EHR coupled with the use of artificial intelligence (AI) and machine learning (ML) can allow the efficient manipulation of large data, thus producing clinically valuable information. It is thus due to the methodologies, applications, perceived challenges, and their future directions of AI and ML in the integration of genomic data with EHRs that this review endeavors to address such gaps that have been emerging. Figure 11.4 shows the market of AI/ML in genomics worldwide, and Figure 11.5 depicts the potential application of AI/ML to the embedding of genomic data into EHRs.

Genomic data is a resource in and of itself comprising complexities. Information of various modalities is generated, such as targeted gene panels, whole-genome sequencing (WGS) and whole-exome sequencing (WES), with variable proportions of structure and high levels of heterogeneity. This was because the active consideration of genomic information in the electronic health records (EHRs) situation required integrated computational workflows that could meet the structural dimensionality and informational diversity. Such needs have seen most processing and normalisation activities going the automated route in an effort to ensure interoperability with current EHR infrastructures. Since such time- and labor-intensive procedures as variant calling, annotation, and normalization are computationally costly, artificial intelligence (AI) and machine learning (ML) algorithms, that is, supervised learning and deep learning algorithms, become invaluable resources.

The AI/ML have found their niche in the clinical setting as the basis of variants analysis and interpretation. Genetic variations are building blocks of genetic information, namely insertions/deletions (indels) and single nucleotide polymorphisms (SNPs). In supervised learning, AI/

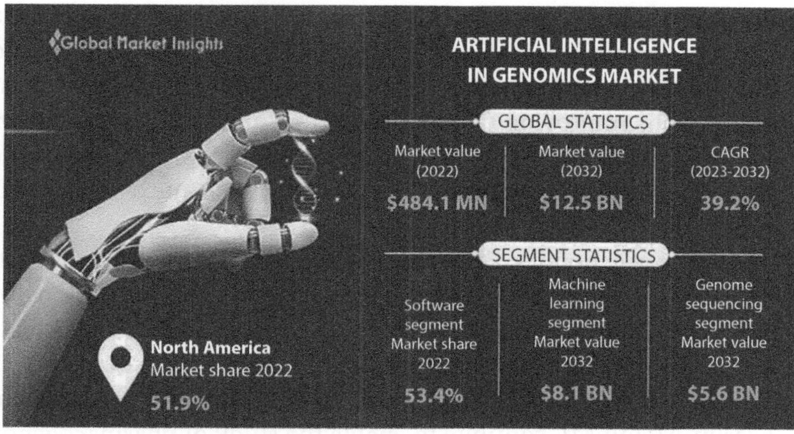

Figure 11.4 AI/ML in genomic market (Ref: Google.com).

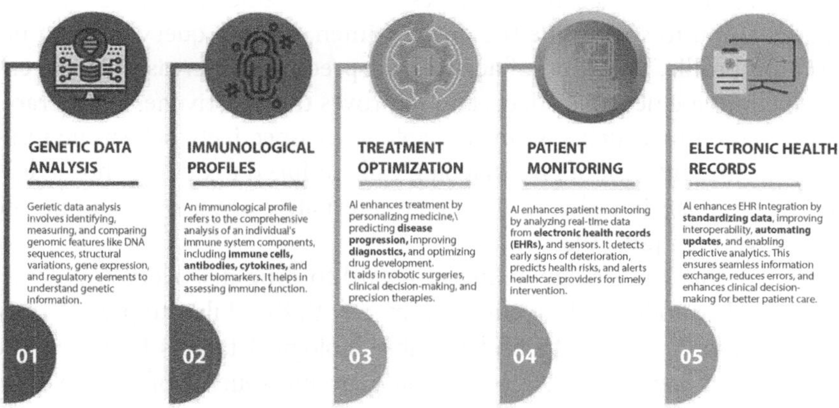

Figure 11.5 AI/ML in integration of genomic data in EHR.

ML systems are trained to forecast pathogenicity of variants using large databases that list genomics loci and their predicated outcome. One of the examples is the construction of models based on gnomAD and ClinVar data to make predictions referring to the phenotypic association of novel variants and inform clinical practice with probabilities. There is also the use of AI-powered analytical pipelines that can prioritize variants based on *in silico* predictions of protein effects and based on the statistical likelihood of causality of specific diseases [27, 28].

At the same time, AI/ML methods enable building precise predictive models of identifying the risk of getting a disease. The AI/ML models can predict if a patient is likely to develop a cancer, cardiovascular disease or neurological condition through a combination of his or her genomic, clinical and demographic data. Such risk estimates may be used in the formulation of customized preventative measures at the early stages of a care continuum, thus improving patient outcomes. Combining individualized risk scores of these models with electronic health records allows clinical practitioners to put the genomic information into context of the wider health profile.

The synergy of genomics, EHRs, and AI/ML, then, is the present key step towards personalized medicine and precision healthcare. Institutionalized research endeavors also proceed to clarify the procedures and advances in relation to this paradigm, its challenges, and its future courses.

Artificial intelligence (AI) and machine learning (ML) Algorithmic advances in AI and ML continue to enable the study of pharmacogenomics, a field focusing on understanding the influence of an individual genetic architecture on a drug response. The computational tools listed above can now be used to predict individual response to medications prescribed to a patient, and to streamline the dose regimen, using a query of germ nal variation profile. This kind of accuracy in prediction increases the potential to eliminate negative effects and improves the effectiveness of therapy. At the same time, such systems are able to prospectively determine individuals with high risk of pharmacologically induced toxicities, thus assisting the clinician in either choosing an alternative agent or changing the dosage regimen [29, 30].

Similar changes are happening in regard to drug discovery and development. The modern AI pipeline is armed with the ability to predict drug effectiveness, as well as to find possible biological targets based on the analysis of comprehensive clinical and genomic sources of information. The overall advantages entail an incentive increase in clinical trial efficacy and accelerated provision of unapproved treatments. At the same time, with the help of AI-powered queries, discoveries of clinically significant bio-markers have been made to track responses to therapy and improve forecasting of patient outcomes. As an example, AI algorithms can explore massive genetic datasets to define the expression-mediated signatures linked with medication responsiveness that may, in the end, result in the establishment of patient-centred treatment protocols.

Facilitating such efforts requires a multidisciplinary approach that brings together physicians, investigators, data scientists and technology developers. The included and in-depth training courses are needed to introduce healthcare professionals to AI-based technologies and include genomic evidence in everyday clinical routines. On the same note, continuous interaction with patients and the larger population is equally important to respond to the related ethical and social implications.

The integration of genomic data sets into electronic health records (EHRs) is fast developing, thus enhancing the achievement of personalised medicine. With the automatisation of data processing, synthesis of genetic variants, approximation of the course of disease development and provision of structured clinical decision support, AI is giving doctors the opportunity to make more specific and evidenced-based treatment decisions.

In using these innovations to their full potential, it will be necessary to pay close attention to data privacy and consistency, methodology standardization and clinical verification. Smooth integration of clinical and genomic data represents a pillar of the future of healthcare delivery, and the essential component that will have to be present in order to fully explore its potential is AI/ML.

11.5 Technical Challenges in Integrating Genetic Data within Current EHRs

The use of genetic information entails profound technical issues regarding intricacy, magnitude, and ever-changing nature of genomic information as it is added to electronic health records (EHRs). Traditional EHRs which have been designed mainly to handle structured data, e.g., laboratory results, diagnosis, prescriptions will not be able to handle the high dimensionality of genetic data. Well-tailored storage systems, workflows, and interpretive systems are accordingly required.

The biggest challenge is due to the sheer amount of information that is produced using high-throughput sequencing. The whole-genome sequence (WGS) alone generates, for example, 100-300 gigabytes per patient, and it is a challenge in logistics to healthcare organizations. Since every genomic data is considered permanent but could be reassessed occasionally with emerging clinical evidence, the demand on its storage and retrieval increases as the patient ages.

There is a complication caused by data formats. In contrast to the codified systems that dominate the sphere of EHR systems (e.g., HL7 or SNOMED CT), the genomic data are usually maintained in such formats as Variant Call Format (VCF), FASTA, and BAM. These niche designs provide robust sequencing metadata but are not well adapted with the incorporation in routine EHR infrastructures. Additionally, unreliable nomenclature also creates the inability to standardize data between datasets: A pathogenic mutation of BRCA1 could be variously reported in ClinVar, OMIM, or dbSNP, thus blocking consistency between institutions and providers.

The aspect that cannot be overlooked is interoperability since the genetic information needs to be shared between laboratories, hospitals, and research organizations. However, the largest EHR companies (such as Epic and Cerner) use proprietary data infrastructure that interferes with the integration of genomic data. Interoperability has been accommodated with the development of emergent standards like HL7 FHIR Genomics, surprisingly, not all health systems have been adopting consistently. The fact that PDF or unstructured reporting format is still used to report genetic findings also reduces the advantage of using such to assist in clinical decision-making. Making the situation more complicated, laboratory-developed tests are usually not consistent based on identical reporting standards, thereby further complicating integration issues.

There exist a number of obstacles to clinical translation in genomics research. Most importantly, it is computationally intensive: analyzing large genetic data requires the use of cloud storage with access to high-performance computing resources. The global healthcare system at the current time has a large percentage of systems that do not contain proper bioinformatics infrastructure and would be unable to process genomics data on a real-time basis. This limitation has an adverse impact on clinicians who need to obtain and transmit genetic testing outcomes in due time. The urgency of fast data processing is depicted in pharmacogenomic recommendations; an example is the adjustment in warfarin dosage as related to VKORC1 and CYPC2C9 variations in which the availability of the data must be at the place of prescription. Genetic data integration cannot be very efficient without provision of adequate computing power.

In conjunction with the computational issues, security and privacy concerns characterize them. In contrast to the medical records that are traditionally not identifiable, genetic data have the inherent property of identifying individuals, as well as sensitive information about ancestry, family relations, and disease risk. A data breach therefore presents severe ethical and legal dilemmas such as the possibility of genetic discrimination and others of such negative impact. Some laws like HIPAA in the United States and GDPR in Europe have strict measures on both storage and exchange of genetic information but it is still hard to comply with them because genome study is global and data sharing is international. As another example, a hospital located in the United States but working with a research institution in the EU will have to work through the intricate provisions of GDPR, which oftentimes includes de-identifying data or having patients consent to its collection and subsequent sharing before it will be permissible to do so.

Genomics is limited by inefficient usability as a clinical tool. In spite of the potential value of genetic information to improve clinical decisions, the majority of the physicians are poorly trained to interpret challenging and complex genetic data. EHRs lack the embedded decision-support tools, further limiting the use; an electrocardiologist can know nothing about the APOE variants which modify the effect of statins, and thus poor treatment selection is made. Additionally, alert fatigue in EHR is a serious problem: uninterrupted and generic alerts may overwhelm clinicians and are a part of burnout. Genetic findings should thus be set with a finer range of alerts to provide useful acts rather than add extra fatigue. In prescribing clopidogrel, an alert after checking body features of a patient would be of greater help to a physician when the body type of the patient is that of a CYP2C19 poor metabolizer.

EHR inclusion of genetic data is a considerable endeavor and requires solutions to data storage, standardization and interoperability problems, computational load, security, and clinical usefulness. The HL7 FHIR Genomics and cloud-based bioinformatics systems are the novelties that create possible pathways; however, extensive implementations and concurrent technical improvements are necessary to introduce scalable genomic medicine to everyday clinical practice. The further development of precision medicine and the successful use of genetic knowledge in the personalized treatment of patients will be impossible without progress in the solution of those problems.

11.6 Conclusion

Finally, this review checks the advantages, potential harms, and approaches to utilising EHR data in secondary research. In order to utilize health data warehouses to their full potential in clinical research, the advancements in EHR data abstraction, biobank combination and inter-institution connections are required. Besides, inter-institutional cooperation and coordination are also required to facilitate the translation of the EHR-based research to make decisions leading to clinical care. Inclusion of genomic data in EHRs is essential to promote precision medicine yet it is a complex challenge. Genomic data are so huge, unstandardized in formats, and cannot be easily integrated. EHRs should be developed to support structured genomics data using such standard systems as HL7 FHIR genomics to provide consistency in healthcare. Moreover, a powerful computational system that can be globally cloud-based is needed to properly store data and process it in real time. The necessity of security and privacy is of the highest importance given the fact that genomic data is very sensitive and strict measures such as the HIPAA and GDPR are required to protect it. It is important that there is a safe interchange of data and at the same time patients remain anonymous so that there can be no misuse. Clinically, the decision support system and physician education need to be enhanced in order to turn genetic knowledge into meaningful treatment decisions. Clinical alerts should be managed cautiously to ensure they do not flood the attendant healthcare providers. To address these barriers, a multidisciplinary collaboration is required by healthcare providers, bioinformaticians, EHR vendors and policy makers. Genomic integration of EHRs has the potential to promote individualized medicine, finer disease prediction, and treatment regimen optimization through smoothing the process of standardization, infrastructure, and usability. By tackling these hurdles, the genomic data will become integrated into daily care of patients, enhancing healthcare outcomes on a global scale.

AI disclosure statement: The author acknowledges the use of AI-based tools for minor language editing or grammatical correction. However, all intellectual contributions, analyses, and conclusions presented in this chapter are entirely the author's own.

References

1. Poissant, L., *et al.*, The impact of electronic health records on time efficiency of physicians and nurses: a systematic review. *J. Am. Med. Inf. Assoc.*, 12, 5, 505–516, 2005.
2. Jha, A.K., *et al.*, Use of electronic health records in US hospitals. *N. Engl. J. Med.*, 360, 16, 1628–1638, 2009.
3. Botsis, T., *et al.*, Secondary use of EHR: data quality issues and informatics opportunities. *Summit Transl. Bioinform.*, 2010, 1, 2010.
4. Lim, H.C., *et al.*, Toward a Learning Health Care System: A Systematic Review and Evidence-Based Conceptual Framework for Implementation of Clinical Analytics in a Digital Hospital. *Appl. Clin. Inf.*, 13, 2, 339–354, 2022.
5. Blumenthal, D. and Tavenner, M., The "meaningful use" regulation for electronic health records. *N. Engl. J. Med.*, 363, 501–504, 2010.
6. Starren, J., Williams, M.S., Bottinger, E.P., Crossing the omic chasm: a time for omic ancillary systems. *JAMA*, 309, 1237–1238, 2013.
7. McCarty, C.A., Chisholm, R.L., Chute, C.G., *et al.*, eMERGE Team. The eMERGE Network: a consortium of biorepositories linked to electronic

medical records data for conducting genomic studies. *BMC Med. Genomics*, 4, 13, 2011.
8. Hooker, G.W., Building an infrastructure to enable delivery of genomic medicine. *Am. J. Med. Genet. C Semin. Med. Genet.*, 187, 1, 95–99, 2021, http://doi.org/10.1002/ajmg.c.31881.
9. Green, E.D., Gunter, C., Biesecker, L.G., *et al.*, Strategic vision for improving human health at the forefront of genomics. *Nature*, 586, 7831, 683–692, 2020, http://doi.org/10.1038/s41586-020-2817-4.
10. Marsolo, K. and Spooner, S.A., Clinical genomics in the world of the electronic health record. *Genet. Med.*, 15, 10, 786–91, 2013 Oct, doi: 10.1038/gim.2013.88. Epub 2013 Jul 11.
11. Carter, A.B., *et al.*, Electronic Health Records and Genomics: Perspectives from the Association for Molecular Pathology Electronic Health Record (EHR) Interoperability for Clinical Genomics Data Working Group. *J. Mol. Diagn.*, 24, 1, 1–17, 2022 Jan, doi: 10.1016/j.jmoldx.2021.09.009.
12. Robertson, A.J., Mallett, A.J., Stark, Z., Sullivan, C., It Is in Our DNA: Bringing Electronic Health Records and Genomic Data Together for Precision Medicine. *JMIR Bioinf. Biotechnol.*, 5, e55632, 2024 Jun 13, doi: 10.2196/55632.
13. Warner, J.L., Jain, S.K., Levy, M.A., Integrating cancer genomic data into electronic health records. *Genome Med.*, 8, 113, 2016, https://doi.org/10.1186/s13073-016-0371-3.
14. Lau-Min, K.S., McKenna, D., Asher, S.B., Bardakjian, T., Wollack, C., Bleznuck, J., Biros, D., Anantharajah, A., Clark, D.F., Condit, C., Ebrahimzadeh, J.E., Long, J.M., Powers, J., Raper, A., Schoenbaum, A., Feldman, M., Steinfeld, L., Tuteja, S., VanZandbergen, C., Domchek, S.M., Ritchie, M.D., Landgraf, J., Chen, J., Nathanson, K.L., Impact of integrating genomic data into the electronic health record on genetics care delivery. *Genet. Med.*, 24, 11, 2338–2350, 2022 Nov, doi: 10.1016/j.gim.2022.08.009. Epub 2022 Sep 15.
15. Ayatollahi, H., Hosseini, S.F., Hemmat, M., Integrating Genetic Data into Electronic Health Records: Medical Geneticists' Perspectives. *Healthc. Inform. Res.*, 25, 4, 289–296, 2019 Oct, doi: 10.4258/hir.2019.25.4.289. Epub 2019 Oct 31. PMID: 31777672; PMCID: PMC6859263.
16. Lander, E.S., Linton, L.M., Birren, B., *et al.*, International Human Genome Sequencing Consortium. Initial sequencing and analysis of the human genome. *Nature*, 409, 860–921, 2001.
17. Sawyers, C.L., Targeted cancer therapy. *Nature*, 432, 7015, 294–297, 2004.
18. Mega, J.L., *et al.*, Cytochrome P-450 polymorphisms and response to clopidogrel. *N. Engl. J. Med.*, 360, 4, 354–362, 2009.
19. Antoniou, A., *et al.*, Average risks of breast and ovarian cancer associated with BRCA1 or BRCA2 mutations detected in case series unselected for family history. *Hum. Mol. Genet.*, 12, 11, 1329–1333, 2003.
20. Wilcken, B., Newborn screening: gaps in the evidence? *Arch. Dis. Child.*, 101, 5, 417–418, 2016.

21. Tarailo-Graovac, M., *et al.*, Exome sequencing and the management of neurometabolic disorders. *N. Engl. J. Med.*, 374, 23, 2246–2255, 2016.
22. Paez, J.G., *et al.*, EGFR mutations in lung cancer: correlation with clinical response to gefitinib. *Science*, 304, 5676, 1497–1500, 2004.
23. Le, D.T., *et al.*, PD-1 blockade in tumors with mismatch-repair deficiency. *N. Engl. J. Med.*, 372, 26, 2509–2520, 2015.
24. Hadfield, J., *et al.*, Nextstrain: real-time tracking of pathogen evolution. *Bioinformatics*, 34, 23, 4121–4123, 2018.
25. Walker, T.M., *et al.*, Whole-genome sequencing for prediction of Mycobacterium tuberculosis drug susceptibility and resistance. *Lancet Infect. Dis.*, 15, 10, 1193–1202, 2015.
26. Evans, J.P., *et al.*, Return of results and the learning health care system. *Genome Med.*, 5, 1, 3, 2013.
27. DeGroat, W., *et al.*, Hygieia: AI/ML pipeline integrating healthcare and genomics data to investigate genes associated with targeted disorders and predict disease. *Software Impacts*, 16, 100493, 2023, ISSN 2665-9638.
28. Quazi, S., Artificial intelligence and machine learning in precision and genomic medicine. *Med. Oncol.*, 39, 8, 120, 2022 Jun 15, doi: 10.1007/s12032-022-01711-1. PMID: 35704152; PMCID: PMC9198206.
29. Chen, Y.M., Hsiao, T.H., Lin, C.H., *et al.*, Unlocking precision medicine: clinical applications of integrating health records, genetics, and immunology through artificial intelligence. *J. Biomed. Sci.*, 32, 16, 2025, https://doi.org/10.1186/s12929-024-01110-w.
30. Ye, J., Woods, D., Jordan, N., Starren, J., The role of artificial intelligence for the application of integrating electronic health records and patient-generated data in clinical decision support. *AMIA Jt Summits Transl. Sci. Proc.*, 2024, 459–467, 2024 May 31, PMID: 38827061; PMCID: PMC11141850.

12

Genomic Counseling: Navigating the Complexities of Genetic Risk and Inheritance

Mudigubba Manoj Kumar[1], Queeni Sharoon Melam[1], Mohammed Azhar Hussain[2], Vishnuvara V. Tummeeti[3] and Akkiraju Sudheer[1]*

[1]Department of Pharmacy Practice, Raghavendra Institute of Pharmaceutical Education and Research, Anantapur, Andhra Pradesh, India
[2]Department of Pharmacy Practice, College of Pharmaceutical Sciences, Dayananda Sagar University, Bangalore, Karnataka State, India
[3]Shoppers Drug Mart, Calgary, Canada

Abstract

Genetic counseling is an important link between the advancing branch of genomic science and customized care for specific genetic predispositions in individuals and families. This chapter explores the possibilities of risk assessment, education, and emotional support for dealing with the genotype of prospective patients and their relatives within the context of genetic counseling. A key component of genetic counseling is knowledge of modes of inheritance of monogenic, polygenic, and multifactorial genes and how they affect health. Counseling involves returning the information gathered from genomic data analysis into understandable results for the patient to use in making decisions about testing, reproductive choices, and disease management. Confidentiality, testing minors, and reproductive choice are basic principles of the discipline, which emphasizes respect for patient values and cultural differences. Technological developments, including next-generation sequencing and artificial intelligence, are enhancing genetic counseling by improving risk assessment and delivering tailored procedures. However, these advancements also raise new issues such as data analysis, equality in data accessibility, and the psychological impact of receiving inconclusive results. Addressing these

**Corresponding author*: sudeerlegend@gmail.com

Swetha Kumari Koduru, Sarad Pawar Naik Bukke, Hussaini Bello, Muhamad Mustafa Mundu and Ayush Madan (eds.) Genomics and Precision Healthcare: Innovations, Applications and Challenges, (373–418) © 2026 Scrivener Publishing LLC

challenges requires a multidisciplinary, patient-oriented approach. The chapter also highlights the growing importance of genetic counseling in areas such as oncology, reproductive medicine, and rare diseases, demonstrating its impact on public health and personalized medicine. Genetic counseling facilitates better decision-making and the incorporation of genomic-based evidence into medical practice. Technology plays a key role in the future deployment of genetic counseling, aiming to enhance patient outcomes while maintaining a human touch. This chapter outlines genetic counseling as an essential component of current healthcare, focusing on decision-making and promoting positive health outcomes even in the presence of hereditary conditions.

***Keywords*:** Genetic counselling, genomic counselling, genetic risk assessment, inheritance patterns, patient autonomy, genetic privacy, emotional support, genetic testing

12.1 Introduction to Genetic Counseling

The term "Genetic Counseling" was coined by Sheldon Reed in 1947 [1]. Since then, genetic counseling and its goals have been defined in many ways. However, considering all aspects of genetic counseling, the National Society of Genetic Counselors task force defines it as the process of helping people understand and adapt to the medical, psychological, and familial implications of the genetic contribution to disease [2].

1. Evaluating the chance of disease occurrence and reoccurrence by interpreting the family and medical histories.
2. i. Providing education on inheritance; ii. managing testing sources; iii. conducting researches.
3. Guidance and support to promote informed choices by counseling and helping people to adapt to the risk or condition.

Genetic counseling seeks to enhance the understanding of genetic conditions within families, facilitating discussions concerning options for disease management, as well as the associated risks and benefits of further testing and alternative strategies [3]. Additionally, it assists in identifying the psychological resources required. The primary objective is to furnish precise and reliable information regarding inherited disorders, empowering individuals affected by such conditions to make informed decisions about future implications [4]. Moreover, it aims to educate the public about inherited disorders and the patterns of inheritance [3, 4].

Who Can Provide Genetic Counseling?
The **NSGC Task Force** (2006) established guidelines; qualified health care professionals who are well educated, trained, and experienced in medical and genetic counseling are eligible for providing genetic counseling [5].

Scope:
Genetic counseling has experienced significant growth alongside advancements in genetic technology [6]. It has yielded favorable outcomes for individuals affected by complex diseases. The genetic counseling profession was originated in 1969 and has since expanded globally, with nearly 7,000 genetic counselors practicing in 27 countries, predominantly in North America, Australia, and New Zealand, as well as a few in South Africa and several Asian countries such as South Korea, Taiwan, Malaysia, and India as of 2007 [7]. This number rose to 10,250 by 2023-24, and now extends across over 45 countries [8]. To establish consistent and appropriate guidelines for genetic counseling, in 1993, the American Board of Genetic Counseling was established to provide training and certification for genetic counselors [9].

In Canada, the graduate-level certifying program commenced in the mid-1980s, initiating the offering of master's degrees from its inception [10]. In 1995, Australia introduced a one-year training diploma, which transitioned to a two-year master's degree by 2008 [11]. The training for genetic counseling in Asia exhibits considerable variability as the field continues to evolve. Many countries, like China, India, Indonesia, Japan, Malaysia, the Philippines, South Korea, and Taiwan, provide graduate-level programs [12]. While some institutions adhere to a two-year master's degree structure, others integrate shorter options within existing programs. For example, Kasturba Medical College in India offers a comprehensive master's degree, whereas the Vellore Institute of Technology includes genetic counseling as part of a broader biomedical genetics curriculum. These diverse training models underscore the region's progressive educational methodologies. In Saudi Arabia, the field evolved in 2015, leading to the establishment of a master's degree program [13]. The demand for genetic counseling is continuously rising worldwide, fueled by increasing public awareness and accessibility of its benefits [14]. The National Society of Genetic Counselors has delineated the clinical scope of genetic counseling, encompassing the following:

A. Medical responsibility like history taking, risk assessment, education on inheritance, natural history, cascade testing, genetic testing, ordering of testing, psychological support involving adaptation assessment, anticipatory guidance and short-term client-centered counseling [15, 16].

B. Delivery of appropriate resources and comprehensive documentation in the preparation of case management [17].

Genetic counselors can also work in industrial settings and laboratory settings.

Industrial Genetic Counselors: Genetic counselors within the industrial sector are being increasingly sought after to mentor students in their thesis projects and to lead research initiatives [18]. This trend is likely driven by the growing awareness of the vast amount of data accessible in industrial contexts, as well as the diverse expertise these professionals contribute to their roles. Furthermore, numerous genetic counselors employed in the industry actively participate in local programs by serving as members of admissions committees and as thesis advisors. They operate as analysts, variant specialists, and leaders in genetic services, and engage with bioinformatics [19]. The field of genetic counseling in industrial environments will continue to advance as it adapts to both emerging challenges and opportunities.

Genetic counselors also expand their roles as social and behavioral scientists, guiding and enhancing outcomes in patient-based research [20]. They have numerous opportunities to engage in social and behavioral research aimed at integrating genomics into precision medicine. The insights they acquire will contribute to improving clinical practices and patient-reported outcomes [21].

Genetic counseling is an extensive and perpetually advancing domain, propelled by significant enhancements in medical genetics, technological innovations, and industry progressions [22]. It assumes a crucial role in personalized medicine, disease prevention, and patient care by aiding individuals in evaluating their genetic risks and making informed health decisions [23]. As genetic research continues to progress, genetic counseling is increasingly becoming indispensable in healthcare, bridging groundbreaking discoveries with practical applications in medicine and society [24].

Beyond clinical practice, the industry significantly contributes by advancing genetic testing, developing innovative therapies, and increasing accessibility to genomic services. Scientists and researchers play a crucial role in identifying new genetic markers, improving diagnostic methods, and enhancing precision medicine strategies [25]. The collaboration between genetic counselors, industry professionals, and scientists ensures that the latest research is effectively translated into practical healthcare solutions, ultimately improving patient outcomes and shaping the future of genetic medicine [26].

Historical Evaluation and Relevance in Modern Medicine

The practice of genetic counseling began nearly five decades ago. The term "Genetic Counseling" was first introduced in 1947 by Sheldon Reed, one of America's pioneering geneticists [1]. Initially, genetic counseling was primarily offered by physicians. While medical geneticists continue to render these services, a notable transformation occurred in 1969 when Sarah Lawrence College established the first master's level program aimed at training a new category of allied health professionals referred to as genetic associates (Marks & Richert, 1976), who are now referred to as genetic counselors [27, 28].

The genetic counseling profession, which is both established and successful, originated from significant genetic discoveries. Notable milestones include Gregor Mendel's principles of inheritance in 1865 [29], the elucidation of the DNA double helix structure in 1953, the identification of the genetic code in 1966 [30], the introduction of the Sanger sequencing method in 1977 [31], the advent of the first genome sequencing method in 1980 [32], the polymerase chain reaction (PCR) technology development in 1983 [33], the launching of the Human Genome Project, the introduction of the first genetic therapy in 1990 [34], the advancement of microarray technology in 1995 [35], the release of the initial draft of the mapped human genome in 2001 [36], and the sequencing of next-generation technologies in 2005 [37]. Collectively, these discoveries have firmly established the genetic counseling profession as a vital component of healthcare, assisting individuals in understanding and managing inherited diseases and conditions.

In contemporary medical practice, genetic counseling constitutes an essential component in the realms of early diagnosis, personalized treatment, and preventive healthcare, ensuring that individuals receive informed and compassionate assistance in their genetic decision-making processes [38]. With the progressive advancements in technology, genetic counseling is poised to continually influence the future of modern medicine, especially within the domains of precision medicine and public health [39, 40].

12.2 Genetic Risk Assessment

a. Tools and methodologies for evaluating genetic risk.
b. Pedigree analysis and family history interpretation.

c. Understanding inheritance patterns: monogenic, polygenic, and multifactorial traits.

a. Tools and Methodologies for Evaluating Genetic Risk

Recent advancements in the field of genetics have enabled the accurate assessment of an individual's risk for hereditary diseases [41]. The evaluation of genetic risk encompasses a diverse array of tools and techniques that facilitate the identification of genetic predispositions, the prediction of disease progression, and the guidance of personalized medical care. These methodologies range from conventional assessments of family history to state-of-the-art genomic sequencing technologies and risk models driven by artificial intelligence [42].

By analyzing genetic variations, chromosomal abnormalities, and inherited patterns, healthcare professionals can provide early diagnosis, preventive strategies, and targeted treatments for genetic conditions [43]. Integrating computational tools, statistical models, and biomarker testing has further enhanced the ability to assess risk and improve patient outcomes [44].

The following are some of the tools:

- Family history analysis
- Genetic testing
- Polygenic risk scores
- Cytogenetic techniques
- Pharmacogenomics
- AI-based predictive models.

These tools play a crucial role in modern precision medicine, helping individuals and healthcare providers make informed decisions about health and disease prevention.

1. **Clinical and Family History Assessment includes:**
   a. **Pedigree Analysis**: To identify the inheritance patterns in various mandelion diseases [45].
   b. **BOADICEA**: The Breast and Ovarian Analysis of Disease Incidence and Carrier Estimation Algorithm is helpful for the prediction of carrying pathogens BRCA1, BRCA2, and other cancer-related genes [46].
   c. **Manchester Scoring System**: Used for the prediction of pathogens BRCA1 and BRCA2 in breast cancer [47].

2. **Genetic testing:**
   a. **Single gene testing**: Examines the particular genes that associates with the disease, e.g., BRCA1/BRCA2 for breast cancer [48].
   b. **Panel testing**: Used for multiple genes disorders and when the gene mutation is not known [49].
   c. **Whole Genome sequencing Testing**: Used to detect the monogenic diseases in neonats [50].
3. **Polygenic Risk scores:** Quantifies risk based on multiple small effect genetic variants [51].
4. **AI-driven risk prediction model:** Machine learning (ML) algorithms to analyse large genomic datasets and identify patterns associated with genetic conditions [52].
5. **AI-powered genomic data analysis:** The application of natural language processing enables the extraction of pertinent genetic information from medical literature and electronic health records and assesses the risk in genetics [53].
6. **Cytogenetic Testing:**
   a. **Conventional Cytogenetic (Karyotyping):** Detects the chromosomal abnormalities by measuring the chromosome number and structure (aneuploidy trisomy in Down syndrome, translocation - philadelphia chromosome in CML, Deletion and duplication - Cri - du - chat syndrome) [54].
   b. **Fluorescence *in situ* Hybridization:** In this technique, the probes bind to specific DNA sequences and detect the microdeletion, duplication, and translocation in some diseases like HER2 amplification in breast cancer. BCR-ABL translocation in chronic myeloid leukemia (CML) [55].
7. **Pharmacogenomics and Pharmacogenetics:**
   a. Next-Generation Sequencing (NGS):
      Whole Genome Sequencing (WGS) & Whole Exome Sequencing (WES): These technologies provide comprehensive insights into genetic variations, including those affecting drug metabolism, drug targets, and disease pathways [56].
   b. Targeted Sequencing (TS): Focuses on specific genes for diagnosis, prognosis, and treatment assessment, improving cost efficiency, reliability, and insurance reimbursement in clinical settings.

c. Single-Cell RNA Sequencing (scRNA-Seq):
   - Enables high-resolution transcriptomic analysis at the individual cell level, allowing the examination of millions of cells within a single study.
   - Facilitates precise classification and differentiation of individual cells based on transcriptomic patterns, aiding in the identification of rare but functionally significant cell types.
   - Plays a critical role in early disease detection, clinical assessment, and understanding drug mechanisms [57].
   - Widely used for analysing gene expression, cellular metabolites, and cell-cell interactions, with applications in embryonic development, tissue and organ research, oncology, and immunology.

Stepwise approach for evaluating the genetic risk:

Detailed Patient History

↓

Family history (Pedigree analysis)

↓

Clinical Examination and biomarkers

↓

Identifying the potential risks based on the information gathered.

↓

Detecting genetic mutations and variations associated with genetic risk involves using genetic testing tools, which may vary depending on the disease.

↓

Analyze genetic data utilizing computational tools to evaluate pathogenicity.

↓

Integrate genetic findings with clinical data collected for risk assessment.

b. Pedigree Analysis and Family History interpretation:
Pedigree analysis is a vital method used in genetic counseling to track the inheritance of traits and genetic conditions across multiple generations. It helps in understanding hereditary patterns, predicting disease risks, and guiding genetic testing and clinical decisions, which eventually aids in genetic counseling. A detailed family history collection is essential for accurate and meaningful pedigree analysis. It provides crucial information about the inheritance of genetics, enhancing the prediction, diagnosis, and personalized treatment of patients [58].

Constructing the Family Tree or Pedigrees
The utilization of clear and standardized symbols significantly enhances the representation of genetic information in a more structured and comprehensible manner as opposed to merely enumerating relatives in a textual format [59]. The compilation of a family history may prove to be a labor-intensive endeavor. Nevertheless, devoting time to the collection and analysis of this data not only facilitates the creation of an accurate pedigree diagram but also yields valuable insights into genetic patterns and health risks. Below are the terminologies employed in pedigree analysis [60].

Family History Importance: Family history holds valuable information about an individual's past and future potential health risks. It provides details about the transmission of disease traits from maternal and paternal lines to offspring, helping to understand relevant disease transmission patterns. It offers accurate diagnostic and genetic testing options based on the family's disease history. Finally, it aids in identifying genetic risks and assists in taking necessary steps for disease prevention by counseling clients on lifestyle modifications and enhancing disease surveillance [61].

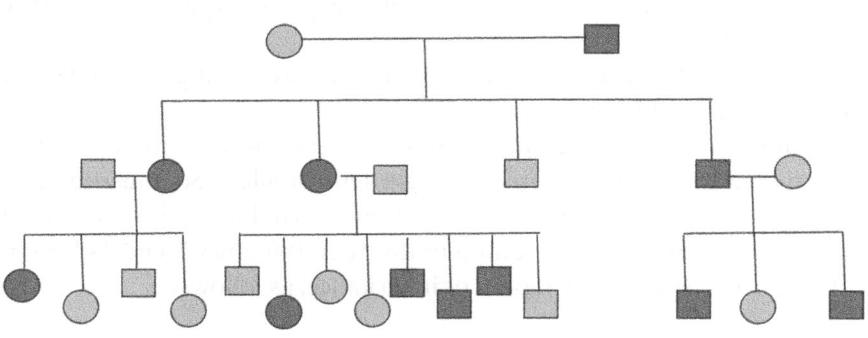

Figure 12.1 Pedigree of a family showing Huntington's disease.

Here is an example of pedigree analysis:

- In this chart, circles represent females, and squares represent males.
- Yellow-coloured individuals are unaffected (genotype: hd hd, meaning they have two recessive alleles).
- Red-coloured individuals are affected (genotype: HD hd, meaning they carry the dominant disease allele).
- The chart is divided into three generations (I, II, III).
- Horizontal lines connecting individuals represent marriages, while vertical lines indicate offspring (Figure 12.1).

Genetic Explanation:

- Huntington's disease follows an autosomal dominant pattern, meaning that an individual with just one copy of the HD allele (HD hd) can develop the disease.
- Non-affected persons have genotype hd hd (homozygous recessive).
- Affected persons have genotype HD hd (heterozygous dominant). Figure 12.2 explaining the pedigree chart symbols for developing the family tree.

c. Understanding Inheritance Patterns: Monogenic, Polygenic, and Multifactorial Traits

Inheritance patterns explain how traits and disorders are transmitted from one generation to subsequent generations. Comprehending these patterns aids in predicting the likelihood of specific traits and genetic conditions. In living organisms, inheritance encompasses the totality of genetic and epigenetic characteristics transmitted from the preceding generation to the progeny [62].

The three main types of inheritance patterns are Monogenic, Polygenic and Multifactorial Inheritance.

Monogenic Inheritance Pattern: In this type of inheritance, the trait is determined by the expression of a single gene or allele. So, mutation to a single gene either in the paternal or maternal germ line [63] results in the expression of the disease condition or disorder. It follows Mendel's laws of inheritance. There are five modes of inheritance, as follows:

1. Autosomal dominant disorders caused by the transfer of an affected gene at the Autosomal Dominant chromosome. Affected parent directly carries the mutated gene to their

offspring. Examples include familial hypercholesterolemia, polycystic kidney disease, Neurofibromatosis type I.

2. Autosomal recessive monogenic disorders are caused by the transfer of a recessive trait of the mutated gene to the offspring. Parents do not get affected but act as carriers. Examples include Sickle cell anemia, cystic fibrosis, Tay–Sachs disease, and Phenylketonuria.
3. X-linked dominant disorders are caused by single gene mutations on the X chromosome. Females are more likely to be affected, but the effects occur in both males and females. Examples include hypophosphatemic rickets (vitamin D-resistant rickets) and ornithine transcarbamylase deficiency.
4. X–linked recessive disorders are caused by a mutation in the X recessive trait. Males are more affected, and affected males often present in each generation. Examples are Hemophilia A and Duchenne muscular dystrophy.
5. Mitochondrial inheritance is caused by the transfer of the mt DNA to the offspring, and both males and females are affected. The traits are passed only through the maternal parent and never by the paternal parent. Disorders like Leber's hereditary optic neuropathy, Kearns-Sayre syndrome.

Y–linked monogenic disorders are caused by a single mutation in the Y chromosome, a defective gene passed from men to their sons. Examples include Y-chromosome infertility and Sweyer syndrome [64].

Polygenic inheritance refers to traits determined by the combined action of many genes, in contrast to monogenic inheritance where a single gene determines the trait. This type of inheritance involves more than one gene, with the dominant alleles exhibiting a cumulative effect, where each dominant allele contributes a part or unit of the trait [65]. The cumulative action of multiple genes results in a significant impact, as every dominant allele expresses a portion of the trait. No individual gene is dominant or recessive over another; their effects are cumulative [66]. Examples of polygenic inheritance in humans include eye color, height, and skin color. Understanding the polygenic inheritance pattern helps to assess polygenic risk scores in certain polygenic diseases such as heart disease, diabetes, and specific cancers.

Multifactorial Inheritance: Genetic traits determined by the inheritance of several genes and environmental predictors. Specific gene implicated in this process is not known. Traits exhibiting multifactorial inheritance

Male Individual	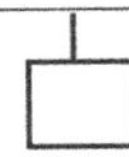
Female Individual	
Unknown Individual	
Affected male individual	
Affected female individual	
Proband male (The person who is affected by a genetic condition or the person who is at risk)	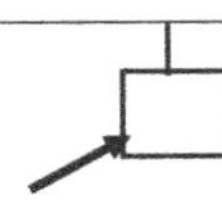
Proband male (The person who is affected by a genetic condition or the person who is at risk)	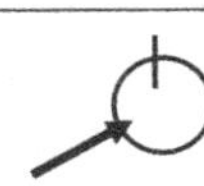
Examined Male	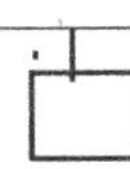
Examined Female	
Diseased Male	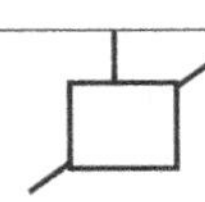
Diseased Female	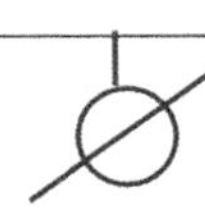

Figure 12.2 Pedigree chart symbols. Pedigree chart [Internet]. Shutterstock. Available from: https://www.shutterstock.com/search/pedigree-chart *(Continued)*

Male fraternal twins	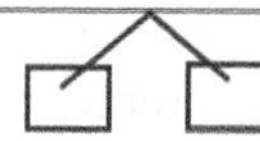
Female fraternal twins	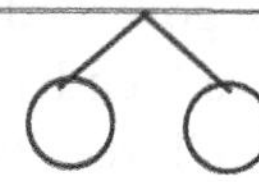
Male identical twins	
Female identical twins	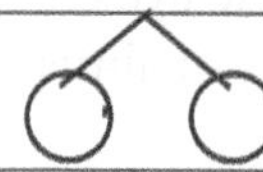
Adopted male	
Adopted male	
Still birth	
Miscarriage	
Pregnancy	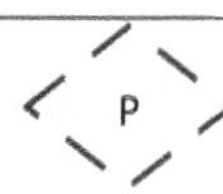
Marriage	
Consanguineous marriage	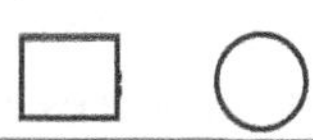
Extra Marital mating	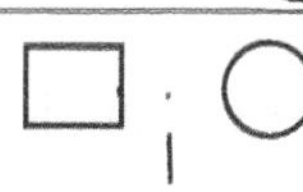
Divorce	
No children	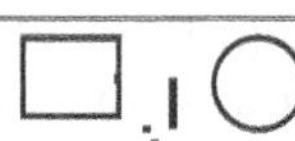

Figure 12.2 (Continued) Pedigree chart symbols. Pedigree chart [Internet]. Shutterstock. Available from: https://www.shutterstock.com/search/pedigree-chart

are typically quite complex, as numerous diverse genetic factors and environmental factors interact to impact their adaptability. Unlike Mendelian traits, which are governed by a single genetic locus, cystic fibrosis is classified as a recessive Mendelian disorder attributed to mutations in the CFTR gene. Although CF results from a single genetic mutation, the severity of the disease varies widely as a result of the impact of the various other environmental factors and genetic factors [67]. Traits that exhibit multifactorial inheritance include eye color, weight, height, and clinical conditions such as type-II diabetes mellitus, bipolar disorder, and sporadic breast cancer.

12.3 The Counseling Process

a. Key stages of genetic counseling (information gathering, risk communication, decision-making). (Figure 12.2) pedigree chart symbols.
b. Role of empathy and psychological support in counseling.
c. Strategies for communicating genetic risk to patients and families.

Key stages of genetic counseling:

1. Enhance the family's comprehension of a hereditary genetic disorder.
2. Evaluate potential strategies for disease management, examining the associated risks and benefits of additional diagnostic testing and alternative interventions.
3. Assist the individual and family in identifying the psychosocial resources necessary to effectively manage potential outcomes.
4. Mitigate the family's anxiety levels.

Stages of genetic counseling:

a. **Pre–counseling:** Prior to the appointment with the client, the genetic counselors can communicate face-to-face, by telephone or in written format [68].
   - Explain or review the purpose of the consultation, describe the consultation process and its objectives, and mention the regular physical examination to address client concerns, if any.

- Identify and provide the necessary client's requirements, such as wheelchair services and attendant services.
- The counselor should also acknowledge any social or cultural issues that may affect the consultation.
- Address any impracticable expectations regarding the consultation.
- Offer emotional assistance to ease pre-clinical anxiety.

b. Preparation for counseling

- Ensure all required information was gathered.
- Cross-verify the necessary health records and clinical-genetics files.
- Consult with your experts, your colleagues, or your supervisor as needed.
- Examine pertinent literature on the concerned topic.
- Disseminate the necessary information, including details about help groups and relevant fact sheets, to be provided to the client [69].

c. Information gathering

- Give genetic file number and relevant hospital unique numbers.
- Gather the demographic details of the client, like name, age, sex, address, and phone number.
- Mention the consultation venue, time, date, and details of the healthcare team involved in the client's health, and the names of the referring health professionals.
- Take the details of each person's present family history, past medical history, and medication history.
- Family history questions encompass the client's health status, the health history of their siblings, details about the family's origins or racial/ethnic background, the age and cause of death for each family member, as well as information on pregnancy outcomes.
- Laboratory tests and reports need to be collected.
- Collect the pedigree for a minimum of three generations, including diagnoses and relevant information that supports genetic inheritance.
- Note every detail and understand the inheritance pattern.
- Collect the information regarding extended family members to know the details of traits [70, 71].

- Based on the gathered information, select the appropriate genetic test to evaluate and present the available options for genetic testing.
- Before choosing the genetic test, ask for the client's consent and clearly counsel them on the risks associated with genetic disorders, including both social and ethical issues.

Diagnosis Verification:
Clinical-geneticists may conduct clinical examinations and necessary tests, as well as review available records and reports to ensure a diagnosis. These records might be considered the report of the tests collected from the patient's family members, even deceased relatives, or examination reports of post-mortem. When collecting information necessitating access to reports of the post-mortem or pathology, records of the registry, or results of the genetic test of concerned relatives, it is imperative to obtain consent from the relevant candidates, next of kin, or a legally responsible person or agency.

Risk Estimation and Communication:
After a family diagnosis is established, individual risk can be determined by analyzing clinical expertise, test results, inheritance patterns, empirical data, and pedigree information. Sharing genetic risk information with the family is not a straightforward process. Communicating genetic risk is often expressed as numerical probability for better understanding [72]. Some key points for risk communication include:

a. Clarity and transparency: information should be clear, accurate, and transparent.
b. Context: Elucidate the context to enhance comprehension of the significance of the risk at hand. Provide a thorough explanation regarding the probability of occurrence, potential consequences, and its interrelation with other associated risks.
c. Confidentiality: Maintain the confidentiality of the client while communicating the risk.
d. Use the information of absolute risk and relative risk: provide both the absolute risk (e.g., the probability of experiencing the cardiovascular event is 10%) and the relative risk (e.g., the risk is twice that of smokers, if the clients are smokers).

e. Contextualize risk: Describe how risk varies with different actions or lifestyle choices. For example, illustrate the absolute risk reduction when an individual quits smoking (e.g., If you quit smoking, your risk of cardiovascular events decreases to 5%).

f. Compare the risk: Compare the individual's risk with the others in different groups or categories (e.g., Your risk is higher than someone of the same age who doesn't smoke).

g. Demonstrate how risk varies over time: Utilize visual aids or programs such as Your Heart Forecast to demonstrate how risk progresses with age and fluctuates based on changes in behaviour or risk factors.

h. Emphasize the influence of behaviour change: Illustrate how modifying behaviour can lower risk. For example, demonstrate how quitting smoking, exercising regularly, or improving diet can mitigate risk.

i. Provide clear and actionable steps: Offer specific guidance on how individuals can reduce their risk. This encompasses lifestyle changes, regular screenings, and preventive measures.

j. Utilize visual aids: Leverage charts, graphs, and other visual tools to assist the individual in understanding their risk and how it can be modified.

k. Support and follow-up: Ensure continuous support and follow-up to help individuals stay motivated and on track with their risk-reducing behaviours.

Incorporating these elements makes risk communication more effective, empowering individuals to understand their risk and take proactive steps to manage it.

Decision-Making

Understanding the Decision Categories:

- Considering Genetic Testing: Deciding to undergo genetic testing involves evaluating the benefits, risks, and potential outcomes of the test.
- Post-Test Decisions: Determining how to utilize the information gained from genetic tests, which may affect treatment options, preventive measures, or reproductive choices.
- Gathering Information: Collect comprehensive medical and family history to understand the genetic risks [73].

- Schedule an appointment with a genetic counselor to explore the purpose, process, and implications of genetic testing.
- Educational Resources: Evaluate the educational materials offered by healthcare professionals to gain a deeper understanding of genetic conditions and testing options.

Weighing Benefits and Risks:

- Benefits of Testing: Consider the potential benefits, such as early detection, informed treatment options, and preventive measures.
- Risks and Limitations: Understand the risks and limitations of genetic testing, including the possibility of uncertain or inconclusive results [74].

Making the Testing Decision:

- Informed Consent: Ensure the patient understands the test and grants informed consent.
- Support System: Include family members or a support network to assist with the decision-making process.
- Emotional Considerations: Consider the emotional impact of testing and potential results.

Post-Test Decision Making:

- Interpreting Results: Collaborate with healthcare professionals to accurately interpret the test results.
- Treatment Options: For patients diagnosed with conditions such as cancer, discuss treatment options based on the results of genetic tests.
- Preventative Measures: For individuals with a family history of conditions like cardiac disease, consider preventative options based on pre-symptomatic testing results.
- Reproductive Choices: For prospective parents, discuss reproductive options based on prenatal testing results, which include potential interventions or alternative reproductive methods.

Role of Empathy and Psychological Support in Genetic Counseling

Empathy and psychological support for clients during genetic counseling are significant elements [75].

The 321 initial codes related to empathy were categorized into five groups: (i) prior context for perceiving empathy, (ii) perceived empathy 1: understanding and consideration, (iii) perceived empathy 2: bedside manner, (iv) affected areas of perceived empathy, and (v) lack of empathy. Categories (ii) and (iii) were recognized as the core categories of perceived empathy.

Understanding and Consideration: Genetic counselors should demonstrate a profound understanding of their clients' thoughts and understandings through their behavior.

Bedside Manner: The integration of elements such as a calm, warm, and friendly demeanor, alongside nonverbal cues, including smiling and fostering a comfortable atmosphere, enables participants to perceive counseling sessions as empathetic. By demonstrating these qualities in their bedside manner, genetic counselors are able to effectively communicate empathy, thereby creating a positive and supportive experience for their clients.

Affected Areas of Perceived Empathy

The empathy shown by genetic counselors can significantly influence clients' decisions and enhance their overall experiences.

Support in Decision-Making: The counselor's insight and encouragement helped the client gain confidence in making this important decision.

Feelings of Support and Trust: Clients should feel that genetic counselors are supportive and trustworthy. The perceived empathy fosters a sense of safety and reliability, encouraging clients to share their concerns and thoughts openly.

Mixed Feelings and Challenges

While empathy is generally beneficial, it can also evoke mixed feelings among clients. According to some studies, the participants have the following concerns:

Positive and Negative Emotions: Participants described both positive and negative feelings following their interactions with empathetic counselors. While many felt supported and valued, some participants experienced feelings of insufficiency or discomfort despite recognizing the counselor's consideration.

Uncomfortable Feelings: Some participants acknowledged that, while the counselor demonstrated empathy, certain aspects of the interaction made them feel uneasy. This highlights the complexity of empathetic communication and the need for counselors to continuously refine their approach to address clients' diverse emotional needs.

Empathy in genetic counseling is a double-edged sword. On one hand, it offers crucial support and builds trust, aiding clients in making informed decisions. On the other hand, it can evoke mixed emotions, highlighting the need for sensitivity and ongoing improvement in empathetic interactions.

Strategies for Communicating Genetic Risk to Patients and Families

Strategies for communicating genetic risk should address challenges such as tailoring genetic counseling, streamlining the process, determining the specific information to reveal, acknowledging that the communication has limited impact, and understanding the effect of context. Developing strategies to tackle these challenges has significant potential to empower people to adopt healthy lifestyles and enhance their health qualitatively.

Customizing communication based on patient profile:
Present risk information of various formats without using qualitative modifiers

Know the format for presenting risk. Some formats are: 1. Percentage 2. Proportions 3. Population comparisons 4. Even gambling odds. More effective communication involves presenting risk figures in multiple numerical formats. You can also communicate by presenting absolute risk and relative risk. Qualitative descriptions, such as high or low risk, may lead to misinterpretation and misunderstanding [76].

Beware of framing biases and present risk information using multiple perspectives

The way the risk figure is framed plays a crucial role in perception. Risks can be presented as gains or losses (positive and negative framing), and these variations can significantly influence the risk interpretation. Based on a study done by Swiss healthcare professionals and patients, the views on how a potential new drug is treated differ when the risk is presented in different ways. The formats presented are survival and mortality proportions, and the relative mortality reduction, or a combination of all three [77].

The views on how a potential new drug is treated differ.
Choose pictures carefully:

Pictographs with 100 boxes are a better way of communicating percentages. And, this method is effective in conveying both specific information and the essence of the message. Drawings represent the percentage in visual units, effectively illustrating the part-to-whole ratio. It also helps to minimise the impact of subjective reports on patients' interpretation of risk [78].

Numerators and Denominators:
Can you envision a room filled with 1,000 people? If the number increases to five digits, it becomes difficult for many individuals to visualize such large quantities [79].

Research on healthcare risk communication has demonstrated that people, especially those with limited numerical skills, encounter difficulties when interpreting large denominators. Therefore, employing smaller denominators facilitates the comprehension process. Furthermore, the representation of risks through ratios is frequently considered.

Consider emotional responses rather than purely factual risk information:
The patient's cultural beliefs, past experiences, acquaintances, and previously perceived risks should be considered over the numerical risk interpretation, as they play an important role in comprehending and applying the information in decision-making.

Online risk assessment tools:
Engaging and interactive sessions with patients enhance their comprehension of disease risks and facilitate behavioral modifications aimed at reducing these risks. Online assessment tools provide optimal risk communication by rapidly processing information regarding patients' lifestyles, family histories, and genetic backgrounds while evaluating genetic risk through the application of risk prediction models [80]. Additionally, these tools empower individuals to input their personal risk factors, participate in the assessment process, and observe the results, thereby fostering an understanding of how specific behavioral changes can impact risk and potential outcomes.

12.4 Applications of Genetic Counseling

a. Prenatal and reproductive counseling.
b. Cancer genetics and familial risk counseling.
c. Pediatric and adult-onset genetic disorders.

a. Prenatal and Reproductive Counseling
Prenatal and reproductive genetic counseling offers essential information for prospective parents concerning the genetic risks that may impact their offspring [81, 88]. Given the significant prevalence of consanguineous

marriages in specific communities, genetic counseling is instrumental in identifying risks associated with inherited disorders such as thalassemia, sickle cell anemia, and congenital metabolic disorders [82, 88].

The primary objective is [81, 88]:

1. To assess genetic conditions
2. Inform parents about the likelihood of transmission
3. Discuss possible reproductive options

One of the most prevalent components of prenatal genetic counseling entails screening and diagnostic assessments. Tests such as non-invasive prenatal testing (NIPT) and biochemical screening (including double, triple, and quadruple marker tests) assist in identifying chromosomal abnormalities, including Down syndrome (trisomy 21), Edwards syndrome (trisomy 18), and Patau syndrome (trisomy 13) [82–84]. Moreover, more invasive techniques, such as chorionic villus sampling (CVS) and amniocentesis, offer definitive genetic diagnoses; however, their utilization is less frequent due to considerations related to accessibility and cost.

Counselors conduct comprehensive assessments of family and medical histories to identify inherited conditions, including cystic fibrosis, sickle cell anemia, and Tay-Sachs disease [85, 90, 91]. Carrier screening tests are employed to ascertain whether one or both parents possess mutations that may lead to genetic disorders in their offspring. For couples at an elevated risk of transmitting a genetic disorder, assisted reproductive technologies (ART), such as preimplantation genetic diagnosis (PGD), offer the opportunity to select embryos devoid of specific genetic conditions prior to implantation during *in vitro* fertilization (IVF) [85, 94, 95].

In addition to genetic risk assessment, prenatal counseling encompasses psychological support. Anticipating potential genetic risks can induce anxiety in parents, and genetic counselors fulfill a vital role in elucidating test results in a comprehensible manner [92, 95]. Furthermore, they offer ethical guidance, ensuring that couples are empowered to make decisions consistent with their personal values and cultural beliefs [86].

Studies indicate that raising awareness and improving access to genetic counseling can significantly enhance informed reproductive choices, ultimately resulting in better maternal and child health outcomes [86, 92].

b. Cancer Genetics and Familial Risk Counseling

Genetic counseling assumes a critical function in the assessment of cancer risk, particularly for individuals possessing a familial history of hereditary cancer syndromes. Numerous cancers, such as breast, ovarian, colorectal, and prostate cancer, exhibit a genetic component. Comprehending one's genetic predisposition facilitates the implementation of targeted surveillance and strategies for risk mitigation [91, 92].

Individuals possessing a substantial family history of cancer are assessed for mutations in high-risk genes, including BRCA1, BRCA2, TP53, and MLH1. Hereditary breast and ovarian cancer syndromes present considerable concern, particularly within urban populations where lifestyle and reproductive factors exacerbate risks. While genetic testing is presently available at premier hospitals and diagnostic centers nationwide, accessibility in rural areas continues to pose significant challenges [91].

A fundamental element of cancer genetic counseling is risk stratification. Counselors undertake comprehensive family histories, develop pedigrees, and evaluate the necessity for genetic testing. Individuals identified as carriers of hereditary mutations may implement preventive measures, which include early and more frequent cancer screenings such as mammograms and colonoscopies [91, 92]. Preventive surgeries, such as mastectomy for BRCA mutation carriers, are less frequently selected due to cultural and social factors; however, awareness is increasing [91].

Genetic counseling additionally addresses the psychological implications of testing, thereby ensuring that patients receive adequate emotional support and ethical guidance. In situations where cancer diagnoses frequently carry a stigma, genetic counselors assist individuals in navigating their risk and effectively communicating genetic information to the members of the client family who may also be vulnerable [89, 95].

For individuals possessing a confirmed genetic predisposition, preventive strategies may encompass enhanced surveillance measures (e.g., early and more frequent mammograms and colonoscopies), chemoprevention, or even prophylactic surgical interventions (e.g., mastectomy or oophorectomy for carriers of the BRCA mutation). Furthermore, genetic counseling offers valuable guidance regarding the implications of genetic findings on family members, who may also derive benefits from testing and early intervention [91, 92, 95].

It is crucial to note that genetic counseling also addresses the psychological and ethical considerations associated with testing. Obtaining a positive genetic test result may induce anxiety and emotional distress. Counselors assist patients in navigating these emotions, ensuring that they are equipped with the necessary knowledge on qualitative health and better lifestyle modifications to enable informed decision-making [86, 95].

With the advancements in precision medicine and the increased affordability of genetic tests, genetic counseling is anticipated to become a fundamental aspect of oncology care.

c. Paediatric and Adult-Onset Genetic Disorders

In the field of pediatrics, genetic counseling plays a pivotal role in the managing conditions such as cystic fibrosis, Duchenne muscular dystrophy, and Fragile X syndrome. Counselors assist parents in comprehending the available options for genetic testing, the associated prognosis, and potential therapeutic interventions. Newborn screening programs frequently identify metabolic and genetic disorders at an early stage, facilitating timely medical interventions. For example, the early diagnosis of phenylketonuria (PKU) permits dietary management that effectively prevents intellectual disabilities [84, 86, 92].

In cases of adult-onset disorders, genetic counseling plays a crucial role in various conditions, including Huntington's disease, Alzheimer's disease, and hereditary cardiovascular disorders. Predictive genetic testing empowers at-risk individuals to evaluate their likelihood of developing a disorder later in life. Another growing area of adult-onset genetic counseling is pharmacogenomics, wherein genetic variations influence individual medication responses. Although a cure may be unavailable for certain conditions, early awareness permits individuals to strategise for their future, engage in clinical trials, and consider lifestyle modifications that might postpone the onset of the disease [86, 87, 92].

In the field of pediatric medicine, genetic counseling plays a critical role in addressing conditions such as thalassemia, Duchenne muscular dystrophy, and Fragile X syndrome. The gradual expansion of newborn screening programs facilitates the early detection of metabolic disorders, including phenylketonuria (PKU) and congenital hypothyroidism, thus enabling timely intervention. Parents are provided with genetic counseling to comprehend the implications of test results and to investigate potential treatment options [92, 94].

In the context of adult-onset disorders, genetic counseling assumes a pivotal role in conditions such as Huntington's disease, hereditary cardiovascular disorders, and neurodegenerative diseases, including Alzheimer's disease [86, 92]. Predictive genetic testing allows individuals at risk to evaluate their probability of developing a disorder in the future. Although awareness regarding predictive testing continues to develop, genetic counselors are positioned to assist individuals in making well-informed decisions pertaining to their health, lifestyle changes, and participation in clinical trials [92, 94].

Pharmacogenomics, an evolving field, is another key aspect of genetic counseling. It helps determine how an individual metabolizes drugs, ensuring that medications are prescribed in a more effective and personalized manner. For instance, genetic testing for CYP2C19 variants can guide the use of antiplatelet therapy in patients with cardiovascular disease, potentially improving treatment outcomes [93, 94].

Genetic counseling for pediatric and adult-onset disorders encompasses various ethical considerations, particularly regarding the testing of minors and managing the psychological impacts. The scope of genetic counseling services is progressively broadening, with institutions such as the Centre for DNA Fingerprinting and Diagnostics (CDFD) and the All India Institute of Medical Sciences (AIIMS) at the forefront of research and awareness initiatives [85, 92].

As genomic medicine continues to evolve, integrating genetic counseling into mainstream healthcare becomes essential for enhancing disease management, facilitating early detection, and developing personalized treatment strategies [92].

Genomic testing, a laboratory technique that examines genetic alterations throughout the entire DNA, encompassing the complete set of genes, within an individual or a specific cell type or tissue [96]. Genomic testing can help identify individuals at risk for developing cancer, cardiovascular diseases, and neurodegenerative conditions [97–99]. By analyzing gene expression, these tests can reveal patients who might not require aggressive treatment, even if they are considered high-risk based on clinical factors. Offering deeper molecular insights, genomic testing aids in making informed treatment decisions, such as whether a patient should undergo active surveillance or receive intensified therapy, ultimately leading to more personalized and effective care strategies [100, 101].

In clinical practice patients exhibit varied responses to drugs, with dosages that are therapeutic for some potentially being ineffective or results in adverse drug reaction (ADR) in others. Pharmacogenomics plays a crucial role in understanding this variability by examining the influence of genetic diversity on drug response. It facilitates a tailored treatment strategy by correlating gene expression patterns and single nucleotide polymorphism (SNP) with pharmacological efficacy and toxicity, enabling a more personalized approach to treatment, optimizing therapeutic outcomes and minimizing adverse effects [102]. Approaches such as PRS-PGx-Bayes demonstrate superior predictive performance over traditional disease Polygenic risk score (PRS) in cardiovascular trial for LDL reduction [103]. The integration of pharmacogenomics can also lead to a reduction in the prescription of drugs with predicted drug interactions compared to standard care [104, 105]. Pharmacogenomics-guided treatment approach can lead to superior therapeutic outcomes when compared to consensus-based treatment [106].

Genomic Sequencing

Genomic sequencing is a molecular method that determines an organism's whole nucleotide sequence, revealing genetic differences that influence disease development [107]. The complete genomic sequencing can be mapped out to help researchers better understand medication responses, evolutionary trends and hereditary disorders. Whole Genome sequencing (WGS) is a comprehensive technique that illustrates the complete DNA sequence of an individual, encompassing both coding (exons) and non-coding (introns, regulatory areas) segments [108]. Enabling rigorous genomic analysis and accurate variant discovery, whole genome sequencing (WGS) is transforming the diagnosis of hereditary illnesses, rare diseases and genetic disorders [109–111]. Additionally, WGS can help in identifying unique or uncommon pharmacogenomics (PGx) indicators that might go unreported with traditional genetic screening techniques [112]. By identifying candidate modifier genes and the pathophysiological pathways they are linked to, the WPS approach can make it easier to understand disease causes and possible treatment targets [113]. By allowing researchers to find mutations unique to cancer cells, WGS has transformed the study of cancer and aided in the development of specific treatments and understandings of tumor evolution [114].

In infectious disease surveillance, WGS provides comprehensive genomic information about organisms, enabling accurate tracking of disease outbreaks and antibiotic resistance [115]. By identifying transmission networks, new resistance patterns and implementing focused public health interventions, this approach can improve disease management and prevention strategies [116].

Integration of AI in risk prediction:

Artificial Intelligence (AI) is transforming genetic counseling by enhancing risk assessment, refining diagnostic precision and individualizing patient management.

1. AI-driven risk prediction models:

AI uses machine learning (ML) algorithms to analyze large genomic datasets and identify patterns associated with genetic conditions. Poirion *et al.* created a hybrid model that combines deep learning techniques to predict patient survival risk in chronic diseases such as cancer [117].

The utilization of machine learning algorithms and bioinformatics approaches can enhance the discovery of biomarkers and genes for different disease states [118, 119]. In the field of microbiology, Antimicrobial resistance plays an important role in the therapeutic outcome. The machine learning-driven approach for predicting anti-microbial resistance (AMR) utilizing whole genome sequencing data has been developed [120–122]. These models demonstrated high predictive performance, identified known resistance-related genes and variants without prior domain knowledge and uncovered a novel potential drug-resistance-associated gene, highlighting its potential for enhancing treatment strategies. Similarly, AI-assisted rapid Whole Genome Sequencing (rWGS) is a diagnostic tool useful for patients who have complex medical histories presenting unknown multi-systemic hereditary conditions [123].

Polygenic risk scores:
AI calculates an individual's genetic risk for diseases like diabetes, cancer and cardiovascular conditions by evaluating multiple genetic markers. Integrating rare variants into an organized genetic risk model leads to greater probability across diverse population and enhances the accuracy of genetic-based risk predictions [124]. In recent years, there have been efforts to substitute linear regression in PRS calculations with more advanced machine learning techniques, which offer the potential for enhanced accuracy owing to their less restrictive modeling assumptions [125, 126]. Deep learning methods have greatly enhanced the accuracy of polygenic risk scores by effectively capturing complicated non-linear relationships among genetic variants [127]. For instance, deep neural networks, part of the extensive domain of deep learning, have been evaluated in predictive risk scores for breast cancer, resulting in enhanced performance relative to alternative statistical and machine learning estimation techniques [128].

Deep learning for variant interpretation:
Deep learning has become a significant tool in genomic variant interpretation, tackling the complexity involved in predicting the impacts of genetic variants. Germline variations have been better classified and prioritized using machine learning techniques, especially penalized logistic regression. These models produce probabilistic scores that improve the resolution of variants of uncertain significance (VUS) more effectively than conventional approach [129]. Using a protein language framework, the deep learning model ESM1b analyzes around 450 million missense variants to improve coding variant prediction [130]. It performs better than current methods for classifying pathogenicity and takes into account

isoform-specific effects, which are essential for interpreting variants. A semi-supervised model for non-coding variants that combines MPRA data with epigenetic features such as chromatin accessibility, methylation and histone modifications can effectively predict the regulatory effects of genetic variants. A semi-supervised model for non-coding variants that combines Massively parallel reporter assay (MPRA) data with epigenetic features such as chromatin accessibility, methylation, and histone modifications can effectively predict the regulatory effects of genetic variants [131]. Additionally, models such as DeepFIGV preducts variant functionality by analyzing epigenetic signals, generating impact scores to refine disease risk and gene expression information [132].

AI-powered genomic data analysis:
Artificial intelligence has the capability to analyze entire genome and exome sequencing data more rapidly and with enhanced accuracy compared to conventional techniques [133]. The application of Natural Language Processing (NLP) is on the rise, enabling the extraction of pertinent genetic information from medical literature and electronic health records (EHRs), thereby enhancing comprehensive risk assessments in genomics [134]. By automating the analysis of enormous volumes of unstructured data, this method improves the diagnostic process overall and makes it possible to identify disease-associated variations more rapidly. A framework utilizing NLP automates the extraction of Human Phemmotype Ontology (HPO) terms with an impressive 92% accuracy, thereby improving gene prioritization for genetic diagnosis [135].

AI models are also being developed to integrate genetic, environmental and lifestyle factors for predicting the onset and progression of genetic disorders [136]. These predictive models utilize extensive datasets and advanced algorithms to uncover patterns that could signal the onset of a disorder, presenting considerable potential for timely intervention.

Personalized treatment and genetic counseling:
Genetic counseling denotes the advice concerning genetic abnormalities offered by a skilled healthcare professional known as a genetic counselor to an individual or family. AI and ML exhibit considerable promise in genomics research including applications such as facial analysis for the diagnosis of genetic disorders, machine learning-based variant categorization and algorithms for risk assessment. The incorporation of chatbot technology in genetic counseling could improve accessibility and serve as a feasible substitute for traditional pre-test counseling approaches. Genetic counselors

can focus on a patient's past experience values and beliefs by using chatbots to communicate genetic information. This approach lets counselors provide patients more personalized advice and help them incorporate genetic knowledge into their lives [136]. Patient satisfaction and understanding in pre-test counseling among cancer patients are equivalent between an automated chatbot-based method and conventional in-person counseling [137]. Large Language Model (LLM) like ChatGPT's ability to support medical professionals and experts will improve with targeted training on rare disorders' clinical presentation, disease development, therapies and family risk assessment [138]. ChatGPT's widespread usage and personalized refining could help identify uncommon condition patients, lowering undetected cases, which may help doctors to examine and make decisions about patients. A considerable proportion of genetic counselors are employing ChatGPT for diverse purposes, primarily for writing-related tasks. Although it has ability to optimize workflow challenges in clinical genetics, it is crucial that counselors have sufficient education and standardized training regarding its limitations to ensure effective and informed utilization [139]. For example, a machine learning (ML) approach using real-world data showed good discrimination between Gaucher's disease (GD) patients and healthy controls, enabling GD identification far earlier than clinical diagnosis [140].

Emerging Tools in personalized pharmacogenomics and pharmacogenetics:

Drug therapy is being revolutionized by personalized pharmacogenomics and pharmacogenetics, which customize therapies according to each patient's unique genetic profile. Numerous innovative tools and technology are transforming this profession, enhancing treatment efficacy, minimizing adverse drug reactions and enhancing overall patient outcomes.

Next-generation sequencing:

Whole Genome Sequencing (WGS) and Whole Exome Sequencing: These technologies provide comprehensive data on genetic variations, including those that influence drug metabolism, drug targets and disease pathways.

Targeted Sequencing:

Targeted sequencing (TS) concentrates on particular genes for diagnosis, prognosis and treatment assessment, hence decreasing expenses while increasing reliability and insurance reimbursement in clinical settings [141].

Single-cell RNA-sequencing:

Single-cell RNA sequencing enabled the detailed analysis of transcriptomes at the individual cell level, allowing for the investigation of millions of cells in a single study. This method facilitates accurate categorization, characterization and differentiation of individual cells according to their transcriptome patterns, thus allowing the identification of rare but significantly functional cell types [142]. Single-cell RNA sequencing (scRNA-Seq) offers significant benefits in understanding the specific cell subset expression at molecular level, identifying diseases at an early stage, assessing clinical conditions, and explaining medication action mechanisms [143]. scRNA-Seq is a crucial tool for analyzing cellular metabolites and gene expression, primarily used to investigate cell-cell interactions and gene regulation. It is widely used in different research fields, including embryonic, tissue and organ development, oncology and immunology [143].

Pharmacogenomics databases:

Pharm GKB:

The pharmacogenomics knowledgebase (PharmGKB: http://www.pharmgkb.org) is a premier global resource for pharmacogenomics information, consistently adapting to progress in genomic medicine. It integrates gene-drug data *via* variant annotations from source literature, enabling streamlined aggregation and analysis. The clinical annotations assist in identifying possibly actionable gene-drug interactions [144].

Clinical Pharmacogenetics Implementation Consortium (CPIC):

The CPIC offers clinical practice recommendations for pharmacogenetic testing and aiding healthcare professionals in making evidence-based individualized treatment decisions informed by genetic data [145].

Drug-gene interaction databases:

These databases document drug-gene interactions, facilitating the prediction of how an individual's genetic profile may alter the drug response. An online resource such as Drug-Gene Interaction Database (DGIdb; [www.dgidb.org] (http://www.dgidb.org)) consolidates about drug-gene interaction and gene druggability [146].

Artificial intelligence and machine learning:

AI techniques evaluate extensive genetic, clinical and demographic datasets to forecast drug responses, adverse reactions and disease progression. A net-based-regression model ranks drugs for specific cancer cell lines using gene expression and drug sensitivity data [147]. This approach enhances scalability and robustness, ensuring consistent performance across drugs while reducing overfitting. Prediction of doses on patient characteristics is made possible by models that incorporate multiple regression and machine learning approaches [148]. Future developments in AI and new laboratory techniques that improve the genotype-phenotype association could result in personalized medicine that can predict adverse drug reactions (ADR) and as a result select the right medication and dosage for each patient [149].

Biomarkers:
Genetic biomarkers have attracted a lot of attention recently due to their potential clinical applications in the field of medication therapy. It has been demonstrated that these biomarkers can partially replace the procedure of monitoring the therapeutic medication level. The combination of genetic biomarkers and drug levels has the potential to further advise optimum dosing among individual patients [150].

Challenges and opportunities:
Gaps in accessibility and workforce training in genetic conseling:
There is a global scarcity of genetic counselors who are trained with a particular emphasis on rural areas and developing countries. The majority of genetic counseling services are predominantly located in urban and wealthy communities, resulting in other people being insufficiently served [151]. The shortage of direct patient care (DPC) genetic counselors can lead to prolonged patient wait times and unequal distribution of services [152]. Genetic counselors employed by the government assume different position titles (nurses or research associates) indicating the improper acknowledgement of genetic counselors [153]. Limited training programs for genetic counselors are one of the biggest challenges for genetic counseling [151]. Training program establishment and maintenance requires large financial resources. Many nations have inadequate financial support for these initiatives, therefore preventing their growth [154].

Addressing Disparities in Genetic Services:
Expanding workforce development and training:
Expand access to genetic counseling knowledge by creating more training programs and making more courses available online. To enhance healthcare provider's capacity to understand genetic data, incorporate genomics education into healthcare professional courses.

Enhancing affordability and insurance coverage:
Raise awareness about the need for more accessible genetic testing by advocating for reimbursement regulations. Developing low-cost-genetic testing models and public funding programs can help marginalized or rural populations.

Development of community-based genetic education programs and culturally competent genetic counseling:
Genetic education programs should be designed to be community-based and tailored to the specific needs of diverse ethnic and linguistic populations. These programs must incorporate culturally competent manner, acknowledging and respecting varying beliefs, traditions and toward genetic testing. These approaches can enhance informed decision-making, promote more access to genetic services and improve public engagement with genetic health information.

Future perspectives in genetic counseling and precision medicine:
Integration of AI in genetic counseling: Risk prediction, variation interpretation and personalized therapy recommendations will all be improved by improved by AI-driven technologies.

Global expansion of genetic services: Partnership and funding initiatives can be implemented to enhance genetic services across low-income and middle-income countries. Collaborative international research in precision healthcare and genetic medicine can results in innovation in the genetics.

Conclusion

Genetic counseling constitutes an emerging discipline within the field of genomics explicitly tailored to individuals and families who are at risk of genetic conditions. This practice facilitates a comprehensive understanding and management of the implications associated with genetic risk factors through educational resources, risk assessment, and the provision of emotional support. It necessitates a profound understanding of inheritance patterns, including monogenic traits, polygenic traits, and multifactorial traits, to convert genomic data into actionable insights. Furthermore, confidentiality and ethical considerations are foundational principles of the discipline.

Technological advancements, including next-generation sequencing and artificial intelligence, are significantly enhancing the efficacy of genetic counseling. These tools facilitate improved risk assessment and provide personalized approaches; however, they also present challenges such as complexities in data analysis, disparities in accessibility, and the psychological implications of inconclusive results. Addressing these challenges necessitates a multidisciplinary and patient-centered approach.

The increasing significance of genetic counseling is evident across multiple domains, including oncology, reproductive medicine, and rare diseases, thereby underscoring its impact on public health and personalized medicine. With advancements in genomic science, the responsibilities of genetic counselors are anticipated to expand further, incorporating genomics-based information into medical practices to foster informed decision-making and promote favorable health outcomes. In order to conduct this genetic counseling smoothly, the genetic counselors play an important role in guiding patients through genetic testing, explaining the results and providing emotional support. They educate healthcare providers, contribute to research, and advocate for policies that improve access to genetic services. The lack of qualified genetic counselors, high costs, and restricted access to genetic services prevent widespread adoption, especially in communities with limited resources. To address these disparities, it is essential to expand educational programs, integrate digital health solutions such as tele-genetics, and advocate for improved insurance coverage and funding support. In collaboration with industry experts, scientists turn the research into real healthcare solutions, enhance patient care, improve access to genetic services, and drive the future of genetic medicine.

In conclusion, genetic counseling represents an indispensable component of contemporary healthcare. By navigating individuals through the complexities associated with genetic risk and inheritance, genetic counselors assume a pivotal role in enhancing patient care and advancing precision medicine. Out of the clinical practice, the future trajectory of genetic counseling will undoubtedly be shaped by ongoing technological advancements, aimed at improving patient outcomes while preserving the essential human connection.

AI disclosure statement: The author confirms that no artificial intelligence (AI)-generated content or AI-assisted writing tools were used in the preparation of this chapter (including summaries and the abstract). All text, analysis, and conclusions are the result of the author's original work.

References

1. Anderson, V.E., Sheldon C. Reed, Ph.D. (November 7, 1910-February 1, 2003): genetic counseling, behavioral genetics. *Am. J. Hum. Genet.*, 73, 1, 1–4, 2003 Jul, doi: 10.1086/376611. PMID: 12875265; PMCID: PMC1180572.
2. National Society of Genetic Counselors' Definition Task Force, Resta, R., Biesecker, B.B., Bennett, R.L., Blum, S., Hahn, S.E., Strecker, M.N., Williams, J.L., A new definition of Genetic Counseling: National Society of Genetic Counselors' Task Force report. *J. Genet. Couns.*, 15, 2, 77–83, 2006 Apr, doi: 10.1007/s10897-005-9014-3. PMID: 16761103.
3. Genetic Alliance and The New York-Mid-Atlantic Consortium for Genetic and Newborn Screening Services, Chapter 5, Genetic Counseling, in: *Understanding Genetics: A New York, Mid-Atlantic Guide for Patients and Health Professionals*, Genetic Alliance, Washington (DC), 2009 Jul 8.
4. Hakim Zada, F., Ahmad Azahari, A.H.S., Wong, S.W., Ali, A., Ismail, N.A.S., Understanding Challenges of Genetic Testing on Neuromuscular Disorders from the Parental Lens. *J. Pers. Med.*, 13, 12, 1652, 2023 Nov 27, doi: 10.3390/jpm13121652. PMID: 38138879; PMCID: PMC10744978.
5. Hooker, G.W., Babu, D., Myers, M.F., Zierhut, H., McAllister, M., Standards for the Reporting of Genetic Counseling Interventions in Research and Other Studies (GCIRS): an NSGC Task Force Report. *J. Genet. Couns.*, 26, 3, 355–360, 2017 Jun, doi: 10.1007/s10897-017-0076-9. Epub 2017 Feb 24. PMID: 28236280; PMCID: PMC5415582.
6. Lilienthal, D. and Cahr, M., Genetic Counseling and Assisted Reproductive Technologies. *Cold Spring Harb. Perspect. Med.*, 10, 11, a036566, 2020 Nov 2, doi: 10.1101/cshperspect.a036566. PMID: 31570374; PMCID: PMC7605230.
7. Abacan M, Alsubaie L, Barlow-Stewart K, Beppy Caanen, Christophe Cordier, Eliza Courtney, Emeline Davoine, Janice Edwards *et al.*, The Global State of the Genetic Counseling Profession. *Eur. J. Hum. Genet.*, 27, 2, 183–197, 2019. doi:10.1038/s41431-018-0252-x.
8. Ormond, K.E., Abad, P.J., MacLeod, R., Nishigaki, M., Wessels, T.M., The global status of genetic counselors in 2023: What has changed in the past 5 years? *Genet. Med. Open.*, 2, Suppl 2, 101887, 2024, doi: 10.1016/j.gimo.2024.101887.
9. Ormond, K.E., Laurino, M.Y., Barlow-Stewart, K., Wessels, T.M., Macaulay, S., Austin, J., Middleton, A., Genetic counseling globally: Where are we now? *Am. J. Med. Genet. C Semin. Med. Genet.*, 178, 1, 98–107, 2018 Mar, doi: 10.1002/ajmg.c.31607. Epub 2018 Mar 25. PMID: 29575600; PMCID: PMC5947883.
10. Roter, D., Ellington, L., Erby, L.H., Larson, S., Dudley, W., The genetic counseling video project (GCVP): models of practice. *Am. J. Med. Genet. C Semin. Med. Genet.*, 142, 4, 209–20, 2006.
11. Wheelahan, L., From old to new: The Australian qualifications framework. *J. Educ. Work*, 24, 323–42, 2011, doi: 10.1080/13639080.2011.584689.

12. Laurino, M.Y., Sternen, D.L., Thompson, J.K., Leppin, K.A., Identifying opportunities for collaboration and growth of genetic counseling services in the Asia Region. *J. Community Genet.*, 8, 243–248, 2017, doi: 10.1007/s12687-017-0307-7.
13. Al-Yousuf, M., Akerele, T.M., Al-Mazrou, Y.Y., Organization of the Saudi health system. *East. Mediterr. Health J.*, 8, 4–5, 645–53, 2002. Available from: https://www.ncbi.nlm.nih.gov/pubmed/15603048.
14. Maio, M., Carrion, P., Yaremco, E., Austin, J.C., Awareness of genetic counseling and perceptions of its purpose: a survey of the Canadian public. *J. Genet. Couns.*, 22, 6, 762–70, 2013 Dec.
15. Institute of Medicine (US) Committee on Assessing Genetic Risks, Andrews, L.B., Fullarton, J.E., Holtzman, N.A., *et al.* (Eds.), 2, Genetic Testing and Assessment, in: *Assessing Genetic Risks: Implications for Health and Social Policy*, National Academies Press (US), Washington (DC), 1994, Available from: https://www.ncbi.nlm.nih.gov/books/NBK236037/.
16. Blystra, Y., Lim, W.K., Kam, S., *et al.*, Family history assessment significantly enhances delivery of precision medicine in the genomics era. *Genome Med.*, 13, 3, 2021, https://doi.org/10.1186/s13073-020-00819-1.
17. Giardino, A.P. and De Jesus, O., Case Management, in: *StatPearls* [Internet], StatPearls Publishing, Treasure Island (FL), 2025 Jan-, [Updated 2023 Aug 14]. Available from: https://www.ncbi.nlm.nih.gov/books/NBK562214/.
18. MacFarlane, I.M. and Zierhut, H., Promoting the integration of genetic counseling education and research across the spectrum of learners at a large academic institution. *J. Genet. Couns.*, 33, 1, 250–4, 2024, doi: 10.1002/jgc4.1810. Epub 2023 Oct 21. PMID: 37864570; PMCID: PMC10923058.
19. Bamshad, M.J., Magoulas, P.L., Dent, K.M., Genetic counselors on the frontline of precision health. *Am. J. Med. Genet. C Semin. Med. Genet.*, 178, 1, 5–9, 2018 Mar, doi: 10.1002/ajmg.c.31610. Epub 2018 Mar 26. PMID: 29582554; PMCID: PMC5910203.
20. Biesecker, B.B., Genetic counselors as social and behavioral scientists in the era of precision medicine. *Am. J. Med. Genet. C Semin. Med. Genet.*, 178, 1, 10–14, 2018 Mar, doi: 10.1002/ajmg.c.31609. PMID: 29675992.
21. Schlesinger, M., Grob, R., Shaller, D., Using Patient-Reported Information to Improve Clinical Practice. *Health Serv. Res.*, 50 Suppl 2, Suppl 2, 2116–54, 2015 Dec, doi: 10.1111/1475-6773.12420. Epub 2015 Nov 17. PMID: 26573890; PMCID: PMC5115180.
22. Weil, J., Genetic counselling in the era of genomic medicine. *EMBO Rep.*, 3, 7, 590–3, 2002 Jul, doi: 10.1093/embo-reports/kvf144. PMID: 12101085; PMCID: PMC1084197.
23. Abul-Husn, N.S. and Kenny, E.E., Personalized Medicine and the Power of Electronic Health Records. *Cell*, 177, 1, 58–69, 2019 Mar 21, doi: 10.1016/j.cell.2019.02.039. PMID: 30901549; PMCID: PMC6921466.
24. Thomas, J., Keels, J., Calzone, K.A., Badzek, L., Dewell, S., Patch, C., Tonkin, E.T., Dwyer, A.A., Current state of genomics in nursing: a scoping review

of healthcare provider oriented (clinical and educational) outcomes (2012-2022). *Genes (Basel)*, 14, 11, 2013, 2023, doi: 10.3390/genes14112013.

25. Hassan, M., Awan, F.M., Naz, A., deAndrés-Galiana, E.J., Alvarez, O., Cernea, A., *et al.*, Innovations in genomics and big data analytics for personalized medicine and health care: a review. *Int. J. Mol. Sci.*, 23, 9, 4645, 2022, doi: 10.3390/ijms23094645.
26. Zierhut, H. and Austin, J., How inclusion of genetic counselors on the research team can benefit translational science. *Sci. Transl. Med.*, 3, 74, 74cm7, 2011, doi: 10.1126/scitranslmed.3001898.
27. Marks, J.H. and Richter, M.L., The genetic associate: a new health professional. *Am. J. Public Health*, 66, 4, 388–90, 1976, doi: 10.2105/ajph.66.4.388.
28. Mertens, T.R., Hendrix, J.R., Kenkel, K.M., The genetic associate: a career option in genetic counseling. *J. Hered.*, 77, 3, 175–8, 1986, doi: 10.1093/oxfordjournals.jhered.a110210. PMID: 3525663.
29. Durmaz, A.A., Karaca, E., Demkow, U., Toruner, G., Schoumans, J., Cogulu, O., Evolution of genetic techniques: past, present, and beyond. *Biomed Res. Int.*, 2015, 461524, 2015, doi: 10.1155/2015/461524. Epub 2015 Mar 22. PMID: 25874212; PMCID: PMC4385642.
30. Tamura, K., The Genetic Code: Francis Crick's Legacy and Beyond. *Life (Basel)*, 6, 3, 36, 2016, doi: 10.3390/life6030036. PMID: 27571106; PMCID: PMC5041012.
31. Eren, K., Taktakoğlu, N., Pirim, I., DNA Sequencing Methods: From Past to Present. *Eurasian J. Med.*, 54, Suppl1, 47–56, 2022 Dec, doi: 10.5152/eurasianjmed.2022.22280. PMID: 36655445; PMCID: PMC11163357.
32. Heather, J.M. and Chain, B., The sequence of sequencers: The history of sequencing DNA. *Genomics*, 107, 1, 1–8, 2016 Jan, doi: 10.1016/j.ygeno.2015.11.003. Epub 2015 Nov 10. PMID: 26554401; PMCID: PMC4727787.
33. Kaunitz, J.D., The Discovery of PCR: ProCuRement of Divine Power. *Dig. Dis. Sci.*, 60, 8, 2230–1, 2015 Aug, doi: 10.1007/s10620-015-3747-0. PMID: 26077976; PMCID: PMC4501591.
34. Collins, F.S. and Fink, L., The Human Genome Project. *Alcohol Health Res. World*, 19, 3, 190–195, 1995. PMID: 31798046; PMCID: PMC6875757.
35. Lenoir, T. and Giannella, E., The emergence and diffusion of DNA microarray technology. *J. Biomed. Discovery Collab.*, 1, 11, 2006 Aug 22, doi: 10.1186/1747-5333-1-11. PMID: 16925816; PMCID: PMC1590052.
36. Naidoo, N., Pawitan, Y., Soong, R., Cooper, D.N., Ku, C.S., Human genetics and genomics a decade after the release of the draft sequence of the human genome. *Hum. Genomics*, 5, 6, 577–622, 2011 Oct, doi: 10.1186/1479-7364-5-6-577. PMID: 22155605; PMCID: PMC3525251.
37. Yang, M. and Kim, J.W., Principles of Genetic Counseling in the Era of Next-Generation Sequencing. *Ann. Lab. Med.*, 38, 4, 291–295, 2018 Jul, doi: 10.3343/alm.2018.38.4.291. PMID: 29611378; PMCID: PMC5895857.

38. White, S., Jacobs, C., Phillips, J., Mainstreaming genetics and genomics: a systematic review of the barriers and facilitators for nurses and physicians in secondary and tertiary care. *Genet. Med.*, 22, 7, 1149–1155, 2020, doi: 10.1038/s41436-020-0785-6.
39. Marques, L., Costa, B., Pereira, M., Silva, A., Santos, J., Saldanha, L., Silva, I., *et al.*, Advancing precision medicine: a review of innovative *in silico* approaches for drug development, clinical pharmacology and personalized healthcare. *Pharmaceutics*, 16, 3, 332, 2024 Feb 27, doi: 10.3390/pharmaceutics16030332. PMID: 38543226; PMCID: PMC10975777.
40. Yadav, P., Sreelakshmi, M., Chauhan, V., Mishra, R., Precision medicine: unveiling progress amidst challenges. *Int. J. Pharm. Investig.* [Internet], 14, 4, 1010–8, 2024 Sep 27. Available from: http://dx.doi.org/10.5530/ijpi.14.4.110.
41. Horton, R.H. and Lucassen, A.M., Recent developments in genetic/genomic medicine. *Clin. Sci. (Lond.)*, 133, 5, 697–708, 2019 Mar 5, doi: 10.1042/CS20180436. PMID: 30837331; PMCID: PMC6399103.
42. Lin, Q., Tam, P.K., Tang, C.S., Artificial intelligence-based approaches for the detection and prioritization of genomic mutations in congenital surgical diseases. *Front. Pediatr.*, 11, 1203289, 2023 Aug 1, doi: 10.3389/fped.2023.1203289. PMID: 37593442; PMCID: PMC10429173.
43. McPherson, E., Genetic diagnosis and testing in clinical practice. *Clin. Med. Res.*, 4, 2, 123–9, 2006 Jun, doi: 10.3121/cmr.4.2.123. PMID: 16809405; PMCID: PMC1483893.
44. Qin, Y., Huo, M., Liu, X., Li, S.C., Biomarkers and computational models for predicting efficacy to tumor ICI immunotherapy. *Front. Immunol.*, 15, 1368749, 2024 Mar 8, doi: 10.3389/fimmu.2024.1368749. Erratum in: Front Immunol. 2024 Jun 04;15:1438587. doi: 10.3389/fimmu.2024.1438587. PMID: 38524135; PMCID: PMC10957591.
45. Genetic Alliance; District of Columbia Department of Health, Appendix B, Classic Mendelian Genetics (Patterns of Inheritance), in: *Understanding Genetics: A District of Columbia Guide for Patients and Health Professionals*, Genetic Alliance, Washington (DC), 2010 Feb 17, Available from: https://www.ncbi.nlm.nih.gov/books/NBK132145/.
46. Petrucelli, N., Daly, M.B., Pal, T., BRCA1- and BRCA2-Associated Hereditary Breast and Ovarian Cancer, in: *GeneReviews®* [Internet], M.P. Adam, J. Feldman, G.M. Mirzaa, *et al.* (Eds.), University of Washington, Seattle, Seattle (WA), 1993-2025 [updated 2023 Sep 21; cited 1998 Sep 4, Available from: https://www.ncbi.nlm.nih.gov/books/NBK1247/.
47. van Wyk, C., Al-Sayegh, H., Al-Kiyumi, S.I., Al-Mamari, G.S., Al Saegh, A.A., The Manchester Scoring System for predicting BRCA1/2 mutations underperforms in Arabic Omani breast cancer patients. *J. Genet. Couns.*, 2024 Jun 18, doi: 10.1002/jgc4.1939. Epub ahead of print. PMID: 38888083.
48. McCarthy, A.M. and Armstrong, K., The role of testing for BRCA1 and BRCA2 mutations in cancer prevention. *JAMA Intern. Med.*, 174, 7, 1023–4,

2014 Jul, doi: 10.1001/jamainternmed.2014.1322. PMID: 24841841; PMCID: PMC4169670.

49. Reid, S. and Pal, T., Update on multi-gene panel testing and communication of genetic test results. *Breast J.*, 26, 8, 1513–1519, 2020 Aug, doi: 10.1111/tbj.13971. Epub 2020 Jul 8. PMID: 32639074; PMCID: PMC7484453.
50. Saunders, C.J., Miller, N.A., Soden, S.E., Dinwiddie, D.L., Noll, A., Alnadi, N.A., *et al.*, Rapid whole-genome sequencing for genetic disease diagnosis in neonatal intensive care units. *Sci. Transl. Med.*, 4, 154, 154ra135, 2012 Oct 3, doi: 10.1126/scitranslmed.3004041. PMID: 23035047; PMCID: PMC4283791.
51. Ayoub, A., McHugh, J., Hayward, J., Rafi, I., Qureshi, N., Polygenic risk scores: improving the prediction of future disease or added complexity? *Br. J. Gen. Pract.*, 72, 721, 396–398, 2022, doi: 10.3399/bjgp22X720437. PMID: 35902257; PMCID: PMC9343049.
52. Quazi, S., Artificial intelligence and machine learning in precision and genomic medicine. *Med. Oncol.*, 39, 8, 120, 2022, doi: 10.1007/s12032-022-01711-1. PMID: 35704152; PMCID: PMC9198206.
53. Dias, R., Torkamani, A., Artificial intelligence in clinical and genomic diagnostics. *Genome Med.*, 11, 70, 2019. https://doi.org/10.1186/s13073-019-0689-8.
54. Genetic Alliance; The New York-Mid-Atlantic Consortium for Genetic and Newborn Screening Services, *Understanding genetics: A New York, Mid-Atlantic guide for patients and health professionals*, Genetic Alliance, Washington (DC), 2009, Available from: https://www.ncbi.nlm.nih.gov/books/NBK115545/.
55. Cui, C., Shu, W., Li, P., Fluorescence *in situ* hybridization: Cell-based genetic diagnostic and research applications. *Front. Cell Dev. Biol.*, 4, 89, 2016, doi: 10.3389/fcell.2016.00089. PMID: 27656642; PMCID: PMC5011256.
56. Schmidt, J.L., Maas, R., Altmeyer, S.R., Genetic counseling for consumer-driven whole exome and whole genome sequencing: A commentary on early experiences. *J. Genet. Couns.*, 28, 2, 449–455, 2019, doi: 10.1002/jgc4.1109. PMID: 30861266.
57. Bustin, S.A. and Jellinger, K.A., Advances in molecular medicine: Unravelling disease complexity and pioneering precision healthcare. *Int. J. Mol. Sci.*, 24, 18, 14168, 2023, doi: 10.3390/ijms241814168. PMID: 37762471; PMCID: PMC10531697.
58. Goetz, L.H. and Schork, N.J., Personalized medicine: motivation, challenges, and progress. *Fertil. Steril.*, 109, 6, 952–963, 2018, doi: 10.1016/j.fertnstert.2018.05.006. PMID: 29935653; PMCID: PMC6366451.
59. Li, M.M., Datto, M., Duncavage, E.J., Kulkarni, S., Lindeman, N.I., Roy, S., Tsimberidou, A.M., Vnencak-Jones, C.L., Wolff, D.J., Younes, A., Nikiforova, M.N., Standards and guidelines for the interpretation and reporting of sequence variants in cancer: A joint consensus recommendation of the Association for Molecular Pathology, American Society of Clinical Oncology,

and College of American Pathologists. *J. Mol. Diagn.*, 19, 1, 4–23, 2017, doi: 10.1016/j.jmoldx.2016.10.002. PMID: 27993330; PMCID: PMC5707196.

60. Genetic Alliance; The New York-Mid-Atlantic Consortium for Genetic and Newborn Screening Services, Chapter 3, Pedigree and family history-taking, in: *Understanding Genetics: A New York, Mid-Atlantic Guide for Patients and Health Professionals*, Genetic Alliance, Washington (DC), 2009 Jul 8, Available from: https://www.ncbi.nlm.nih.gov/books/NBK115557/.
61. Cragun, D. and Zierhut, H., Development of FOCUS-GC: Framework for Outcomes of Clinical Communication Services in Genetic Counseling. *J. Genet. Couns.*, 27, 1, 33–58, 2018, doi: 10.1007/s10897-017-0145-0. PMID: 29022205; PMCID: PMC5796847.
62. Nelson, V.R. and Nadeau, J.H., Transgenerational genetic effects. *Epigenomics*, 2, 6, 797–806, 2010, doi: 10.2217/epi.10.57. PMID: 22122083; PMCID: PMC3720237.
63. Mohiuddin, M., Kooy, R.F., Pearson, C.E., De novo mutations, genetic mosaicism and human disease. *Front. Genet.*, 13, 983668, 2022, doi: 10.3389/fgene.2022.983668. PMID: 36226191; PMCID: PMC9550265.
64. Houston, B.J., Riera-Escamilla, A., Wyrwoll, M.J., Salas-Huetos, A., Xavier, M.J., Nagirnaja, L., Friedrich, C., Conrad, D.F., Aston, K.I., Krausz, C., Tüttelmann, F., O'Bryan, M.K., Veltman, J.A., Oud, M.S., A systematic review of the validated monogenic causes of human male infertility: 2020 update and a discussion of emerging gene-disease relationships. *Hum. Reprod. Update*, 28, 1, 15–29, 2021, doi: 10.1093/humupd/dmab030. PMID: 34498060; PMCID: PMC8730311.
65. Sinha, D., Maurya, A.K., Abdi, G., Majeed, M., Agarwal, R., Mukherjee, R., Ganguly, S., Aziz, R., Bhatia, M., Majgaonkar, A., Seal, S., Das, M., Banerjee, S., Chowdhury, S., Adeyemi, S.B., Chen, J.T., Integrated genomic selection for accelerating breeding programs of climate-smart cereals. *Genes (Basel)*, 14, 7, 1484, 2023, doi: 10.3390/genes14071484. PMID: 37510388; PMCID: PMC10380062.
66. Kingdom, R. and Wright, C.F., Incomplete penetrance and variable expressivity: From clinical studies to population cohorts. *Front. Genet.*, 13, 920390, 2022, doi: 10.3389/fgene.2022.920390. PMID: 35983412; PMCID: PMC9380816.
67. Drumm, M.L., Ziady, A.G., Davis, P.B., Genetic variation and clinical heterogeneity in cystic fibrosis. *Annu. Rev. Pathol.*, 7, 267–282, 2012, doi: 10.1146/annurev-pathol-011811-120900. PMID: 22017581; PMCID: PMC4029837.
68. Pestoff, R., Ingvoldstad, C., Skirton, H., Genetic counsellors in Sweden: their role and added value in the clinical setting. *Eur. J. Hum. Genet.*, 24, 3, 350–355, 2016, doi: 10.1038/ejhg.2015.110. PMID: 26014428; PMCID: PMC4755363.
69. Biesecker, B., Genetic counseling and the central tenets of practice. *Cold Spring Harb. Perspect. Med.*, 10, 3, a038968, 2020, doi: 10.1101/cshperspect.a038968. PMID: 31570379; PMCID: PMC7050579.

70. Genetic Alliance; The New York-Mid-Atlantic Consortium for Genetic and Newborn Screening Services, *Understanding genetics: A New York, Mid-Atlantic guide for patients and health professionals*, Genetic Alliance, Washington (DC), 2009, Available from: https://www.ncbi.nlm.nih.gov/books/NBK115552/.
71. Clarke, A.J. and Wallgren-Pettersson, C., Ethics in genetic counselling. *J. Community Genet.*, 10, 1, 3–33, 2019, doi: 10.1007/s12687-018-0371-7. PMID: 29949066; PMCID: PMC6325035.
72. Lautenbach, D.M., Christensen, K.D., Sparks, J.A., Green, R.C., Communicating genetic risk information for common disorders in the era of genomic medicine. *Annu. Rev. Genomics Hum. Genet.*, 14, 491–513, 2013, doi: 10.1146/annurev-genom-092010-110722. PMID: 24003856; PMCID: PMC3862080.
73. Salema, D., Townsend, A., Austin, J., Patient decision-making and the role of the prenatal genetic counselor: An exploratory study. *J. Genet. Couns.*, 28, 1, 155–163, 2019, doi: 10.1002/jgc4.1049. PMID: 30629775.
74. Shaw, T., Fok, R., Courtney, E., Li, S.T., Chiang, J., Ngeow, J., Missed diagnosis or misdiagnosis: Common pitfalls in genetic testing. *Singapore Med. J.*, 64, 1, 67–73, 2023, doi: 10.4103/singaporemedj.SMJ-2021-467. PMID: 36722519; PMCID: PMC9979802.
75. Tomozawa, C., Kaneko, M., Sasaki, M., Miyake, H., Clients' and genetic counselors' perceptions of empathy in Japan: A pilot study of simulated consultations of genetic counseling. *PLoS One*, 18, 7, e0288881, 2023, doi: 10.1371/journal.pone.0288881. PMID: 37467241; PMCID: PMC10355444.
76. Austin, J.C., Re-conceptualizing risk in genetic counseling: implications for clinical practice. *J. Genet. Couns.*, 19, 3, 228–234, 2010, doi: 10.1007/s10897-010-9279-z. PMID: 20119700; PMCID: PMC3706327.
77. Lazarin, G.A., Detweiler, S., Nazareth, S.B., Ashkinadze, E., Genetic counselors' perspectives and practices regarding expanded carrier screening after initial clinical availability. *J. Genet. Couns.*, 25, 2, 395–404, 2016, doi: 10.1007/s10897-015-9881-1. PMID: 26354338; PMCID: PMC4799270.
78. Smerecnik, C.M., Mesters, I., Verweij, E., de Vries, N.K., de Vries, H., A systematic review of the impact of genetic counseling on risk perception accuracy. *J. Genet. Couns.*, 18, 3, 217–228, 2009, doi: 10.1007/s10897-008-9210-z. PMID: 19291376; PMCID: PMC7451018.
79. Davis, K.W., Roter, D.L., Schmidlen, T., Scheinfeldt, L.B., Klein, W.M.P., Testing a best practices risk result format to communicate genetic risks. *Patient Educ. Couns.*, 104, 5, 936–943, 2021, doi: 10.1016/j.pec.2020.10.021. Availablefrom:https://www.sciencedirect.com/science/article/pii/S0738399120305589.
80. Shickh, S., Rafferty, S.A., Clausen, M., Kodida, R., Mighton, C., Panchal, S., Lorentz, J., Ward, T., Watkins, N., Elser, C., Eisen, A., Carroll, J.C., Glogowski, E., Schrader, K.A., Lerner-Ellis, J., Kim, R.H., Chitayat, D., Shuman, C., Bombard, Y., Incidental Genomics Study Team. The role of digital tools in

the delivery of genomic medicine: enhancing patient-centered care. *Genet. Med.*, 23, 6, 1086–1094, 2021, doi: 10.1038/s41436-021-01112-1. PMID: 33654192; PMCID: PMC9306004.
81. Indian Council of Medical Research (ICMR), Guidelines on genetic disorders and prenatal screening.
82. Tata Memorial Centre (TMC), Cancer genetics and genetic counselling in India.
83. All India Institute of Medical Sciences (AIIMS), Genetic testing and counselling services.
84. Centre for DNA Fingerprinting and Diagnostics (CDFD), Research in human genetics and inherited diseases.
85. National Institute of Mental Health and Neurosciences (NIMHANS), Genetic counselling for neurodevelopmental and psychiatric disorders.
86. National Centre for Biotechnology Information (NCBI), Research articles on genetic counselling in India.
87. GenomeIndia Project, An initiative to map genetic diversity and improve precision medicine in India.
88. Alkilani, H.M., El-Akouri, K., Farooq, A., Shi, Z., Al-Shafai, M., Stotland, M., Khodjet-El-Khil, H., Parental knowledge and attitudes toward genetic counseling and childhood genetic testing for congenital anomalies in Qatar. *J. Genet. Couns.*, 34, 4, e70096, 2025 Aug. doi: 10.1002/jgc4.70096. PMID: 40833203; PMCID: PMC12366482.
89. Wiley, *Counseling About Cancer: Strategies for Genetic Counseling*, 4th Edition.
90. Wiley, *A Guide to Genetic Counseling*, 3rd Edition, Wiley.
91. Barnes & Noble, *A Guide to Genetic Counseling*, Paperback - Barnes & Noble.
92. American Board of Genetic Counseling, Studying for the Genetic Counseling Board Exam: A Look at Prenatal Study Resources.
93. Taylor & Francis, *Harper's Practical Genetic Counselling*, Eighth Edition.
94. Oxford Academic, *Foundations of Perinatal Genetic Counseling: A Guide for Counselors.*
95. Parker, M., Lucassen, A., Using a genetic test result in the care of family members: How does the duty of confidentiality apply? *Eur. J. Hum. Genet.*, 26, 7, 955–959, 2018 Jul. doi: 10.1038/s41431-018-0138-y. Epub 2018 Apr 27. PMID: 29700390; PMCID: PMC6018806.
96. Definition of genomic testing - NCI Dictionary of Cancer Terms, NCI, February 2, 2011, Accessed February 1, 2025. https://www.cancer.gov/publications/dictionaries/cancer-terms/def/genomic-testing.
97. Thodeson, D.M. and Park, J.Y., Genomic testing in pediatric epilepsy. *Cold Spring Harb. Mol. Case Stud.*, 5, 4, a004135, 2019, doi: 10.1101/mcs.a004135.
98. K LB, S W, M H, E M., Genomic testing in patients with renal disease. *Br. J. Hosp. Med. (Lond. England 2005)*, 84, 7, 2023, doi: 10.12968/hmed.2023.0085.

99. Fogel, B.L., Genetic and genomic testing for neurologic disease in clinical practice. *Handb. Clin. Neurol.*, 147, 11–22, 2018, doi: 10.1016/B978-0-444-63233-3.00002-6.
100. *Molecular Genetics of Prostate Cancer and Role of Genomic Testing*, PubMed, Accessed February 1, 2025. https://pubmed.ncbi.nlm.nih.gov/36344179/.
101. Cardoso, F., van't Veer, L.J., Bogaerts, J., *et al.*, 70-Gene Signature as an Aid to Treatment Decisions in Early-Stage Breast Cancer. *N. Engl. J. Med.*, 375, 8, 717–729, 2016, doi: 10.1056/NEJMoa1602253.
102. Roden, D.M., McLeod, H.L., Relling, M.V., *et al.*, Pharmacogenomics. *Lancet*, 394, 10197, 521–532, 2019, doi: 10.1016/S0140-6736(19)31276-0.
103. Zhai, S., Zhang, H., Mehrotra, D.V., Shen, J., Pharmacogenomics polygenic risk score for drug response prediction using PRS-PGx methods. *Nat. Commun.*, 13, 1, 5278, 2022, doi: 10.1038/s41467-022-32407-9.
104. Oslin, D.W., Lynch, K.G., Shih, M.C., *et al.*, Effect of Pharmacogenomic Testing for Drug-Gene Interactions on Medication Selection and Remission of Symptoms in Major Depressive Disorder: The PRIME Care Randomized Clinical Trial. *JAMA*, 328, 2, 151–161, 2022, doi: 10.1001/jama.2022.9805.
105. Thase, M.E., Parikh, S.V., Rothschild, A.J., *et al.*, Impact of Pharmacogenomics on Clinical Outcomes for Patients Taking Medications With Gene-Drug Interactions in a Randomized Controlled Trial. *J. Clin. Psychiatry*, 80, 6, 19m12910, 2019, doi: 10.4088/JCP.19m12910.
106. Wang, A., Mbikyo, M.B., Zhang, J., Cui, N., Li, Z., Treatment for Resistant Hypertension Under the Guidance of Pharmacogenomics: A Randomised Controlled Open-Label Trial. *J. Coll. Physicians Surg. Pak.*, 34, 4, 383–389, 2024, doi: 10.29271/jcpsp.2024.04.383.
107. Definition of genomic sequencing - NCI Dictionary of Cancer Terms, NCI, February 2, 2011, Accessed February 2, 2025. https://www.cancer.gov/publications/dictionaries/cancer-terms/def/genomic-sequencing.
108. Whole Genome Sequencing - an overview, ScienceDirect Topics, Accessed February 2, 2025. https://www.sciencedirect.com/topics/neuroscience/whole-genome-sequencing#definition.
109. French, C.E., Delon, I., Dolling, H., *et al.*, Whole genome sequencing reveals that genetic conditions are frequent in intensively ill children. *Intensive Care Med.*, 45, 5, 627–636, 2019, doi: 10.1007/s00134-019-05552-x.
110. Stranneheim, H., Lagerstedt-Robinson, K., Magnusson, M., *et al.*, Integration of whole genome sequencing into a healthcare setting: high diagnostic rates across multiple clinical entities in 3219 rare disease patients. *Genome Med.*, 13, 1, 40, 2021, doi: 10.1186/s13073-021-00855-5.
111. Freson, K. and Turro, E., High-throughput sequencing approaches for diagnosing hereditary bleeding and platelet disorders. *J. Thromb. Haemost.*, 15, 7, 1262–1272, 2017, doi: 10.1111/jth.13681.
112. Katsila, T. and Patrinos, G.P., Whole genome sequencing in pharmacogenomics. *Front. Pharmacol.*, 6, 61, 2015, doi: 10.3389/fphar.2015.00061.

113. Thami, P.K., Choga, W.T., Dandara, C., *et al.*, Whole genome sequencing reveals population diversity and variation in HIV-1 specific host genes. *Front. Genet.*, 14, 1290624, 2023, doi: 10.3389/fgene.2023.1290624.
114. Priestley, P., Baber, J., Lolkema, M.P., *et al.*, Pan-cancer whole-genome analyses of metastatic solid tumours. *Nature*, 575, 7781, 210–216, 2019, doi: 10.1038/s41586-019-1689-y.
115. Köser, C.U., Ellington, M.J., Cartwright, E.J.P., *et al.*, Routine Use of Microbial Whole Genome Sequencing in Diagnostic and Public Health Microbiology. *PLoS Pathog.*, 8, 8, e1002824, 2012, doi: 10.1371/journal.ppat.1002824.
116. NIHR Global Health Research Unit on Genomic Surveillance of AMR, Whole-genome sequencing as part of national and international surveillance programmes for antimicrobial resistance: a roadmap. *BMJ Glob. Health*, 5, 11, e002244, 2020, doi: 10.1136/bmjgh-2019-002244.
117. Poirion, O.B., Jing, Z., Chaudhary, K., Huang, S., Garmire, L.X., DeepProg: an ensemble of deep-learning and machine-learning models for prognosis prediction using multi-omics data. *Genome Med.*, 13, 112, 2021, doi: 10.1186/s13073-021-00930-x.
118. Tang, X., Zhou, Y., Chen, Z., *et al.*, Identification of key biomarkers for predicting CAD progression in inflammatory bowel disease *via* machine-learning and bioinformatics strategies. *J. Cell. Mol. Med.*, 28, 6, e18175, 2024, doi: 10.1111/jcmm.18175.
119. Liu, C., Zhou, Y., Zhou, Y., Tang, X., Tang, L., Wang, J., Identification of crucial genes for predicting the risk of atherosclerosis with system lupus erythematosus based on comprehensive bioinformatics analysis and machine learning. *Comput. Biol. Med.*, 152, 106388, 2023, doi: 10.1016/j.compbiomed.2022.106388.
120. Jiang, Z., Lu, Y., Liu, Z., *et al.*, Drug resistance prediction and resistance genes identification in Mycobacterium tuberculosis based on a hierarchical attentive neural network utilizing genome-wide variants. *Brief. Bioinform.*, 23, 3, bbac041, 2022, doi: 10.1093/bib/bbac041.
121. Ren, Y., Chakraborty, T., Doijad, S., *et al.*, Prediction of antimicrobial resistance based on whole-genome sequencing and machine learning. *Bioinformatics*, 38, 2, 325–334, 2021, doi: 10.1093/bioinformatics/btab681.
122. Lüftinger, L., Májek, P., Beisken, S., Rattei, T., Posch, A.E., Learning From Limited Data: Towards Best Practice Techniques for Antimicrobial Resistance Prediction From Whole Genome Sequencing Data. *Front. Cell. Infect. Microbiol.*, 11, 610348, 2021, doi: 10.3389/fcimb.2021.610348.
123. Ye, G.X., Ontiveros, E., Ivander, A., Velinov, M., Simotas, C., Autosomal Recessive Infantile Hyaline Fibromatosis Identified Using Artificial Intelligence-Assisted Rapid Whole Genome Sequencing: A Rare, Multisystemic, Hereditary Disorder. *Cureus*, 16, 6, e62037, 2024, doi: 10.7759/cureus.62037.

124. Fiziev, P.P., McRae, J., Ulirsch, J.C., *et al.*, Rare penetrant mutations confer severe risk of common diseases. *Science*, 380, 6648, eabo1131, 2023, doi: 10.1126/science.abo1131.
125. Ho, D.S.W., Schierding, W., Wake, M., Saffery, R., O'Sullivan, J., Machine Learning SNP Based Prediction for Precision Medicine. *Front. Genet.*, 10, 267, 2019, doi: 10.3389/fgene.2019.00267.
126. Elgart, M., Lyons, G., Romero-Brufau, S., *et al.*, Non-linear machine learning models incorporating SNPs and PRS improve polygenic prediction in diverse human populations. *Commun. Biol.*, 5, 1, 856, 2022, doi: 10.1038/s42003-022-03812-z.
127. Elliott, J., Bodinier, B., Bond, T.A., *et al.*, Predictive Accuracy of a Polygenic Risk Score-Enhanced Prediction Model vs a Clinical Risk Score for Coronary Artery Disease. *JAMA*, 323, 7, 636–645, 2020, doi: 10.1001/jama.2019.22241.
128. Badré, A., Zhang, L., Muchero, W., Reynolds, J.C., Pan, C., Deep neural network improves the estimation of polygenic risk scores for breast cancer. *J. Hum. Genet.*, 66, 4, 359–369, 2021, doi: 10.1038/s10038-020-00832-7.
129. Nicora, G., Zucca, S., Limongelli, I., Bellazzi, R., Magni, P., A machine learning approach based on ACMG/AMP guidelines for genomic variant classification and prioritization. *Sci. Rep.*, 12, 1, 2517, 2022, doi: 10.1038/s41598-022-06547-3.
130. Brandes, N., Goldman, G., Wang, C.H., Ye, C.J., Ntranos, V., Genome-wide prediction of disease variant effects with a deep protein language model. *Nat. Genet.*, 55, 9, 1512–1522, 2023, doi: 10.1038/s41588-023-01465-0.
131. Peña-Martínez, E.G. and Rodríguez-Martínez, J.A., Decoding Non-coding Variants: Recent Approaches to Studying Their Role in Gene Regulation and Human Diseases. *Front. Biosci. (Schol. Ed.)*, 16, 1, 4, 2024, doi: 10.31083/j.fbs1601004.
132. Hoffman, G.E., Bendl, J., Girdhar, K., Schadt, E.E., Roussos, P., Functional interpretation of genetic variants using deep learning predicts impact on chromatin accessibility and histone modification. *Nucleic Acids Res.*, 47, 20, 10597–10611, 2019, doi: 10.1093/nar/gkz808.
133. Vilhekar, R.S. and Rawekar, A., Artificial Intelligence in Genetics. *Cureus*, 16, 1, e52035, doi: 10.7759/cureus.52035.
134. Juhn, Y. and Liu, H., Artificial intelligence approaches using natural language processing to advance EHR-based clinical research in Allergy, Asthma, and Immunology. *J. Allergy Clin. Immunol.*, 145, 2, 463–469, 2020, doi: 10.1016/j.jaci.2019.12.897.
135. Parikh, J.R., Genetti, C.A., Aykanat, A., *et al.*, A data-driven architecture using natural language processing to improve phenotyping efficiency and accelerate genetic diagnoses of rare disorders. *HGG Adv.*, 2, 3, 100035, 2021, doi: 10.1016/j.xhgg.2021.100035.
136. Nazareth, S., Nussbaum, R.L., Siglen, E., Wicklund, C.A., Chatbots & artificial intelligence to scale genetic information delivery. *J. Genet. Couns.*, 30, 1, 7–10, 2021, doi: 10.1002/jgc4.1359.

137. Al-Hilli, Z., Noss, R., Dickard, J., *et al.*, A Randomized Trial Comparing the Effectiveness of Pre-test Genetic Counseling Using an Artificial Intelligence Automated Chatbot and Traditional In-person Genetic Counseling in Women Newly Diagnosed with Breast Cancer. *Ann. Surg. Oncol.*, 30, 10, 5990–5996, 2023, doi: 10.1245/s10434-023-13888-4.
138. Zampatti, S., Peconi, C., Megalizzi, D., *et al.*, Innovations in Medicine: Exploring ChatGPT's Impact on Rare Disorder Management. *Genes (Basel)*, 15, 4, 421, 2024, doi: 10.3390/genes15040421.
139. Ahimaz, P., Bergner, A.L., Florido, M.E., Harkavy, N., Bhattacharyya, S., Genetic counselors' utilization of ChatGPT in professional practice: A cross-sectional study. *Am. J. Med. Genet. A*, 194, 4, e63493, 2024, doi: 10.1002/ajmg.a.63493.
140. Tenenbaum, A., Revel-Vilk, S., Gazit, S., *et al.*, A machine learning model for early diagnosis of type 1 Gaucher disease using real-life data. *J. Clin. Epidemiol.*, 175, 111517, 2024, doi: 10.1016/j.jclinepi.2024.111517.
141. Bewicke-Copley, F., Arjun Kumar, E., Palladino, G., Korfi, K., Wang, J., Applications and analysis of targeted genomic sequencing in cancer studies. *Comput. Struct. Biotechnol. J.*, 17, 1348–1359, 2019, doi: 10.1016/j.csbj.2019.10.004.
142. Jovic, D., Liang, X., Zeng, H., Lin, L., Xu, F., Luo, Y., Single-cell RNA sequencing technologies and applications: A brief overview. *Clin. Transl. Med.*, 12, 3, e694, 2022, doi: 10.1002/ctm2.694.
143. Wang, S., Sun, S.T., Zhang, X.Y., *et al.*, The Evolution of Single-Cell RNA Sequencing Technology and Application: Progress and Perspectives. *Int. J. Mol. Sci.*, 24, 3, 2943, 2023, doi: 10.3390/ijms24032943.
144. Barbarino, J.M., Whirl-Carrillo, M., Altman, R.B., Klein, T.E., PharmGKB: A worldwide resource for pharmacogenomic information. *Wiley Interdiscip. Rev. Syst. Biol. Med.*, 10, 4, e1417, 2018, doi: 10.1002/wsbm.1417.
145. Relling, M.V., Klein, T.E., Gammal, R.S., Whirl-Carrillo, M., Hoffman, J.M., Caudle, K.E., The Clinical Pharmacogenetics Implementation Consortium: 10 Years Later. *Clin. Pharmacol. Ther.*, 107, 1, 171–175, 2020, doi: 10.1002/cpt.1651.
146. Drug Gene Interaction database (DGIdb), PubChem Data Source, Accessed February 4, 2025. https://pubchem.ncbi.nlm.nih.gov/source/15679.
147. Sotudian, S. and Paschalidis, I.C., Machine Learning for Pharmacogenomics and Personalized Medicine: A Ranking Model for Drug Sensitivity Prediction. *IEEE/ACM Trans. Comput. Biol. Bioinform.*, 19, 4, 2324–2333, 2022, doi: 10.1109/TCBB.2021.3084562.
148. Ahn, S., Building and analyzing machine learning-based warfarin dose prediction models using scikit-learn. *Transl. Clin. Pharmacol.*, 30, 4, 172–181, 2022, doi: 10.12793/tcp.2022.30.e22.
149. Micaglio, E., Locati, E.T., Monasky, M.M., Romani, F., Heilbron, F., Pappone, C., Role of Pharmacogenetics in Adverse Drug Reactions: An Update towards

Personalized Medicine. *Front. Pharmacol.*, 12, 651720, 2021, doi: 10.3389/fphar.2021.651720.

150. Sadee, W., Wang, D., Hartmann, K., Toland, A.E., Pharmacogenomics: Driving Personalized Medicine. *Pharmacol. Rev.*, 75, 4, 789–814, 2023, doi: 10.1124/pharmrev.122.000810.
151. Aizawa, Y., Watanabe, A., Kato, K., Institutional and Social Issues Surrounding Genetic Counselors in Japan: Current Challenges and Implications for the Global Community. *Front. Genet.*, 12, 646177, 2021, doi: 10.3389/fgene.2021.646177.
152. Dawson, C., Syverson, E., Chelius, T.H., *et al.*, Does Supply Equal Demand? The Workforce of Direct Patient Care Genetic Counselors in Wisconsin. *WMJ*, 119, 3, 158–164, 2020.
153. Abad, P.J., Tumulak, M.A.J., Guerbo, R., *et al.*, Landscape of genetic counseling in the Philippines. *J. Genet. Couns.*, 33, 5, 934–942, 2024, doi: 10.1002/jgc4.1804.
154. Kim, N., Kong, S.Y., Yoo, J., Kim, D.H., Seo, S.H., Kim, J., Current Issues, Challenges, and Future Perspectives of Genetic Counseling in Korea. *Ann. Lab. Med.*, 42, 3, 314–320, 2022, doi: 10.3343/alm.2022.42.3.314.

13

Healthcare Policy and Genomics: Shaping the Future of Health Care Systems

A. Revathi, B. Princy, K. Janani, M. Divya, B. Lokeshwari and P. Saranraj*

Department of Microbiology, Sacred Heart College (Autonomous), Tirupattur, Tamil Nadu, India

Abstract

Genomics forms the backbone of precision medicine and allows for individualised treatment based on genomics. Genomic medicine has been implemented for early disease detection and targeted therapies. Some genomic research and clinical applications are provided by national and international organisations like the World Health Organization (WHO), the U.S. Food and Drug Administration (FDA), and the European Medicines Agency (EMA). Ethical and social concerns continue to dominate genomic healthcare. Policies and ethical guidelines are needed due to genetic discrimination and ethical considerations regarding gene editing technologies such as CRISPR. Many advances in public health genomics could transform disease prevention practices through early identification of genetic predisposition and applying preventive measures. This involves utilising whole-genome sequencing (WGS) as part of routine care, enabling therapies through pharmacogenomics, and improving cancer treatment. Healthcare can improve public health and wellness through targeted lifestyle interventions and disease prevention. The COVID-19 pandemic showed the importance of genomic surveillance in identifying viral mutations and vaccine development. Genomics and policy developments show the future of health systems. Artificial Intelligence (AI) in genomic research, gene therapy, and bioinformatics requires collaboration from policymakers, researchers, and healthcare providers to realise the full potential of genomics in an efficient and ethical healthcare environment. Genomics is for improved patient outcomes in global health systems. This study examines healthcare policy and genomics for adopting and regulating genomic technologies.

**Corresponding author*: microsaranraj@gmail.com

Swetha Kumari Koduru, Sarad Pawar Naik Bukke, Hussaini Bello, Muhamad Mustafa Mundu and Ayush Madan (eds.) Genomics and Precision Healthcare: Innovations, Applications and Challenges, (419–440) © 2026 Scrivener Publishing LLC

Keywords: Genomics, healthcare policy, precision medicine, personalized healthcare, pharmacogenomics and gene therapy

13.1 Introduction

Precision medicine and treating individuals according to their genes are rapidly relying on genomic research. It is anticipated to be an even more effective, personalised piece of medicine. Genomic medicine offers significant advantages by enabling early disease detection and effective therapy [1]. In order to keep healthcare costs down, genomic technologies are becoming widely available and affordable. The full potential of genetic medicine could not be realised without sufficient education and technology assistance that would be used in healthcare delivery. Scientific advancements in precision medicine, personalised care, and diagnostic revolutions benefit patient rights by safeguarding against concerns about consent, genetic counselling, and the moral implications of novel gene editing technologies like Clustered Regularly Interspaced Short Palindromic Repeats (CRISPR) [2]. Biotechnology is generating a more accurate and efficient healthcare system for patients, and the human genome is used to create focused medicines and artificial intelligence to analyse data [3]. In addition to improving treatment and potentially revolutionising the entire healthcare system, biotechnology breakthroughs have a global impact that extends far beyond health issues. The advancement of information technology (IT) and the remarkable remote monitoring of health have considerably helped healthcare services. Electronic health records, telemedicine, and patient portals are all examples of information and communication technologies (ICT) that work well [4]. With the advent of these technologies, the healthcare industry has been able to tailor patient treatment, transforming the conventional system into a more individualised one. AI (artificial intelligence) systems should promote responsibility, specificity, and equity. In order to implement its principles at the national and regional levels, the World Health Organisation (WHO) provides technical advice and support [5]. By improving disease preventive strategies, enabling precision treatment, and simplifying public health initiatives that require policy changes, the integration of genetics into healthcare policy revolutionises modern healthcare systems [6]. Finding genetic variations connected to complex disorders has been made possible by the rapid development of genomic technologies like Next-Generation Sequencing (NGS), Whole-Genome Sequencing (WGS), and Genome-Wide Association Studies (GWAS). This has led to the creation of treatment plans that significantly improve clinical outcomes.

Healthcare policy must address issues of data governance, privacy, fair access, and specific frameworks to enable various populations to take advantage of genetic innovations due to its extensive use [7].

Effective legal protections like the Genetic Information Nondiscrimination Act (GINA) in the United States and comparable laws in other countries are necessary to ensure that health policy structures consider the ethical aspects of genomic testing, including informed consent, genetic discrimination, and potential misuse of genetic information in insurance or workplace settings [8]. Additionally, disparities in the representation of genomic research are necessary because most genetic data comes from populations of European ancestry. This means that global efforts to increase diversity in genomic analysis are needed to ensure that precision medicine is accessible to all races. The incorporation of genomic data into Electronic Health Records (EHRs) has additionally enabled the implementation of population-specific early disease detection screening programs, predictive modelling through the use of AI and ML, and extensive biobanking initiatives [9]. Innovation and sustainability must be balanced in the economic aspects of genomic medicine, such as the cost-effectiveness of genetic testing and long-term medical treatment resulting from early disease identification [10]. With gene-altering technologies like Clustered Regularly Interspaced Short Palindromic Repeats-associated protein 9 (CRISPR-Cas9), healthcare policy and genomics may have a bright future, but there are also moral dilemmas in healthcare policy, combining genomics with proteomics, metabolomics, and transcriptomics, and integrating multi-omics technologies.

Co-regulation, self-regulation, and adaptive regulation are some of the novel and inclusive approaches to AI for health regulation that the WHO supports since they can balance the risks and benefits of AI for the public and the healthcare industry [11].

13.2 Genomic Public Health Care

Genomic science has advanced rapidly in recent decades of precision medicine and the worldwide transformation of healthcare systems. The advancements in primary and secondary disease prevention, pre-symptomatic diagnosis, and precision treatment in the public healthcare system are based on the human genome that can effectively balance innovation with ethics, data security, access equity, and regulations [12]. Public health systems have the potential to transform individuals, and healthcare providers and policymakers need to address the challenges to health delivery systems. Genomic medicine marks the transition into personalised healthcare in treatments that are based on genetics rather than general population differences. The genomic data in public healthcare will require critical issues, including genetic privacy, ethical concerns, inequalities in access to genomic medicine and regulation of novel genetic technologies, including CRISPR/gene editing [13].

Genomic public healthcare is an innovative field utilising genomics data in public health policy to improve precision medicine, epidemiology, and systemic health at the population level [14]. In recent years, new genomic sequencing technologies in Next-Generation Sequencing (NGS) have significantly lowered the cost and time needed to complete whole-genome and whole-exome sequencing, dramatically increasing the utility of large-scale genomics in public health [15]. The genes of Pathogen sequencing are one of the fastest techniques for emerging infectious diseases like SARS-CoV-2, influenza, and bacterial outbreaks, which help develop more timely and effective public health interventions, vaccines, and antimicrobial stewardship programs.

Genomics has been used to benefit newborn screening programs to diagnose genetic disorders and metabolic disorders early on while intervening in a matter that improves the modifiable aspects of outcomes and saves money long-term in health costs. Advances in screening and reproductive genomics have provided prospective parents with information about inherited conditions that allow for reproductive choices and improved neonatal health [16]. Artificial intelligence (AI) models use extensive datasets to estimate disease risk and improve care delivery. These advances also came with challenges to introducing genomic medicine in public health systems, including ethical, legal and social implications (ELSI), access to genomic services, and concerns surrounding genetic privacy and data security [17].

The evolution of genomic public health in the new gene editing technologies like CRISPR-Cas9 that offer new treatment options for genetic diseases shows ethical concerns related to many ethical consequences [18]. The application of blockchain technology for secure storage and sharing of genomics data and federated learning models for genomics in public health. Genomics, environmental determinants and society intersect in the precision public health interventions as genetic makeup interacts with diet, lifestyle behaviours, and exposure to environmental toxins in influencing disease risk. Genomic public healthcare must be integrated into more traditional healthcare systems [1]. Pharmacogenomics is the field of genomic medicine that improves treatment efficacy by an individual's genetic profile to affect their response to drugs. The healthcare policies enable the widespread integration of genomic testing into clinical practices. Genomic public healthcare has excellent potential with significant ethical and legal implications. The advances in genomics may involve the existing healthcare and the benefits of precision medicine [19].

13.3 COVID-19 Pandemic Healthcare

The COVID-19 pandemic has been in the realm of viral sequencing and surveillance. Genomic sequencing allowed these methods to quickly determine the genetic code for SARS-CoV-2 in their mutations and the spread of new variants. The differences in global genomic infrastructure exposed inequalities in pandemics [20]. The large-scale genomic surveillance in low- and middle-income countries was used to identify novel variants and apply new interventions. Healthcare policies need to be designed to rectify through the genomic capabilities across the globe in sequencing technology so that all countries can swiftly respond to infectious diseases. Genomic surveillance policies will be pivotal in future pandemics [21].

Genomics should also play a role in new SARS-CoV-2 vaccines. Traditional vaccine development can result in a vaccine inequity in regard to the pandemic and its impacts on populations. Healthcare policies must protect against genomic research and ensure that genomic-based healthcare solutions are available in many pharmaceutical industries [22]. Genomics in pandemic healthcare develop moderate illness while others suffer severe complications, and it was the observation of these genetic variations that led researchers to conclude that it must have played a decisive role. Pharmacogenomics also influenced the efficacy of COVID-19 treatments due to genetic variation in how patients metabolised certain drugs. Genomic data in electronic health records (EHRs) quickly became vital tools during the COVID-19 response to monitor patient responses to vaccines and therapeutics [23].

The COVID-19 pandemic also intersects with genomics and social impacts on health. Due to genetic, environmental and socioeconomic factors, some populations, formed by race and ethnicity minorities, were identified with more significant mortality due to COVID-19. The COVID-19 pandemic illness shows the role of genomics as a transformative force in developing healthcare policy and future health systems. Genomic sequencing, mRNA vaccine development, personalised medicine and data-driven public health strategies were key to dealing with the crisis. On the other hand, the pandemic revealed inequalities in access to genomic technologies, ethical issues related to the use of genetic data, and the importance of advancing health policy to ensure the use of genomic innovations [24].

13.4 Impact of Genomics on Personalized Medicine

Personalised medicine, precise diagnosis, focused treatments, and tailored care of their molecular makeup and environmental interactions have all been transformed by genomics. The availability of Next-Generation Sequencing (NGS), Whole-Genome Sequencing (WGS), and Whole-Exome Sequencing (WES) has made it easier to identify genetic variations associated with drug metabolism, illness susceptibility, and response to treatment, improving precision healthcare [25]. The ability to metabolise antiplatelet medications is impacted by genetic variations in the CYP2C19 gene, which can alter how well these medications work to prevent cardiovascular events or inflammation. In contrast, leukaemia patients' medication tolerance is influenced by the variation in the Thiopurine Methyltransferase (TPMT) gene. Developments in liquid biopsy methods for extracellular vesicles and circulating tumour DNA (ctDNA) provide non-invasive early cancer detection and diagnosis, supporting precision oncology [26]. Genomics has advanced personalised medicine, preventive care, treatment effectiveness, and diagnostic precision across a range of medical specialities.

Advances in Next-Generation Sequencing (NGS), Whole-Genome Sequencing (WGS), Whole-Exome Sequencing (WES), and Genome-Wide Association Studies (GWAS) have made it easier to identify genetic variants linked to disease, giving rise to precision-driven interventions that go beyond standard treatment modalities that are one-size-fits-all [27]. In oncology, cancer genomics has produced tailored treatments based on mutations unique to a patient's tumour, which is used in personalised medicine. Variations in the genes of the enzymes break down popular medications like clopidogrel, antidepressants, and thiopurines (CYP2C19, CYP2D6, and TPMT, respectively). By merging genomic data with electronic health records (EHRs), artificial intelligence (AI), and machine learning algorithms, genomics and personalised health monitoring will result in precision diagnosis, predictive modelling, and tailored care. The genome can be therapeutically altered to produce harmful mutations thanks to gene-editing tools like CRISPR-Cas9 [28]. The integration of genomic and imaging data to ascertain further cancer classification, disease progression, and individualised treatment plans is made easier by AI-based radio genomics. In addition to cancer, genomics has revolutionised the treatment of rare genetic diseases with genome-editing technologies like CRISPR-Cas9 and gene therapy, which help fix disease-causing mutations at the DNA level. Due to the lack of effective conventional medicines, gene therapies have become promising treatment alternatives for otherwise rare disorders such as sickle cell disease (SCD), Duchenne muscular dystrophy (DMD), and spinal muscular atrophy (SMA) [29].

Customised probiotics and faecal microbiota transplantation (FMT) are two examples of microbiome-directed therapeutics being investigated for the treatment of conditions like metabolic syndrome and inflammatory bowel disease (IBD). Dietary regimens are customised for each person depending on their risk of metabolic disease, hypersensitivity, or micronutrient deficiencies using precision nutrition based on nutrigenomics [30].

13.5 Genomics in Disease Prevention and Public Health

The use of genomic technologies, such as whole-genome sequencing (WGS), whole-exome sequencing (WES), and genome-wide association studies (GWAS), has led to the identification of genetic variants associated with a number of diseases, the ability to predict risk early in life, new preventive measures, and targeted screening and applications [31]. This makes it possible for health systems to transition from a reactive care paradigm to a proactive, preventative one that uses genomic data at the individual and population levels to optimise health outcomes and reduce the burden of disease. By identifying high-risk individuals for hereditary disorders like Lynch syndrome, which increases the risk of colorectal and other cancers due to mutations in the Mismatch Repair (MMR) genes, or Hereditary Breast and Ovarian Cancer (HBOC) syndrome, which is caused by mutations in BRCA1 and BRCA2, genomics may help prevent disease. Numerous large-scale sequencing projects helped track viral variations, guide vaccine development, and enhance treatment techniques during the SARS-CoV-2 pandemic, highlighting the significance of genomics in controlling infectious disease epidemics [32]. Human Leukocyte Antigen (HLA) type and other genomic markers provide tailored vaccination strategies that increase vaccine effectiveness while reducing side effects.

By studying the gut, skin, and respiratory microbiota and their significance for immunological competency and disease risk, genomics and microbiome research are combined to improve public health treatments [33].

Probiotics, prebiotics, and faecal microbiota transplantation (FMT) are examples of microbiome-modulating therapies that have emerged as a result of the association between microbial diversity and disorders like metabolic syndrome, autoimmune diseases, and inflammatory bowel disease (IBD). In order to ensure that genomic breakthroughs are included into equitable and successful public health interventions, synthetic biology and gene editing tools are becoming more and more potent tools for developing eradication programs using biologically based interventions [34].

13.6 Integration of Genomics into Healthcare Systems

As medical practice changes, healthcare systems now give patients the choice to use genetics to improve patient care through focused medicines, precision diagnosis, and personalised disease preventive programs [35].

You can identify genetic variations linked to complex diseases, family disorders, and pharmacogenomic reactions for early intervention and individualised treatment protocols by using Next-Generation Sequencing (NGS), Whole-Genome Sequencing (WGS), Whole-Exome Sequencing (WES), and Genome-Wide Association Studies (GWAS). The creation of tailored medicines as a result of genomic profiling of cancers with particular genetic alterations has had a significant impact on oncology. After thorough genomic analysis, EGFR inhibitors for non-small cell lung cancer, HER2 therapies for breast cancer, and BRAF inhibitors for melanoma were developed. These treatments are more efficient and significantly less harmful. PRS can be used to group patients into risk groups in cardiovascular medicine, enabling personalised decision-making to prevent conditions including atrial fibrillation, hypertension, and coronary artery disease by medication, lifestyle modifications, and monitoring [36]. For example, CYP2C19 polymorphisms that result in variations in clopidogrel metabolism, TPMT variants that determine thiopurine tolerance in leukaemia patients, or SLCO1B1 mutations that predispose to the development of statin-induced myopathy are examples of pharmacogenomics in the genetic differences that influence drug-metabolizing enzymes, transporters, and receptors. Pharmacogenomics can maximise drug effectiveness while minimising side effects. Today, genomic technologies are being used to diagnose rare genetic diseases, to detect them early through newborn screening programs, and to develop gene-directed treatments such as RNA therapies for spinal muscular atrophy (SMA), enzyme replacement therapy for lysosomal storage diseases, and gene editing techniques like CRISPR-Cas9 for sickle cell disease symptoms and beta-thalassemia [38]. As demonstrated by the quick-genomic characterisation of SARS-CoV-2, which aided in the development of vaccines and antiviral medications, genomic sequencing has revolutionised pathogen surveillance, outbreak monitoring, and immunisation in infectious disease management. With ongoing studies on *in vivo* gene editing in genetic diseases, gene therapy for inherited retinal diseases, and precision oncology applications involving T-cell engineered therapies like Chimeric Antigen Receptor (CAR) T-cell therapy, emerging therapies employing synthetic biology and gene editing technologies offer hope for disease prevention and treatment. For the use of genomic medicine in healthcare systems, interdisciplinary cooperation among geneticists, practitioners, bioinformaticians, legislators, and patient advocates is necessary to address obstacles and train medical professionals in genomic innovations [39]. Additionally, cost and turnaround time remain major barriers in clinical adoption of genetic testing [37].

13.7 Genomic Research for Humans

Human genomics improves patient outcomes, removes pharmacological adverse effects, and applies precision medical treatment to an individual's DNA. Some of the most serious hereditary diseases that can cause a lifetime of illness are revealed by genomic studies. Scientists are now able to find genetic markers for hereditary diseases like Huntington's disease, sickle cell anaemia, and cystic fibrosis thanks to advancements in genome sequencing [40]. Genetic screening's ability to identify these disorders early on has completely changed how medical practitioners treat and prevent these diseases. For incurable diseases, genome-editing technologies like CRISPR-Cas9 have made it possible to fix genetic errors before the disease manifests. The Human Genome Project (HGP), the first complete reference genome, was completed in 2003. It helped to develop and make Whole-Genome Sequencing (WGS), Whole-Generation Sequencing (NGS), Whole-Exome Sequencing (WES), and Genome-Wide Association Studies (GWAS) easier. These techniques allow for the identification of genetic variants that cause complex traits, uncommon conditions, and common disorders like diabetes, cardiovascular disease, and cancer [41].

Particularly in areas like oncology, psychiatry, and cardiology, these genomic studies have made a substantial contribution to personalised medicine. For example, pharmacogenomics of an individual is used to predict drug responses, enabling treatment to meet individual needs, reducing drug side effects, and improving therapeutic effectiveness. To better understand human health and disease development, this entails combining genomic data with multi-omics techniques (transcriptomics, proteomics, epigenomics, and metabolomics) to gain a deeper understanding of gene regulation, cellular pathways, and disease mechanisms [42]. Health disparities, ancestry-informed illness risk, and access to genomic medicine are all significantly impacted by population genomics, which includes evolutionary adaptations, genetic drift, migratory routes, human ancestry, admixture events, and ethnic group diversity [43]. Driver mutations, changes to tumour suppressor genes, and epigenetic modifications that promote carcinogenesis have all been identified by cancer genomics.

This cleared the path for the creation of targeted treatments such as immune checkpoint inhibitors for tumor-microenvironment high microsatellite instability (MSI), chimeric antigen receptor (CAR)-T cell therapies for haematogenous tumours, and tyrosine kinase inhibitors (TKIs) for EGFR-mutant lung cancer [44]. The discovery of numerous biomarkers for the early diagnosis, prognosis, and assessment of cancer treatment response has also been made feasible by genomics studies. Utilising liquid biopsy techniques also enables low residual disease and non-invasive monitoring of Circulating Tumour DNA (ctDNA) to track the progression of the disease. In addition to the development of various gene targeting therapies based on either RNA interference (RNAi) or gene replacement strategies, the widespread use of genomics for rare disease diagnosis has led to an increase in the identification of pathogenic variants for monogenetic disorders like Huntington's disease, Duchenne muscular dystrophy, and cystic fibrosis that preclude late-stage treatment by enabling genetic counselling and early-onset medical management. Through improved reproductive genetic testing, such as prenatal screening, carrier testing, and *In Vitro* Fertilisation (IVF) technologies through Preimplantation Genetic Testing (PGT), reproductive genomics advancements can improve genetic disease prevention by enabling the selection of embryos free from genetic disorders that are either life-threatening or severely debilitating. This is not only beneficial for families, but also for the prevention of disease in future generations [45]. In infectious disease research, pathogen genomics has transformed vaccine design, AMR surveillance, and outbreak identification. Within weeks of the COVID-19 pandemic, SARS-CoV-2 variants were being isolated and genomically described for public health and vaccine adaptation. Regarding environmental interactions and genomic research, a person's genetic predisposition to a multifactorial disease can be predicted using PRS, which are multiple genetic variants linked to a disease or trait that are calculated for each individual. This allows for precision prevention of the disease, lifestyle changes, and targeted interventions. Precision medicine, drug repurposing, and the identification of novel gene-disease correlations have all been accelerated by the application of artificial intelligence (AI) and machine learning (ML) in genomics [46].

Precision medicine uses medications made to target certain genetic abnormalities, while conventional chemotherapy destroys both cancerous and healthy cells. A genetic-based approach to medication therapy has also been illustrated by the rise of pharmacogenomic medicine, which is provided by chemists [47]. Hospitalization-causing adverse drug reactions can be reduced with personalised medication prescriptions based on genomic profiles. In addition to the health benefits of genomic research, genomics is essential to public health, especially in the fight against common complex diseases like diabetes, heart disease, and neurological disorders. Disease prevention may be revolutionised by polygenic risk scores, which measure a disease's risk based on several genetic variations [48].

13.8 AI in Healthcare

The healthcare and medicine industry is being transformed by the infusion of AI technology that helps in diagnostics, treatment, drug discovery, and personalised medicine. The advancement of AI-driven technologies and their integration with genomics is shaping the future of healthcare systems for precise and efficient patient care [49]. Artificial intelligence-led genomic analysis has enhanced cancer therapy with mutation discovery in tumours and recommending customised treatment. AI is also key in predicting inherited risks for diseases like Alzheimer's, diabetes, and heart conditions, leading to earlier intervention. AI has revolutionised medical imaging and diagnostics, leading to more accurate disease detection. AI applications in pathology and radiology have improved the efficiency of diagnosing neurological disorders, cardiovascular diseases, and genetic conditions [50].

AI is indispensable in drug discovery and development, and this process has been long and expensive for a new drug. AI-linked genome analytics in the detection of the virus and optimal vaccine formulations. Healthcare policies should promote the ethical use of AI in drug development and regulations around drug development for their safety and efficacy. AI algorithms can also be used in Electronic Health Records (EHR) and genomics to identify people at high disease risk. They play a crucial role in enabling healthcare providers to intervene early, providing lifestyle modulations, screenings, and personalised treatment to prevent the progression of diseases. AI-based healthcare may provide preventive care through vital signs and real-time alerts for abnormalities. AI in healthcare has tremendous potential; it has ethical and legal implications under the regulation of healthcare policymakers [51]. Figure 13.1 illustrates the integration of digital technologies such as AI, EHRs, robotics, telemedicine, laboratory automation, pharmaceuticals, and clinical services within modern healthcare infrastructure, underscoring how digitization drives interconnected healthcare delivery.

AI has the potential for healthcare by providing advanced diagnostic and treatment solutions. AI-propelled healthcare solutions will be concentrated in rich countries, leaving low- and middle-income countries in the dark [52]. At high prices, AI-based genomic testing and precision medicine are still a significant barrier to many populations. Healthcare

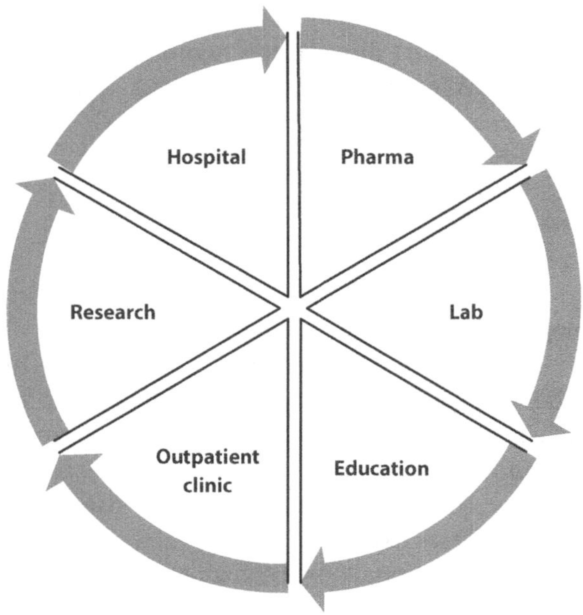

Figure 13.1 Digital technologies in healthcare.

policies must promote initiatives to make AI-powered healthcare technologies affordable and accessible globally. Governments, research institutions, and private companies can work collaboratively around the world. Genomics and AI combined have great potential to advance public health and transform disease prevention and treatment. The implementation of wide-ranging policies in AI protects data and privacy and ensures AI algorithms are transparent and the usage of AI in healthcare [53].

13.9 AI Applications in Healthcare

13.9.1 Medical Imaging Analysis

AI imaging analysis methods Convolutional Neural Networks (CNNs) have revolutionised the field of radiology by improving image segmentation and classification and reducing human error and variability. Deep learning algorithms have large datasets that enable the detection of pulmonary nodules, breast lesions, and brain tumours with their accuracy. Real-time AI methods in MRI and CT reduce the scanning time and radiation exposure significantly by adjusting low-quality images through denoising and super-resolution techniques to improve diagnostic performance. AI-powered echocardiography analysis facilitates the automated assessment of left ventricular function, myocardial infarction detection, and coronary artery disease risk estimation, enhancing patient management. Automating imaging-based interventions like robotic-assisted surgical navigation and AI-optimized radiotherapy planning increases procedural precision and patient success rates [54].

Pathology imaging powered by AI aids in automating the categorisation of histopathological slides, resulting in prompt diagnosis of cancer while also alleviating the workload of a pathologist. AI integration with multimodal imaging may improve disease characterisation by associating structural metabolic data [55]. They help to enhance transparency and interpretability in medical imaging by employing Explainable Artificial Intelligence (XAI) techniques like maps and attention-based mechanisms for AI-based diagnostics. Recent advancements in the fields of quantum computing, self-supervised learning, and multimodal AI promise to make further refinements in medical imaging analysis, leading to more precise, fast, and available diagnostic capabilities for all global communities [56].

13.9.2 Drug Discovery

Drug discovery remains a complicated and resource-consuming process, and cutting-edge developments in Artificial Intelligence (AI), Computational Biology, and Molecular Modelling have been transforming and speeding up the process of finding new therapeutic agents [57]. Drug discovery proceeds through target identification, discovery, lead optimisation, preclinical validation, and clinical trials. Generative models (GAEs), as well as variants such as Variational Autoencoders (VAEs) and Generative Adversarial Networks (GANs), utilise AI to enable *de novo* molecular design by generating potential drug-like compounds that may have favourable Pharmacokinetic (PK) and Pharmacodynamic (PD) profiles. Natural Language Processing (NLP) allows for the optimisation of knowledge-based drug discovery by mining patents and clinical trials in AI-driven High-Throughput Virtual Screening (HTVS), accelerates the screening process by evaluating vast amounts of chemical libraries *in silico,* enabling researchers to narrow down potential drug discovery [58].

AI-enabled omics analysis, encompassing genomics, transcriptomics, proteomics, and metabolomics, pinpoints disease-associated biomarkers and therapeutic targets, paving the way for personalised medicine strategies [59]. Quantum computing improves molecular simulation, enabling precise modelling of complex biological interactions and polypharmacological effects. AI-based network pharmacology helps with drug repurposing that relies on identifying the therapeutic repositioning in discovering treatments for COVID-19. AI-powered synthesis in the laboratories accelerates drug development by optimising syntheses. AI-based toxicity prediction models help evaluate the safety profile of drug candidates in the early stages of the late-stage clinical trials [60].

13.9.3 Precision Medicine

Precision medicine is a global pioneering approach in disease diagnosis, treatment, and prevention; it utilises the cross-disciplinary evolution of Genomics, Bioinformatics, Artificial Intelligence (AI), and systems biology to facilitate personalised medical intervention based on an individual's genetic, molecular, environmental, and lifestyle factors [61]. Precision medicine adopts a patient-specific and response–response-specific approach to therapeutic strategies. It integrates multi-omics data such as genomics, transcriptomics, proteomics, metabolomics, and epigenomics to assess disease-associated biomarkers and personalised drug responses [62]. Next-generation sequencing (NGS) technologies like Whole-Genome Sequencing (WGS) and Whole-Exome Sequencing (WES) enable the identification of pathogenic mutations and genetic variations associated with complex diseases, including cancer, cardiovascular disorders, and neurodegenerative conditions. In oncology, precision medicine has transformed the use of targeted therapies, with the advent of Tyrosine Kinase Inhibitors (TKIs) in Estimated Glomerular Filtration Rate (EGFR)-mutant lung cancer, V-Raf Murine Sarcoma Viral Oncogene Homolog B (BRAF) inhibitors in melanoma, and HER2-targeted monoclonal antibodies in breast cancer [63].

The integration of pharmacogenomics, a key component of precision medicine, utilises genetic variations in drug metabolism pathways to the appropriate drug to minimise adverse effects or maximise drug sensitivity. Precision medicine in rare genetic disorders has resulted in revolutionary gene-editing therapies and mutations using CRISPR-Cas9 technologies in diseases such as sickle cell anaemia and Duchenne muscular dystrophy [64]. AI-enhanced digital health solutions can monitor real-time physiological data and facilitate personalised disease management in conditions like chronic hypertension and atrial fibrillation. AI models provide multi-institutional data-sharing. Standardising precision medicine guidelines and regulatory frameworks is essential to ensure clinical validity and reproducibility to provide precision medicine and lead to increasingly ultra-personalized therapeutics globally [65].

13.10 Challenges Posed by the Use of AI in Healthcare

Artificial intelligence (AI) in healthcare has transformed diagnostics, treatment, patient management, and medical research to be ethical, safe, and effective [66]. AI applications assist in diagnosis, treatment plans, and even robotic surgery; these terms may also need to be addressed under algorithmic bias and fairness, especially since biases in training datasets can lead to disparities in diagnostic accuracy and treatment recommendations across demographic groups. AI models primarily trained on population-related data may perform less well in underrepresented communities, increasing health disparities. The absence of universal guidelines for assessing AI models and their limited applicability to varying clinical environments complicates integrating these tools into practice. Convolutional Neural Networks (CNNs) and transformer-based approaches work as a black box over which practitioners cannot interpret AI-predicted class labels [67].

Adopting Explainable AI (XAI) techniques like mechanisms and interpretable machine learning approaches is essential to developing clinical uses in AI practical use and ensuring that AI systems use these aspects of evidence-based medical practice. Implementing AI in healthcare often requires substantial expenditure on infrastructure, data storage, and computation resources, leading to economic barriers in low-resource healthcare applications, resulting in a lack of equitable access to AI-directed innovation and solutions [68]. The potential for overdiagnosis and overtreatment induced by AI is another challenge, as overly sensitive AI models can identify - symptomatic abnormalities in an irrelevant pathology that does not require treatment and patient anxiety. The use of AI for clinical decision support can lead to clinician overreliance on AI algorithms, where healthcare providers place too much dependence on the AI tools without doing an adequate critical assessment of patient safety. AI could enhance AI in healthcare, but overcoming the fundamental challenges of bias, explainability, regulation, and ethical deployment will be essential for the potential to improve healthcare worldwide [69].

13.11 Future Challenges in Healthcare

Rapid evolution in medicine, technology, and genomics are shaping the future of healthcare. These innovations hold great potential to alter disease prevention, treatment and patient care, but they also create healthcare systems worldwide [70]. In addition, the regulations should be unambiguous on the appropriate use and the ethics of gene editing technologies such as CRISPR that can cure

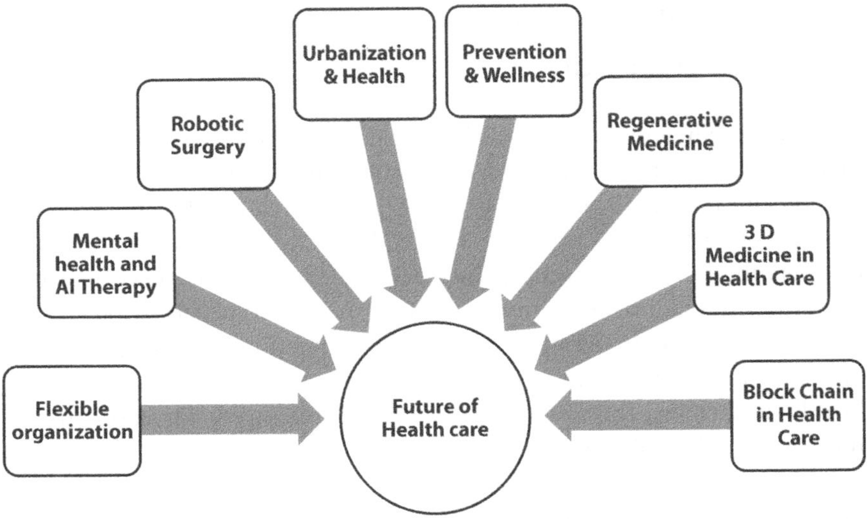

Figure 13.2 Future of healthcare.

genetic diseases but may lead to risks such as mutations and ethical guidelines. AI systems must be effectively regulated to avoid biases, mistakes, and ethical guidelines [71]. The growing use of electronic health records (EHRs), telemedicine, and genetic databases is making healthcare institutions. Figure 13.2 demonstrates the future dimensions of healthcare transformation, including genomic medicine, regenerative medicine, robotic surgery, blockchain-secured health data, urbanization–health dynamics, mental-health AI therapy, and adaptable organizational health models.

The COVID-19 pandemic revealed in healthcare systems an uneven vaccine distribution. New pathogens, antibiotic-resistant bacteria, may be responsible for future pandemics [73]. The potential for the future of healthcare and precision medicine is set to transform health systems, but the scale raises ethical transformation, affordability challenges, data privacy concerns, and health crises. Solutions offered here are necessary for the evolved state of healthcare systems that will be inclusive, efficient, and ethically responsible institutions in the genomics and medical advancements to benefit society. They are based on political integrity and technological accountability worldwide. Further, microbiome-informed prognostics such as lung microbial signatures are emerging as predictors of patient outcomes [72].

13.12 Conclusion

The future of healthcare systems is being reshaped by the incorporation of genetics into health policy, which supports personalised therapy, improves disease prevention, and maximises public health initiatives. As genomic research advances, policymakers must address ethical, legal, and economic concerns to support data privacy, equal access, and innovative regulatory laws. By promoting cooperation between researchers, physicians, and legislators, we may create a more efficient, patient-centered healthcare system that can address current and upcoming health concerns. Policies based on genetics have the potential to revolutionise healthcare by improving patient outcomes and guiding the development of a more egalitarian and sustainable healthcare system in the future.

AI disclosure statement: The author confirms that no artificial intelligence (AI)-generated content or AI-assisted writing tools were used in the preparation of this chapter (including summaries and the abstract). All text, analysis, and conclusions are the result of the author's original work.

References

1. Mattick, J.S., Dziadek, M.A., Terrill, B.N., The impact of genomics on the future of medicine and health. *Med. J. Aust.*, 201, 17–20, 2014.
2. Coiera, E., *Guide to Health Informatics*, 3rd ed., CRC Press Imprint, London, UK, 2015.
3. Siemieniuk, R.A.C., Harris, I.A., Agoritsas, T., Arthroscopic surgery for degenerative knee arthritis and meniscal tears: A clinical practice guideline. *BMJ*, 357, j1982, 2017.
4. Emery, D.J., Shojania, K.G., Forster, A.J., Overuse of magnetic resonance imaging. *JAMA Intern. Med.*, 173, 823–825, 2013.
5. Amalberti, R., Nicklin, W., Braithwaite, J., Preparing national health systems to cope with the impending tsunami of ageing and its associated complexities: Towards more sustainable health care. *Int. J. Qual. Health Care*, 28, 412–414, 2016.
6. Burke, W., Matheny Antommaria, A.H., Bennett, R., Botkin, J., Clayton, E.W., Henderson, G.E., Recommendations for returning genomic incidental findings? We need to talk! *Genet. Med.*, 15, 11, 854–859, 2013.
7. Green, R.C., Berg, J.S., Grody, W.W., Kalia, S.S., Korf, B.R., Martin, C.L., ACMG recommendations for reporting incidental findings in clinical exome and genome sequencing. *Genet. Med.*, 15, 7, 565–574, 2013.
8. Evans, B.J., Minimizing liability risks under the ACMG recommendations for reporting incidental findings in clinical exome and genome sequencing. *Genet. Med.*, 15, 12, 915–920, 2013.
9. Townsend, A., Adam, S., Birch, P.H., Friedman, J.M., Paternalism and the ACMG recommendations on genomic incidental findings: Patients seen but not heard. *Genet. Med.*, 15, 9, 751–752, 2013.

10. Berg, J.S., Amendola, L.M., Eng, C., Allen, E.V., Gray, S.W., Wagle, N., Processes and preliminary outputs for identification of actionable genes as incidental findings in genomic sequence data in the Clinical Sequencing Exploratory Research Consortium. *Genet. Med.*, 15, 11, 860–867, 2013.
11. Stanton, M.W., Expanding patient-centred care to empower patients and assist providers, vol. 5, Agency Healthc. Res. Qual., pp. 1–12, 2002.
12. Chawla, N.V. and Davis, D.A., Bringing big data to personalised healthcare: A patient-centered framework. *J. Gen. Intern. Med.*, 28, Suppl 3, S660–S665, 2013.
13. Sclafani, F., Gullo, G., Sheahan, K., Crown, J., BRAF mutations in melanoma and colorectal cancer: A single oncogenic mutation with different tumour phenotypes and clinical implications. *Crit. Rev. Oncol. Hematol.*, 87, 1, 55–68, 2013.
14. Mao, M., Tian, F., Mariadason, J.M., Tsao, C.C., Lemos, R., Jr, Dayyani, F., Resistance to BRAF inhibition in BRAF-mutant colon cancer can be overcome with PI3K inhibition or demethylating agents. *Clin. Cancer Res.*, 19, 3, 657–667, 2013.
15. Nutbeam, D. and Muscat, D.M., Health promotion glossary 2021. *Health Promot. Int.*, 36, 6, 1578–1598, 2021.
16. Jena, A.B., Malpractice risk according to physician speciality. *N. Engl. J. Med.*, 365, 629–630, 2011.
17. Vanderpool, D., The Standard of Care. *Innov. Clin. Neurosci.*, 18, 50–51, 2021.
18. Lander, E.S., Linton, L.M., Birren, B., Initial sequencing and analysis of the human genome. *Nature*, 409, 860–921, 2001.
19. Venter, J.C., Adams, M.D., Myers, E.W., The human genome sequence. *Science*, 291, 1304, 2001.
20. Australian Institute of Health and Welfare, Aboriginal and Torres Strait Islander Health Performance Framework Summary Report 2023, AIHW, Canberra (AU), 2023, Available from: https://www.indigenoushpf.gov.au/report-overview/overview/summary-report. Accessed: February 2024.
21. Buntin, M.B., Burke, M.F., Hoaglin, M.C., Blumenthal, D., The benefits of health information technology: A review of the recent literature shows predominantly positive results. *Health Aff.*, 30, 3, 464–471, 2011.
22. Omotayo, O., Maduka, C.P., Muonde, M., Olorunsogo, T.O., Ogugua, J.O., The rise of non-communicable diseases: A global health review of challenges and prevention strategies. *Int. Med. Sci. Res. J.*, 4, 1, 74–88, 2024.
23. Dowell, S.F., Blazes, D., Desmond-Hellman, S., Four steps to precision public health. *Nature*, 540, 189–191, 2016.
24. Guthrie, J.L., Strudwick, L., Roberts, B., Allen, M., McFadzen, J., Roth, D., Comparison of routine field epidemiology and whole genome sequencing to identify tuberculosis transmission in a remote setting. *Epidemiol. Infect.*, 148, e15, 2020.
25. Fauci, A.S., Lane, H.C., Redfield, R.R., COVID-19: Navigating the uncharted. *N. Engl. J. Med.*, 382, 13, 1268–1269, 2020.

26. Calzone, K., Kirk, M., Tonkin, E., Badzek, L., Benjamin, C., Middleton, A., The global landscape of nursing and genomics. *J. Nurs. Scholarsh.*, 50, 249–256, 2018.
27. Association of Schools and Programs of Public Health, Areas of Study. Published 2020. Accessed May 17, 2021.
28. Hamid, J.S., Hu, P., Roslin, N.M., Ling, V., Greenwood, C.M., Beyene, J., Data integration in genetics and genomics: Methods and challenges. *Hum. Genomics Proteomics*, 2009.
29. Cogin, J.A., Ng, J.L., Lee, I., Controlling healthcare professionals: How human resource management influences job attitudes and operational efficiency. *Hum. Resour. Health*, 14, 1, 1–8, 2016.
30. Khoury, M.J., Engelgau, M., Chambers, D.A., Mensah, G.A., Beyond Public Health Genomics: Can Big Data and Predictive Analytics Deliver Precision Public Health? *Public Health Genomics*, 21, 5–6, 244–250, 2018.
31. Frieden, T.R., The Future of Public Health. *N. Engl. J. Med.*, 373, 18, 1748–1754, 2015.
32. Dolley, S., Big data's role in precision public health. *Front. Public Health*, 2018.
33. Lazarus, D., "Big data" could mean big problems for people's healthcare privacy, L.A. Times Online, Oct 11, 2016.
34. Murdoch, T.B. and Detsky, A.S., The inevitable application of big data to health care. *JAMA*, 309, 13, 1351–1352, 2013.
35. Schultz, D., As patients' records go digital, theft and hacking problems grow, News KH: KHN, June 3, 2012.
36. Newcomb, A., Massive Equifax Data Breach Could Affect Half of the U.S. Population, NBC News, Sept 10, 2017.
37. National Library of Medicine: What is the cost of genetic testing, and how long does it take to get the results? (link unavailable), 2017.
38. Ramsey, L., I tried 23andMe's new genetics test — and now I know why the company caused such a stir, Bus. Insider, Dec 23, 2015.
39. National Library of Medicine, What is direct-to-consumer genetic testing? (link unavailable), 2017.
40. Mahmood, S., Hasan, K., Colder Carras, M., Labrique, A., Global preparedness against COVID-19: we must leverage the power of digital health. *JMIR Public Health Surveill.*, 6, e18980, 2020.
41. Polap, D., Srivastava, G., Yu, K., Agent architecture of an intelligent medical system based on federated learning and blockchain technology. *J. Inf. Secur. Appl.*, 58, 102748, 2021.
42. Royal College of General Practitioners, Position statement on Direct-to-Consumer Genomic Testing. (link unavailable), 2019.
43. Sayer, A., *Why things matter to people*, Cambridge University Press, 2011.
44. Gref, A., Merid, S.K., Gruzieva, O., Ballereau, S., Becker, A., Bellander, T., Genome-wide interaction analysis of air pollution exposure and childhood asthma with functional follow-up. *Am. J. Respir. Crit. Care Med.*, 195, 10, 1373–1383, 2017.

45. Prosperi, M., Min, J.S., Bian, J., Modave, F., Big data hurdles in precision medicine and public health. *BMC Med. Inform. Decis. Mak.*, 18, 1, 139, 2018.
46. Rasmussen, L.V., Smith, M.E., Almaraz, F., Persell, S.D., Rasmussen-Torvik, L.J., Pacheco, J.A., Chisholm, R.L., Christensen, C., Herr, T.M., Wehbe, F.H., Starren, J.B., An ancillary genomics system to support the return of pharmacogenomic results. *J. Am. Med. Inform. Assoc.*, 26, 4, 306–310, 2019.
47. National Academies of Sciences, Engineering, and medicine, Action Collaboratives, Genomics and Population Health—A Precision Health Activity. (link unavailable), 2018.
48. Jianxun, C., Arkorful, V.E., Shuliang, Z., Electronic health records adoption: Do institutional pressures and organisational culture matter. *Technol. Soc.*, 65, 101531, 2021.
49. Correa, V.C., Lugo-Agudelo, L.H., Aguirre-Acevedo, D.C., Contreras, J.A.P., Borrero, A.M.P., Patiño-Lugo, D.F., Valencia, D.A.C., Individual, health system, and contextual barriers and facilitators for the implementation of clinical practice guidelines: A systematic metareview. *Health Res. Policy Syst.*, 18, 1–11, 2020.
50. Korjian, S. and Gibson, C.M., Digital technologies and the democratisation of clinical research: Social media, wearables, and artificial intelligence. *Contemp. Clin. Trials*, 117, 106767, 2022.
51. Winter, J.S. and Davidson, E., Harmonizing regulatory regimes for the governance of patient-generated health data. *Telecommun. Policy*, 46, 5, 102285, 2022.
52. Colloud, S., Metcalfe, T., Askin, S., Belachew, S., Ammann, J., Bos, E., Cerreta, F., Evolving regulatory perspectives on digital health technologies for medicinal product development. *npj Digit. Med.*, 6, 1, 56, 2023.
53. Zhou, Y., Wang, F., Tang, J., Nussinov, R., Cheng, F., Artificial intelligence in COVID-19 drug repurposing. *Lancet Digit. Health*, 2, 12, e667–e676, 2020.
54. Mohanty, S., Rashid, M.H.A., Mridul, M., Mohanty, C., Swayamsiddha, S., Application of Artificial Intelligence in COVID-19 drug repurposing. *Diabetes Metab. Syndr. Clin. Res. Rev.*, 14, 5, 1027–1031, 2020.
55. Floresta, G., Zagni, C., Gentile, D., Patamia, V., Rescifina, A., Artificial Intelligence Technologies for COVID-19 De Novo Drug Design. *Int. J. Mol. Sci.*, 23, 6, 3261, 2022.
56. Zhu, Y., Bi, D., Saunders, M., Prediction of chronic kidney disease progression using recurrent neural network and electronic health records. *Sci. Rep.*, 13, 22091, 2023.
57. Schena, F.P., Anelli, V.W., Abbrescia, D.I., Prediction of chronic kidney disease and its progression by artificial intelligence algorithms. *J. Nephrol.*, 35, 1953–1971, 2022.
58. Sayani, A., Health equity in national cancer control plans: an analysis of the Ontario cancer plan. *Int. J. Health Policy Manag.*, 8, 9, 550–556, 2019.
59. Aarts, J., Histories of the electronic medical record in The Netherlands 1970-2015, 2023.

60. Deasfernandes, S.L., Telehealth efforts to reduce hospital readmissions of congestive heart failure patients, Walden University, 2020.
61. Cresswell, K. and Sheikh, A., Organizational issues in the implementation and adoption of health information technology innovations: An interpretative review. *Int. J. Med. Inform.*, 82, 5, e73–e86, 2013.
62. Carter, A.B., Abruzzo, L.V., Hirschhorn, J.W., Jones, D., Jordan, D.C., Nassiri, M., Temple-Smolkin, R.L., Electronic health records and genomics: Perspectives from the association for molecular pathology electronic health record (EHR) interoperability for clinical genomics data working group. *J. Mol. Diagn.*, 24, 1, 1–17, 2022.
63. Suzuki, T., Hotta, J., Kuwabara, T., Possibility of introducing telemedicine services in Asian and African countries. *Health Policy Technol.*, 9, 1, 13–22, 2020.
64. Jaffee, E.M., Dang, C.V., Agus, D.B., Alexander, B.M., Anderson, K.C., Ashworth, A., Future cancer research priorities in the USA: A Lancet Oncology Commission. *Lancet Oncol.*, 18, e653–e706, 2017.
65. Spira, A., Yurgelun, M.B., Alexandrov, L., Rao, A., Bejar, R., Polyak, K., Precancer atlas to drive precision prevention trials. *Cancer Res.*, 77, 1510–1541, 2017.
66. Yurgelun, M.B., Chenevix-Trench, G., Lippman, S.M., Translating germline cancer risk into precision prevention. *Cell*, 168, 566–570, 2017.
67. Sorensen, K., Health literacy is an emerging strategic priority in national cancer control plans in the EU. *J. Cancer Policy*, 26, 100255, 2020.
68. Spira, A., Disis, M.L., Schiller, J.T., Vilar, E., Rebbeck, T.R., Bejar, R., Leveraging premalignant biology for immune-based cancer prevention. *Proc. Natl. Acad. Sci. U.S.A.*, 113, 10750–10758, 2016.
69. Braithwaite, J., Matsuyama, Y., Mannion, R., How to do better health reform: A snapshot of change and improvement initiatives in the health systems of 30 countries. *Int. J. Qual. Health Care*, 28, 843–846, 2016.
70. Coiera, E. and Hovenga, E.J., Building a sustainable health system. *Yearb. Med. Inform.*, 11–18, 2007.
71. Teutsch, S.M. and Fielding, J.E., Rediscovering the core of public health. *Annu. Rev. Public Health*, 34, 287–299, 2013.
72. Dickson, R.P., Schultz, M.J., van der Poll, T., Schouten, L.R., Falkowski, N.R., Luth, J.E., Lung microbiota predict clinical outcomes in critically ill patients. *Am. J. Respir. Crit. Care Med.*, 201, 555–563, 2020.
73. Asnicar, F., Berry, S.E., Valdes, A.M., Nguyen, L.H., Piccinno, G., Drew, D.A., Microbiome connections with host metabolism and habitual diet from 1,098 deeply phenotype individuals. *Nat. Med.*, 27, 321–332, 2021.

14

Patient Perspectives and Experiences with Genomic Testing

Safa A. Abdalla[1,2]

[1]*Department of Clinical Pharmacy and Pharmacy Practice, School of Pharmacy, Kampala International University - Western Campus, Ishaka, Bushenyi, Uganda*
[2]*Department of Pharmacology - Faculty of Pharmacy, University of Khartoum, Khartoum, Sudan*

Abstract
Genomics testing is increasingly being exploited in healthcare for disease diagnosis and treatment. However, there is a challenge to patients' perspectives and experiences with genomics testing which brings about setbacks. This chapter aims to harmonize the existing reports on patients' perspectives and experiences with genomics testing to improve the exposure to the benefits and challenges of genomics testing from the patient perspective. The patients' views about genomics testing, with some reporting benefits, such as increased diagnostic certainty and personalized treatment options, were considered. Also, challenges such as difficulty understanding test results, concerns about regulations about patients' privacy, and discrimination are discussed. The importance of patients' perspectives and experiences considered with genomics testing is seen as a boost to provide clear and supportive communication to patients undergoing genomics testing, and to address concerns about privacy and discrimination.

***Keywords*:** Patient perspectives, patient experiences, genetic diagnosis, patient communication, health literacy, genetic risk perception, barriers to genomic testing, patient trust

Email: safaamohamed@kiu.ac.ug; safaarahim.sr@gmail.com

Swetha Kumari Koduru, Sarad Pawar Naik Bukke, Hussaini Bello, Muhamad Mustafa Mundu and Ayush Madan (eds.) Genomics and Precision Healthcare: Innovations, Applications and Challenges, (441–462) © 2026 Scrivener Publishing LLC

14.1 Introduction

In the current decade, there has been a clear increase in genomic testing (also known as genetic testing) utilization. This could be due to increased knowledge about genomes and genetic testing [1]. Next-generation sequencing (NGS) like whole genome/exome sequencing has become available and affordable, encouraging clinicians and researchers to ask patients to perform such tests. These tests help understand the etiology and prevention of different diseases and could be used as a diagnostic tool for some diseases [2]. These diseases include various types of cancers and autoimmune diseases [3]. On the other hand, pharmacogenomics studies clarify variations in drug response due to genetic variability. Pharmacogenomics, defined as the study of the whole range of genes related to drug safety and efficacy, reduces the number of errors in prescribing [4]. Pharmacogenomics findings help select personalized or precision medicine, which is targeted treatment that combines pharmacogenomics findings with the patient's medical history. Personalized medicine decreases the development of side effects and improves the efficacy of different drugs [3]. Still, there are unresolved issues concerning genetic testing as a diagnostic tool, including standards for informed consent and format for results reporting. Also, different studies revealed different gaps limiting the use of genetic testing, including the education and training of health professionals to improve the application of genetic testing and interpretation of results, and the level of patients' understanding of genetics and genetic testing [1, 5].

Although Exome and Genome sequencing are considered powerful diagnostic tools, the probability of genetic variation being the absolute cause of genetic diseases is still extremely low. The 1000 Human Genome Project stated that the human genome differs from the reference genomes in 4.1 to 5 million sites, and most of these variations are benign [6]. A very small number of variants can cause diseases either to the individual or their children. Although genomic testing succeeded in identifying all the variants, the result of genomic testing needs filtering to classify variants according to their pathogenicity. Unfortunately, the interpretation and classification of variants were found to be different in different labs. This can lead to an incorrect diagnosis of patients with genetic diseases [7].

14.2 The Use of Genetic Testing to Screen for Diseases

Traditionally, screening uses specific markers and surveys the population to identify individuals with a high risk of particular disorders after determining the screening cut-off parameter. Genetic screening is a survey of the population to identify individuals having a high risk of developing or transmitting a genetic disease [8].

Screening programs for patients with increased risk of genetic diseases started as early as 1960 by screening asymptomatic newborns with phenylketonuria. Then, in 1968, "Principles and Practice of Screening for a Disease" was published. It is a guideline that evaluates the benefits of screening for different diseases [9]. It stated the criteria for the success of screening programs include the following: the disease must be an important health problem, the biochemical basis of the disease was known, having a sufficient prevalence in the population, a cost-effective and reliable screening test was developed, effective treatment was available, moreover, genetic screening must be important for the patients. Two diseases in which these criteria were met at that time were Tay-Sachs and sickle-cell diseases [8, 10].

Tests used for screening should be accurate, specific, simple, safe, rapid, and inexpensive. The test should be specific to avoid false positive results, and accurate to screen every carrier and avoid false negative results [10]. However, in most instances, there are no genetic markers of many diseases, and even when they are present, they are insensitive and nonspecific for definite diseases. Additionally, the presence of a mutation might not lead to the development of diseases. Nonetheless, positive genetic screening results might affect the carriers and their families psychologically [11].

14.3 Patient Understanding and Knowledge

Health literacy refers to the knowledge and ability of individuals to find, understand, assess, and use health information to make rational decisions about their health in various contexts of their lives [12]. Health literacy includes the ability of the patients to read and understand written information, the skills to use quantitative data, and the capability to communicate verbally [12]. Healthcare providers should know how to introduce medical information to patients to help them make different decisions [12]. People misunderstand various genetic concepts like carrier status and homozygosity versus heterozygosity, besides genetic testing mechanisms and statistics [13]. With the introduction of different types of genetic testing, the term genetic literacy has emerged. Genetic literacy is considered one of the causes of negative patients' perceptions of genetic testing. Research concerning genetic literacy is limited. The Rapid Estimate of Adult Literacy in Genetics (REAL-G) assesses patients' fluency by giving them a list containing different medical terms that differ in complexity [14]. This scoring aims to predict individuals' ability to memorize transferred information during genetic counselling [14]. Individuals who scored sixth-grade level and above could recall more communicated information after simulated genetic counseling, meaningfully. However, individuals with a score lower than sixth grade were found to overestimate the impact of genetic results [14].

A meta-analysis was conducted to identify the requirements and preferences of patients and healthcare providers to understand pharmacogenomics tests. The study included 36 articles; most of the included studies were done in the USA and published between 2004 and 2019. The data from these studies were collected from either patients/general public, healthcare providers, or both. The results found that the general public used terms like "genetics and genes" inaccurately, and they failed to explain them. Moreover, they used the terms "genetics and genomics" interchangeably. Individuals reported that they had heard about genetic testing; however, they had not heard about genomics and how these tests help in drug selection for patients. The study revealed that only genotyped patients knew about pharmacogenetic testing; however, they were unsure about what these tests are [12].

Several studies clarified that public understanding of genetics was the concept of heredity rather than molecular genetics. Terms like DNA, chromosomes, genome, traits within the family, and first-degree relative are sometimes misinterpreted by the public [15]. Public understanding of patterns of inheritance is determined by the social relationship. Studies assessing knowledge of the public about the pattern of inheritance found that most of the participants reported that children shared 50% of their genes with their fathers, while genetic closeness is lower between siblings. However, the public showed uncertainty about the sex-related dimension of inheritance as they do not report that a child inherits 50% of their genes from each parent in a random fashion. Another area in which there is misunderstanding by the patients is the recessive genes; people do not usually understand that two copies of a recessive gene are needed to develop a disease. The term "skipping generation" is given to individuals who do not have the disease, though they are still carriers [15]. Personal experience is considered one of the factors influencing genetic knowledge; people who have a family history of a disease have a better understanding of genetics than people who developed a disease without a known family history. The way of predicting the development of hereditary disease in the family depends on the physical resemblance between family members. Physical resemblance is considered a risk factor for the development of hereditary diseases [13].

Genetic education for patients with genetic diseases is of great importance. Usually, education sessions are facilitated by a geneticist, genetic counselor, or a nurse trained in genetics. The education sessions are face-to-face sessions to provide information about inheritance patterns, different family members' expected risk of having genetic disease, various options for genetic testing, symptoms, treatment, and prognosis of the disease [16]. This will help the patients to gain knowledge, which may improve the outcomes concerning their perception of risk, coping with stress and anxiety, and making informed decisions. All these improve genetic services and reduce burden of genetic diseases [16]. However, due to the scarcity of time and personnel, the genetic personnel may be unable to deliver thorough information to the patients [12].

Calsbeek *et al.* [17] assessed knowledge and attitudes of patients with diabetes mellitus, asthma, and cardiovascular diseases toward genetic testing. The study identified that increased knowledge about genetics and genetic testing was associated with a higher level of education and a younger age. The study found that patients of older ages were related to lower levels of knowledge [17].

14.4 Perception of Patients towards Genetic Testing

Perception about genetic testing determines test utilization and builds theories of health behavior and decision-making. Television programmes overstress the benefits of genetic testing, especially for forensic purposes, while the media also overstate the benefits of genetic testing and ignore its limitations [18]. Generally, public opinion toward genetic testing is favorable. Genetic tests for the detection and screening of diseases are considered valuable. Even when these tests were used for diseases with limited clinical utility, like psychiatric diseases, public opinion was positive. However, biotechnologies such as genetically modified food, human cloning, and gene therapy, which change natural processes, are not as popular. There are additional factors that govern the preference of the public towards genetic factors. The first factor to be considered is age; studies have found that individuals at a young age are always open to new technologies. The second factor is gender; although males have more positive attitudes toward genetic testing, women are more knowledgeable about genetic testing. The last factor is education level, as the education level becomes higher, some individuals favor genetic testing, though it can also lead to negative expectations about the benefits of genetic tests [19].

A 2010 study [19] using a questionnaire to assess public perceptions toward genetic testing revealed that respondents expected genetic tests to be more common. They also expected that knowing the genetic background of the disease would lengthen people's lives. On the other hand, this study showed that people became curious about their genetic makeup, although there were some concerns reported by highly educated individuals and younger respondents [19].

In a study conducted by Chin and Tham [16] to assess the knowledge, awareness, and perception of Malaysians in Klang Valley about genetic testing for hereditary diseases, 79.8% of the participants were aware of genetic testing, and the source of this information was the internet. However, 71.1% of the participants knew that genetic testing is usually used to screen for genetic diseases, while 57.2% were uncertain about the utility of genetic testing in reducing hereditary diseases [16]. Regarding the use of genetic testing for screening of different types of cancer, 58.3% were unaware of it. Half of the participants (50.2%) were willing to experience genetic testing. The participants reported that genetic testing should be conducted in hospitals under physician supervision. They were against selling genetic kits in stores or through the internet [16]. The participants had a good perception of genetic testing and agreed to encourage pregnant women with a history of hereditary diseases to undergo prenatal testing. Barriers to genetic testing reported by a minority of respondents include religious belief that opposes genetic testing. Lack of education about genetic testing increases the fear of the participants of unethical procedures in the practice of genetic testing. Regarding this issue, participants recommend that the safety and efficiency of genetic testing would be better ensured through government-established policies and regulations [16].

Another study conducted by McGowan *et al.* [2] describes the understanding and prospects of genetic patients and their caregivers towards genetic testing using semi-structured interviews with the patients, their spouses, and parents of patients under 18 in three genetics clinics in Cleveland, Ohio, USA. The study assessed participants' views on clinical genetic testing by asking them about their interest in genetic testing, preferences for running tests and receiving results, apparent benefits and risks, and communicating results with others [2]. People who experienced pharmacogenomics testing felt more confident that the medications selected based on these tests would be harmless and would recover their health. Some patients reported that these tests validated their need for dose increments when they claimed the medications were not working. They felt that physicians do not always believe them [2].

Evaluation of genetic testing services by patients who experienced genetic testing, although the testing was considered controversial, the presentation of these services in a good way was associated with good outcomes in clinical parts, i.e., increased patient compliance to preventive measures and treatment, and better consequences [20]. Patients always evaluated features like how the caregiver communicated with them, the sensitivity of the clinical staff to patient needs, and the discussion between different providers in the clinical process [20]. The utility of predictive genomics could be evaluated by clinical measures such as the rate of hospitalization and the rate of death, initiating prophylactic treatment, and quality of life from a health perspective. The importance of evaluating the clinical value of genetic testing is to provide patients with appropriate information [20].

Perception of the public towards causes of diseases varies according to different studies. The public tends to believe that causes of diseases like hypertension, cancer, and heart diseases are neither genetic nor environmental alone. Moreover, they believe that age can determine whether the cause is genetic or environmental, with the incidence of diseases at younger ages indicating genetic inheritance [15]. On the other hand, public perception of genetic testing revealed a fear of discrimination against individuals who carry a higher risk of genetic diseases, and individuals could be labeled as "good" or "bad" accordingly [21].

Another challenge affecting genetic testing is return of genetic results. Many studies showed that return of genetic results of monogenetic diseases to carriers of risk genes was associated with strong negative emotions such as worry, anxiety, general distress, depression, disease-related distress, hopelessness, poorer general wellbeing, and decreased quality of life. Fortunately, these negative symptoms might probably disappear after six months of return of genetic results [22].

On the other hand, the return of the genetic results of diseases with polygenic inheritance was not associated with strong negative emotions, which might be because of the uncertain and small effect of the genes associated with such diseases. Still, carriers of risk genes were found to increase in reporting of healthy behaviors; carriers of the APOE gene for Alzheimer's disease started to positively change their diet, and carriers of melanoma genes used sunscreens to protect from skin cancer [22].

14.5 Barriers against Pharmacogenomics Testing

Although the providers of pharmacogenomics tests know the significance of these tests in fields like the prevention of some diseases and personalized medicine selection, they have many concerns about these tests in clinical practice. The most reported concern is the cost of these tests and whether the insurance companies will cover these costs or not. Moreover, extra time is spent on the incorporation of test results, evidence supporting the pharmacogenomics relationship, and conflict with the clinical information. Additionally, they reported that pharmacogenomics information

Figure 14.1 Barriers encountered by genetic testing providers [23].

should be easily accessible by the providers and easily integrated into workflows. The providers stated that lack of knowledge, lack of comprehensive guidelines for use of pharmacogenomics testing, and absence of pharmacogenomics data for most drugs are the barriers that prevent them from asking for pharmacogenomics testing [23] (Figure 14.1).

Lack of formal genetic training for healthcare providers can lead to challenges in genetic testing results. These interpretations include a misconception of a harmless variant as pathogenic, a misconception of a risk variant as benign, and a variant of unknown impact as either benign or pathogenic [24] (Figure 14.2).

The perception of low-income communities or minorities on genetic testing is understudied. These communities were not covered by health insurance. Moreover, these groups are underserved, especially for genetic testing. In the U.S., studies identified that only a few black and Hispanic females were sent for genetic diagnosis for breast cancer compared with white females. Reasons behind the low rate of genetic testing for minorities were deficiency of knowledge and consciousness of genetic testing and

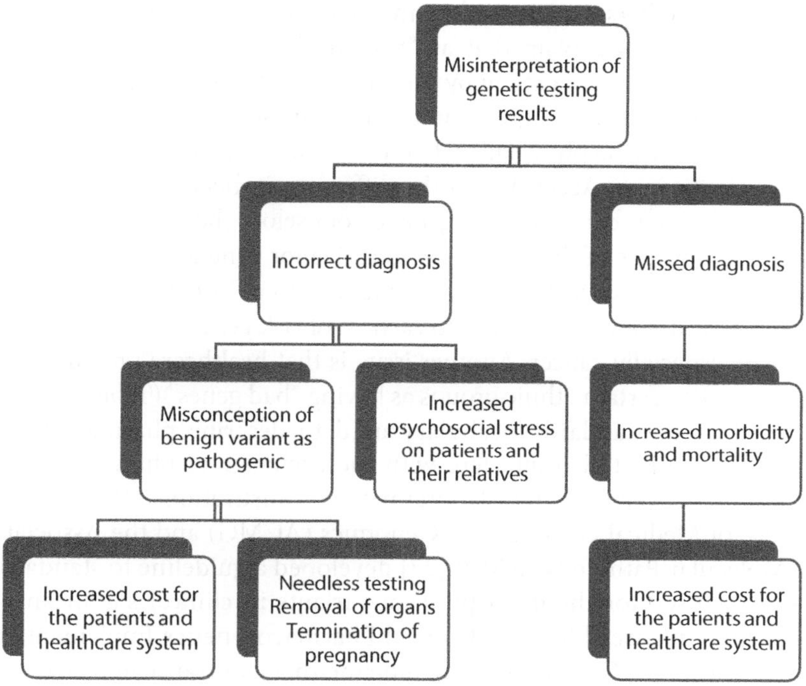

Figure 14.2 Consequences of misinterpretation of genetic testing results [24].

available services, language barrier, stigma associated with cancer gene carriers, bad views about cancer, expectation of negative emotions, uncertainty about information provided, and mistrust in how data would be used [21]. These results agreed with the results of a study in the Latino community in the U.S. Latinos comprised around 17% of the U.S. community. They did not have total health insurance coverage, although they suffered from a lot of cancer types, which prevented them from early screening for cancer. Only 4.2% of Latino females were screened for BRCA1/2 genes associated with breast and ovarian cancer [25].

Additionally, cultural and religious beliefs might prevent individuals with a risk of genetic diseases from conducting genetic testing. A study conducted on Australians of Chinese origin revealed that this population did not believe cancer can be hereditary or that mutations in certain genes can lead to cancer development in some members of the same family. They also did not believe that the reason behind some diseases could be genetic unless there was an extensive family history of the disease. Members of the population in the study did not know the meaning of mutations exactly. Some of them explained its emergence by what the prior generation had done.

Christian members explained the emergence of genetic disease as "God's will", while others explained it as "bad luck". The Chinese community's pedigree, according to a study by Eisenbruch *et al.*, was different from the Western community pedigree due to the Chinese patrilineal system. This change may confuse physicians when they assess the degree of inheritance of genetic diseases. According to the difference in beliefs and conceptions between different communities, genetic counselors should learn about the culture and beliefs of the families who arrive for genetic counseling before their arrival at the centers providing genetic testing. Nonetheless, physicians should consider care to be given to carriers of risk genes of different genetic diseases, especially cancer. Another issue is that healthcare providers must not stigmatize certain ethnic groups as having "bad genes" (Table 14.1) [26].

The lack of standard vocabulary used to describe pharmacogenomics findings might also interfere with the exchange of pharmacogenomics results between different laboratories and institutions. The American College of Medical Genetics and Genomics (ACMG) and the Association for Molecular Pathology (AMP) [27] developed a guideline to standardize the terms used for the description of variants in clinical significance in Mendelian diseases [28]. The Clinical Pharmacogenetics Implementation Consortium (CPIC) developed a guideline that can translate pharmacogenomic results into actionable decisions regarding the prescribed drug. They used the Delphi survey method, which is an approach used to seek expert agreement on a specific topic [28]. The study revealed that the use of the term "extensive metabolizer" to describe the status of individual metabolism was confusing to the clinicians and they preferred to use the term "normal metabolizers". Another issue raised by this study was the varying opinions of the experts about terms describing alleles or functions when there is no data or conflicting data supporting the results. Clinicians used two terms to describe the aforementioned situations: "no literature describing function" and "conflicting data" [28].

Another issue is the confidentiality of individual persons when pursuing familial genetic testing. In clinical practice, guidelines put measures in place to keep the confidentiality of family results as a whole, but no guidelines discuss the confidentiality of individuals within the family who do not want others to know about their results. This could lead to disruption of family dynamics and decrease trust in the health services [7].

The lack of absolute treatment for genetic diseases decreases the utility of these tests by individuals or families with genetic diseases. Many of the individuals refused to receive genetic counseling or genetic testing because there was no available treatment for their conditions if they were carriers of genetic diseases [7, 29].

The presence of strong emotional responses like anxiety, interest, pride, sorrow, and tenseness can drive seeking or avoidance behaviors and may also trigger early medical decisions made about genetics [30]. Emotions related to engagement, like interest, decreased in six weeks to six months post-disclosure of genetic results. This may challenge the healthcare team to convince the patients to take action related to medical recommendations before their motivation decreases [30]. Low psychological impact might result from overcorrection of genetic results framing to limit the emotional response of the patients. The degree of emotional reactions after disclosure of genetic results differs between patients seeking genetic testing, and patients who undergo genetic testing for cancer genes suffer more from negative emotions compared to the cardiovascular disease patients [30].

14.6 Incidental Findings

Whole-genome and exome tests became available to be used for different clinical indications. This sequencing process yields results related to the purpose of the sequencing as well as incidental findings or secondary results that are not related to the indication of the sequencing. However, incidental findings may be of medical importance for the clinician and the patients. The American College of Medical Genetics and Genomics (ACMG) recommends the return of these incidental findings as a set of actionable genes [31]. ACMG mentioned that the issue of these incidental findings should be considered in pre- and post-sequencing counseling with the patients [31]. Dorschner *et al.* [32] screened 1,000 randomly selected individuals from the National Heart, Lung, and Blood Institute (NHLBI) Exome Sequencing Project for variants in a list of 114 actionable genes. These actionable genes are associated with genetic conditions that might remain undiagnosed in adults. The list used in this study includes 52 genes out of 56 genes recommended by the ACMG to be returned to the patients. The study found that only 23 individuals showed pathogenic or likely pathogenic variants. Thus, the allele frequency of these mutations is very low. Moreover, the study revealed a deficit in genetic literature on variants for non-European participants. The study mentioned that current databases do not report the pathogenicity of many variants. As a result, the study suggests developing more consistent criteria for reporting incidental findings [32].

Another study assessed the psychological impact on the patients upon receiving secondary results of genetic testing. The study found that the negative psychological impact was limited. The study revealed that there is a decrease in negative emotions after six months. Return of secondary genetic results for patients who were tested for a specific genetic disease can help in the discovery of other genetic diseases or proposed carriers at early stages [30].

14.7 Perceptions towards Prenatal Genetic Testing

The term "prenatal" means from pregnancy till birth. Pre-natal screening has been established for several decades as screening for cystic fibrosis carriers. Nowadays, the availability of genetic testing at almost affordable prices, besides the appearance of what is called "Direct to Consumer" marketing, increases the awareness and the desire of patients to do genetic testing. The new biotechnological techniques make screening for cancer in patients with a family history of malignancy attractive to gynecologists. The American College of Obstetricians and Gynecologists Committee of Genetics confirms the need for the knowledge and expertise of obstetricians and gynecologists or another healthcare team who offer these tests, in addition to the time and resources required for counseling of the patients [33]. Pre-test considerations include appropriate selection of the population and the tests to be used, especially if the patient is pregnant. In the case of screening for fetal aneuploidy, the committee recommends testing all pregnant women in the early stages of pregnancy using standardized methods to avoid false-negative tests. However, this raises ethical issues as these tests might be ordered as routine tests. There is also a fear of putting parents under pressure to do these tests, and some of the parents might decide to terminate the pregnancy in response to the results of genetic testing [7]. Pre-test counseling should provide information about the type of tests, expected results, benefits, risks, and limitations. Post-test counseling informs the patients of the significance of the results, and whether the patients may need further genetic testing or further follow-up should be discussed with the patients. Additionally, the discovery of "variants of uncertain significance" should be addressed with the patients [33].

14.8 Perception towards Genetic Testing for Cancer

The demand for genetic testing for cancer has increased in the last two decades. This resulted from the discovery of different mutations associated with cancer. Previous studies and surveys revealed positive attitudes of physicians towards genetic testing for hereditary types of cancer, such as breast cancer. Some of the physicians in the early 2000s reported that they recommended genetic testing for patients only upon their request. Physicians also preferred to allocate money for the development of better diagnostic and treatment protocols. Primary care physicians did not prefer to play a role in genetic counseling or follow-up of families with a known risk of hereditary cancer. Reasons behind that included a lack of knowledge, as they did not have training on conducting genetic testing and interpreting the results, besides the limited time assigned for these tasks [21]. However, primary care physicians with good knowledge of genetics appropriately assessed their patients. There is a positive association between a few years of practice and enhancement in family history taking. Specialty also had a role in the quality of history taking, as in the case of breast cancer, obstetricians and genetic counsellors seem to be more knowledgeable than other healthcare professionals [21].

Still, oncology was considered one of the leading specialities that apply genomic tests as routine tests to screen the KRAS gene for colorectal cancer [1], the BRAF gene for melanoma [1], and multiple genes associated with breast cancer/ovarian cancer [1]. Genetic studies confirmed the association of mutations in BRCA1 and BRCA2 with breast, ovarian, and prostate cancers. Families with a history of hereditary cancer should receive counseling and testing to identify personal cancer risk. This will help physicians to decide whether these patients should attend regular screening or go for surgery to decrease the burden of cancer [34]. Education of families with a history of hereditary cancer and referral of these families for genetic counseling might help in the avoidance of cancer development. In addition, the availability of centers that provide genetic testing services will encourage families to do genetic testing. Education plans for families should focus on the correction of negative beliefs especially in ethnic minority groups like removing cancer stigma and mistrust in healthcare providers which can affect patients' psychological well-being and may discourage cancer patients from seeking help [34].

A study conducted by Ford *et al.* [29] assessed the perception of women with a family history of breast/ovarian cancers toward breast cancer genetic counseling. The study involved three groups of women: Group One involved African-American women who were referred to genetic counseling and approved of taking the session, Group Two consisted of African-American women who were offered genetic counseling and refused to receive it, and Group Three involved Caucasian women who were offered genetic counseling and refused to receive it [29]. All these women had the costs of genetic counseling waived. All women involved in the study reported they were afraid of the word "breast cancer." Women engaged in Group Two reported that they knew a lot about breast cancer, BRCA1 and BRCA2, and the hereditary tendency of this type of cancer from the internet and television shows. This result was in contrast with other studies, which indicated that the knowledge of African-Americans was less than that of Caucasians concerning genetic testing and counseling. On another note, women who received genetic counseling became careful about having mammograms every year, doing self-exams, and enlightening their relatives about the importance of breast cancer screening [29]. Reasons behind the reluctance of the other two groups to receive genetic counseling were fear of getting breast cancer, which was synonymous with death in their opinion, fear of different treatment by others, and fear of only having one breast if they got a mastectomy, which would be embarrassing. In addition to that, Caucasian women reported they would like to be cured of breast cancer, which decreases the need for prophylactic mastectomy [29]. The two groups who refused to go for genetic counseling believed in the importance of early cancer screening, and hence, they had mammograms [29]. However, they refused because there is no cure for cancer, and they were afraid of prophylactic mastectomy. Other issues that prevent women from receiving genetic counseling mentioned by the participants included the cost of genetic counseling, fear of sharing secret family health information with healthcare workers and insurers, financial barriers to the treatment of cancer, and mistrust in the ability of doctors to remove all cancer cells during surgery [29] (Table 14.1). Some of the participants did not know that their father's genes may have a role in increased risk for breast cancer [29].

Heshka *et al.* [35] conducted a systematic review to assess the psychological and behavioral impacts of genetic testing, including studies that evaluated hereditary nonpolyposis colorectal carcinoma, hereditary breast, ovarian cancer, and Alzheimer's disease [35]. After reviewing thirty studies, the review found there was no impact of genetic testing on negative psychological responses such as anxiety, distress, or depression, either in carriers of the risk genes or in the noncarriers, although some studies included in the review reported a short-term (around four months) increase in some psychological aspects among carriers. Regarding prophylactic surgeries

Table 14.1 Barriers encountered by the patients or patients' relatives against genetic testing.

Barriers encountered by the patients/patients' relatives
• Lack of knowledge about genetic testing. • Stigma associated with risk variant carriers. • Confidentiality concerns. • Religious and cultural beliefs. • Cost of genetic testing and insurance company coverage.

for carriers, the studies showing an increase in female response to oophorectomy compared to mastectomy. Post-genetic testing screening showed that the proportion of breast cancer screening was similar in carriers and non-carriers; however, the ovarian and colorectal screening proportion was higher in carriers compared to non-carriers. This might be due to the invasive techniques used for the screening of both ovarian and colorectal cancers. Moreover, the review identified no difference in risk perception among carriers and non-carriers after twelve months of genetic testing. At this time, the risk perception was lower compared to the baseline [35].

14.9 Perception towards Genetic Testing for Neurological Degenerative and Psychiatric Diseases

In a study conducted by Christensen *et al.* [18] to measure the change in perception of first-degree relatives of Alzheimer's disease patients before and after doing genetic testing of the APOE gene, participants filled out a questionnaire in which they had to rate the reasons why someone would do genetic testing and why someone would not do genetic tests before conducting the test and after one-year follow-up. The study found that participants rated reasons for taking genetic testing more than reasons for not taking genetic testing (Table 14.2). The APOE gene test is not a routine test because there are no available prevention measures for Alzheimer's disease; still, participants thought they would like to be part of research concerning Alzheimer's disease and know more about preventative measures especially those who identified as carriers of risk variants. On another note, the study revealed that participants rated the reason "to give information about my children's possible risk for AD" lower at follow-up than at baseline [18]. This change could be due to the feeling of guilt for carriers of risk variants of transferring risk genes to the new generation [18].

14.10 Perception towards Genetic Testing for Other Diseases

The general perception of the public towards genetic testing increased as the majority of the participants in a study conducted by Ganne *et al.* [37] reported they would do genetic testing if they had a genetic eye disease. They felt genetic testing should be carried out even if there is no available treatment for the genetic disease. They felt these could lead to a better

Table 14.2 Reasons reported by patients/patients' relatives participated in different studies behind taking/not taking genetic testing.

Study	Disease	Reasons for taking genetic testing	Reasons for not taking genetic testing
[18]	Alzheimer's disease	"To seek information about preventative measures" [18] "To know about risk in case better treatments become available" [18] "To make arrangements for long-term care" [18] "to start doing things sooner" [18]	"The results could affect my employment," [18] "The results could affect my health insurance" [18] "The results could change how people look or act toward me" [18] "Lack of a cure or prevention" [18]
[2]	Different genetic diseases	"I think knowing something is so much better than not knowing…then at least you can plan for something you know" [2] "I also look at the people that have children that they go years and years wondering, knowing as a mother that something isn't quite right with your child, and maybe finally finding out what that is, and just knowing so that you can treat it in a better way than not knowing, and just spending years and years, and how much money testing, trying to figure something out. You can actually target all your efforts towards that one diagnosis once you really kind of know." [2]	"Well, I don't think anyone would want to know…I mean, it's not something you would sign up and say 'please tell me', but you start having symptoms, you want to know why. Why is this happening? So if you had something, symptoms, ok do the test and just tell me what's going on." [2] "I wouldn't want to unduly stress out or worry family members over something that perhaps is not that … strongly correlated and, you know, cause that kind of concern over some- thing that may not be set in stone in a concrete diagnosis." [2]

(Continued)

Table 14.2 Reasons reported by patients/patients' relatives participated in different studies behind taking/not taking genetic testing. (*Continued*)

Study	Disease	Reasons for taking genetic testing	Reasons for not taking genetic testing
[29]	Breast cancer	"To rule you out. Give you that satisfaction, and some sense of security" "The more information you have, the more likely you are to go" [29] "Prevention; the earlier, the better" [29]	"Fear... I guess lack of knowledge, too..." "The fears of finding out you are a carrier [of the gene]" [29] "If I took that genetic testing, that's definitely going to scare me even more" [29]
[36]	Direct-to-consumer genetic testing/ Alzheimer's disease	"I was happy because it did provide me with information about myself based on my genetics—which itself was fascinating to look at and dig through. But, the results were disappointing because they didn't really provide me with much insight into myself. . . . I secretly hoped that there would be more. More detail, more specifics, more certainty. When I think about risk and how I understood risk—before and after testing, my expectations were not at all in the realm of what the test could offer, but instead, were informed by what I wanted the test to be able to do" [36]	"My relatives took part because they thought it would be 'interesting' and being a generally open bunch thought nothing of sharing information more widely. But would information about my mother's genetic Alzheimer predisposition not be better and exclusively kept in her confidential health records? That way, I could only find out about my genotype if I had wanted a test myself or got specific consent from both my parents to examine their genotypes. I did neither" [36]

knowledge of the disease and prediction of the future of their children [37]. On the other hand, around 5% of the participants in the Ganne *et al.* study reported that genetic testing should be prohibited due to many reasons, including the emotional stress of the discovery of the genetic risk, high cost, unavailability of treatment, and ethical concerns [37]. The study also assessed public perception toward gene therapy, for which the majority of the participants agreed with gene therapy for genetic eye problems. Only 19% of the participants felt that the information provided on the internet about genetic eye diseases is reliable. The study also identified poor awareness of the participants about the location and availability of the genetic facilities [37].

14.11 Conclusion

Rapid advance in health genetics increases the opportunities to use these tests for many purposes, like a diagnosis of monogenic diseases in presymptomatic individuals, among other uses [38]. However, many challenges are still facing genetic testing from both the providers' and the patients' sides. Lack of knowledge of providers about the guidelines of genetics testing leads to avoidance of asking the patients to do genetic testing [23]. Moreover, misinterpretation of genetic results can lead to misdiagnosis with either the implementation of unnecessary treatment or increased morbidity and mortality due to an unidentified risk variant [24]. On another note, lack of knowledge of patients and patients' relatives about genetic testing and services provided can lead to wrong decisions such as not taking the genetic testing or it can lead to higher expectations in genetic results which may need correction in genetic counselling session [36]. Another issue is genetic literacy; many studies have shown that the public misunderstands terms like inheritance, carrier status, and physical resemblance with the presence of risk genes [13].

Abbreviations

ACMG	The American College of Medical Genetics and Genomics
AMP	Association for Molecular Pathology
APOE	Apolipoprotein E
BRAF	Gene encodes a protein called B-Raf
BRCA1/2	Tumor suppressor genes
CPIC Clinical	Pharmacogenetics Implementation Consortium

KRAS	A gene that controls cell growth
NGS	Next-Generation Sequencing
NHLBI	National Heart, Lung, and Blood Institute
REAL-G	Rapid Estimate of Adult Literacy in Genetics
WHO	World Health Organization

AI disclosure statement: The author acknowledges the use of AI-based tools for minor language editing or grammatical correction. However, all intellectual contributions, analyses, and conclusions presented in this chapter are entirely the author's own.

References

1. Hall, M.J., Forman, A., Montgomery, S., Rainey, K., Daly, M.B., Understanding patient and provider perceptions and expectations of genomic medicine. *J. Surg. Oncol.*, 111, 1, 9–17, 2016.
2. Mcgowan, M.L., Glinka, A., Asaad, G., Genetics patients' perspectives on clinical genomic testing. *Pers. Med.*, 10, 4, 339–347, 2013 Jun 1, doi: 10.2217/pme.13.32.
3. Mishra, V., Chanda, P., Tambuwala, M.M., Suttee, A., Personalized medicine: An overview. *Int. J. Pharm. Qual. Assur.*, 10, 2, 290–4, 2019.
4. Shukla, R., Pharmacogenomics: Overview, Applications, and Recent Developments, in: *Drug Des - Nov Adv Omi F Appl.*, 2021.
5. Frueh, F.W. and Gurwitz, D., From pharmacogenetics to personalized medicine: A vital need for educating health professionals and the community. *Pharmacogenomics*, 5, 5, 571–9, 2004.
6. Project T 1000 G. ARTICLE A global reference for human genetic variation. Open, 2015.
7. Horton, R.H. and Lucassen, A.M., Recent developments in genetic/genomic medicine. *Clin. Sci. (Lond.)*, 133, 5, 697–708, 2019 Mar 5, doi: 10.1042/CS20180436. Print 2019 Mar 15.
8. Traeger-Synodinos, J. and Rousseau, F., Introduction to Genetic Screening. *OBM Genet.*, 3, 3, 1–7, 2019.
9. Cornel, M.C., Rigter, T., Jansen ME, H.L., Neonatal and carrier screening for rare diseases: How innovation challenges screening criteria worldwide. *J. Community Genet.*, 257–65, 2021.
10. Philip, N., Screening for genetic disorders. *Childs Nerv. Syst.*, 19, 7–8, 436–9, 2003.
11. Wald, N.J., What is Genetic Screening Anyway? *J. Med. Screen.*, 3, 2, 57, 1996.
12. Veilleux, S., Bouffard, M., Bourque Bouliane, M., Patient and Health Care Provider Needs and Preferences in Understanding Pharmacogenomic and

Genomic Testing: A Meta-Data Analysis. *Qual. Health Res.*, 30, 1, 43–59, 2020.

13. Klitzman, R.L., Misunderstandings concerning genetics among patients confronting genetic disease. *J. Genet. Couns.*, 19, 5, 430–46, 2010.
14. Hooker, G.W., Peay, H., Erby, L., Bayless, T., Biesecker, B.B., Roter, D.L., Genetic literacy and patient perceptions of IBD testing utility and disease control: A randomized vignette study of genetic testing. *Inflamm. Bowel Dis.*, 20, 5, 901–8, 2014.
15. Condit, C., Public understandings of genetics and health. *Clin. Genet.*, 1–9, 2010.
16. Chin, J.J. and Tham, H.W., Knowledge, Awareness, and Perception of Genetic Testing for Hereditary Disorders Among Malaysians in Klang Valley. *Front. Genet.*, 11, December, 1–10, 2020.
17. Calsbeek, H., Morren, M., Bensing, J., Rijken, M., Knowledge and attitudes towards genetic testing: A two year follow-up study in patients with asthma, diabetes mellitus and cardiovascular disease. *J. Genet. Couns.*, 16, 4, 493–504, 2007.
18. Christensen, K.D., Roberts, J.S., Uhlmann, W.R., Green, R.C., Changes to perceptions of the pros and cons of genetic susceptibility testing after APOE genotyping for Alzheimer disease risk. *Genet. Med.* [Internet], 13, 5, 409–14, 2011, Available from: https://doi.org/10.1097/GIM.0b013e3182076bf1.
19. Henneman, L., Vermeulen, E., Van El, C.G., Claassen, L., Timmermans, D.R.M., Cornel, M.C., Public attitudes towards genetic testing revisited: Comparing opinions between 2002 and 2010. *Eur. J. Hum. Genet.*, 21, 8, 793–9, 2013.
20. D'Andrea, E., Lagerberg, T., De Vito, C., Pitini, E., Marzuillo, C., Massimi, A., *et al.*, Patient experience and utility of genetic information: A cross-sectional study among patients tested for cancer susceptibility and thrombophilia. *Eur. J. Hum. Genet.* [Internet], 26, 4, 518–26, 2018, Available from: http://dx.doi.org/10.1038/s41431-017-0083-1.
21. Pichert, G., Dietrich, D., Moosmann, P., Zwahlen, M., Stahel, R.A., Sappino, A.P., Swiss primary care physicians' knowledge, attitudes and perception towards genetic testing for hereditary breast cancer. *Fam. Cancer*, 2, 3–4, 153–8, 2003.
22. Frieser, M.J., Wilson, S., Vrieze, S., Behavioral Impact of Return of Genetic Test Results for Complex Disease: Systematic Review and Meta-Analysis. *Health Psychol.*, 37, 12, 1134–44, 2018.
23. Borden, B.A., Galecki, P., Wellmann, R., Danahey, K., Mee, S., Patrick-Miller, L., *et al.*, Assessment of provider-perceived barriers to clinical use of pharmacogenomics during participation in an institutional implementation study. *Pharmacogenet. Genom.*, 31–8, 2019, doi: 10.1097/FPC.0000000000000362.
24. Donohue, K.E., Gooch, C., Katz, A., Wakelee, J., Slavotinek, A., Bruce R., Korf. Pitfalls and challenges in genetic test interpretation: An exploration of

genetic professionals experience with interpretation of results. *Clin. Genet.*, 99, 5, 638–49, 2021.
25. Hamilton, J.G., Shuk, E., Arniella, G., González, C.J., Gold, G.S., Gany, F., *et al.*, Genetic Testing Awareness and Attitudes among Latinos: Exploring Shared Perceptions and Gender-Based Differences. *Public Health Genomics*, 19, 1, 34–46, 2016.
26. Eisenbruch, M., Yeo, S.S., Meiser, B., Goldstein, D., Tucker, K., Barlow-Stewart, K., Optimising clinical practice in cancer genetics with cultural competence: Lessons to be learned from ethnographic research with Chinese-Australians. *Soc. Sci. Med.*, 59, 2, 235–48, 2004.
27. Bergquist, T., Stenton, S.L., Nadeau, E.A.W., ABB, S. M, Greenblatt, Harrison, S.M., Tavtigian, S.V., *et al.*, Calibration of additional computational tools expands ClinGen recommendation options for variant classification with PP3/BP4 criteria. 1–17, 2024.
28. Caudle, K.E., Dunnenberger, H.M., Freimuth, R.R., Peterson, J.F., Burlison, J.D., Whirl-carrillo, M., *et al.*, Open Standardizing terms for clinical pharmacogenetic test results: Consensus terms from the Clinical Pharmacogenetics Implementation Consortium (CPIC). *Genet. Med.*, 19, 2, 215–23, 2017.
29. Ford, M.E., Alford, S.H., Britton, D., McClary, B., Gordon, H.S., Factors influencing perceptions of breast cancer genetic counseling among women in an urban health care system. *J. Genet. Couns.*, 16, 6, 735–53, 2007.
30. McCormick, C.Z., Yu, K.D., Johns, A., Campbell-Salome, G., Hallquist, M.L.G., Sturm, A.C., *et al.*, Investigating Psychological Impact after Receiving Genetic Risk Results—A Survey of Participants in a Population Genomic Screening Program. *J. Pers. Med.*, 12, 12, 2022.
31. Green, R.C., Berg, J.S., Grody, W.W., Kalia, S.S., Korf, B.R., Martin, C.L., McGuire, A., Nussbaum, R.L., O'Daniel, J.M., ACMG Recommendations for Reporting of Incidental Findings in Clinical Exome and Genome Sequencing. *Genet. Med.*, 15, 7, 565–74, 2013.
32. Dorschner, M.O., Amendola, L.M., Turner, E.H., Robertson, P.D., Shirts, B.H., Gallego, C.J., *et al.*, Actionable, pathogenic incidental findings in 1,000 participants' exomes. *Am. J. Hum. Genet.* [Internet], 93, 4, 631–40, 2013, Available from: http://dx.doi.org/10.1016/j.ajhg.2013.08.006.
33. Committee on Genetics, Counselling About Genetic Testing and Communication of Genetic Test Results: Committee Opinion Number 693. *Obstet. Gynecol.*, 130, 2, e210, 2017.
34. Hann, K.E.J., Freeman, M., Fraser, L., Waller, J., Sanderson, S.C., Rahman, B., *et al.*, Awareness, knowledge, perceptions, and attitudes towards genetic testing for cancer risk among ethnic minority groups: A systematic review. *BMC Public Health*, 17, 1, 1–30, 2017.
35. Heshka, J.T., Palleschi, C., Howley, H., Wilson, B., Wells, P.S., A systematic review of perceived risks, psychological and behavioral impacts of genetic testing. *Genet. Med.* [Internet], 10, 1, 19–32, 2008, Available from: https://doi.org/10.1097/GIM.0b013e31815f524f.

36. Anderson, E.E., Wasson, K., Inquiry, N., Anderson, E.E., Wasson, K., Personal Narratives of Genetic Testing: Expectations, Emotions, and Impact on Self and Family Commentary Personal Narratives of Genetic Testing: Expectations, Emotions, and Impact on Self and Family. *Narrat. Inq. Bioeth.*, 5, 3, 229–35, 2018.
37. Ganne, P., Garrioch, R., Votruba, M., Perceptions and understanding of genetics and genetic eye disease and attitudes to genetic testing and gene therapy in a primary eye care setting. *Ophthalmic Genet.*, 36, 1, 50–7, 2015.
38. Goldsmith, L., Jackson, L., Connor, A.O., Skirton, H., Direct-to-consumer genomic testing: Systematic review of the literature on user perspectives. *Eur. J. Hum. Genet.*, 44, October 2011, 811–6, 2012.

15
Global Prospectives on Genomic Healthcare: Opportunities and Challenges

Shaik Harun Rasheed[1]*, Kondapuram Parameshwar[1], G. Buela Priyanka[2], Laith Hamza Samein[3] and Ravi Kumar Vakkalagadda[1]

[1]*Department of Pharmaceutics, School of Pharmacy, Guru Nanak Institutions Technical Campus, Hyderabad, Telangana, India*
[2]*Department of Pharmaceutical Chemistry, School of Pharmacy, Guru Nanak Institutions Technical Campus, Hyderabad, Telangana, India*
[3]*Department of Pharmaceutics, College of Pharmacy, University of Mashreq, Baghdad, Iraq*

Abstract

Genomic healthcare has emerged as a transformative approach in modern medicine, offering unprecedented opportunities for personalized care and precision treatments. This chapter explores the global landscape of genomic healthcare, highlighting its potential to revolutionize diagnostics, therapeutics, and preventive medicine. By integrating genomic data with advanced technologies such as artificial intelligence, healthcare systems increasingly embrace a patient-centric model. The chapter delves into the opportunities presented by genomic advancements, such as tailored interventions for rare and complex diseases, enhanced drug discovery processes, and the identification of genetic predispositions. Case studies from different regions illustrate the progress and challenges in implementing genomic healthcare, including ethical dilemmas, data privacy concerns, and equitable access to these innovations. Furthermore, this chapter discusses collaborative frameworks between nations to promote research, data sharing, and capacity building, emphasizing the importance of addressing disparities in resource-limited settings. The future of genomic healthcare relies on fostering a global vision

**Corresponding author*: shaikharunrasheed@gmail.com; ORCID: 0000-0002-3698-3331
Kondapuram Parameshwar: ORCID: 0000-0001-7174-0426
G. Buela Priyanka: ORCID: 0009-0005-4759-1615
Laith Hamza Samein: ORCID: 0000-0002-3531-9324
Ravi kumar Vakkalagadda: ORCID: 0000-0003-0643-1103

Swetha Kumari Koduru, Sarad Pawar Naik Bukke, Hussaini Bello, Muhamad Mustafa Mundu and Ayush Madan (eds.) Genomics and Precision Healthcare: Innovations, Applications and Challenges, (463–488)

that combines scientific innovation, ethical integrity, and health equity. By examining the diverse opportunities and challenges, this chapter provides a comprehensive perspective on the pivotal role of genomics in shaping the future of global healthcare systems.

Keywords: Genomic healthcare, personalized medicine, precision treatments, global health equity, genomic data integration

Abbreviations

AI	Artificial Intelligence
EHR	Electronic Health Records
NGS	Next-Generation Sequencing
CRISPR	Clustered Regularly Interspaced Short Palindromic Repeats
GWAS	Genome-Wide Association Study
FDA	Food and Drug Administration
HIPAA	Health Insurance Portability and Accountability Act

15.1 Introduction

Genomic healthcare, also called genomic medicine, precision medicine, or personalized medicine, is an emerging paradigm that applies genomic information to improve health outcomes and patient care. Genomic healthcare comprises the entire spectrum from research to implementation within a population or healthcare system [1]. It includes the healthcare applications of non-clinical genomic services and direct-to-consumer tests. Genomic healthcare is currently the most debated and fastest-growing topic in medicine. It has far-reaching applications in improving clinical care pathways, clinical trials, public health surveillance, and population health management. One in five genomic healthcare services is implemented or adopted in clinical care, research, and health system workflow. With an additional 306 services in the developmental pipeline, genomic healthcare could impact 37% of all medical services by 2027. Genomic healthcare presently spans 30 medical fields, with oncology (38%), rare and infectious diseases (21%), pharmacology (14%), obstetrics (8%), and cardiology (6%) being the most impacted areas of care. Genomic healthcare services transform existing treatment protocols and create novel pathways for diagnosis, prevention of disease, and health management [2]. Ultimately, genomics could facilitate a paradigm shift in healthcare from reactive and symptom-based to proactive and precautionary care. Genomic research advancements impact 36 different aspects of clinical care, with the majority relating to diagnosis (48%), treatment selection (23%), and disease risk assessment (9%). Of concern is that 31 genomic technologies developed to impact clinical care are non-commercial. Twenty-three recently adopted genomic technologies in healthcare systems democratize access to care and disease management across 13 medical fields. The technologies' most commonly reported impacts are on public health surveillance (30%) and access to care (20%). Necessitating consideration, 64 policies are currently implemented to regulate the impacts of 47 genomic technologies on clinical care, public health, and healthcare systems [3].

15.1.1 Scope

"Genomic healthcare" is a comprehensive health approach encompassing genomic testing, diagnostics, and therapeutic interventions. Genomic testing includes a variety of applications ranging from the screening of germline genomic aberrations to somatic genomic profiling of tumors or infectious agents. Genomic diagnostics refers to the interpretation of genomic data to inform clinical decision-making. In contrast, genomic therapeutics uses genomics as a target or an adjunct to devise new medical therapies. Genomic healthcare is an interdisciplinary field that integrates genomics with clinical practice, public health, and research. Genomic healthcare is not limited to individual patient care. However, it extends to population health and public health, as well as one-off interventions of care and disease management, as well as preventive strategies informed by genomic data [4]. Efforts to translate genomic discoveries into genomic applications in medicine were initially focused on individual reimbursement-driven services to patients and healthcare providers. However, consideration of payers led to the recognition of added value for population-based services. Furthermore, healthcare systems grappling with high costs associated with rising numbers of genome-based precision medicines have started to consider how to deploy genomics for screening, risk stratification, and monitoring in the equal access to care pathways.

The ongoing molecular technological and scientific breakthroughs directly impact genomic healthcare. For instance, next-generation sequencing has revolutionized the landscape of genomic research, transforming once highly specialized endeavors into routine screenings in research and healthcare environments. High-performance bioinformatics pipelines are now the norm, and access to vast biomedical databases curated and enriched in phenotypic and clinical information is a critical enabler to genomic research, testing, and diagnostics, represented in Figure 15.1.

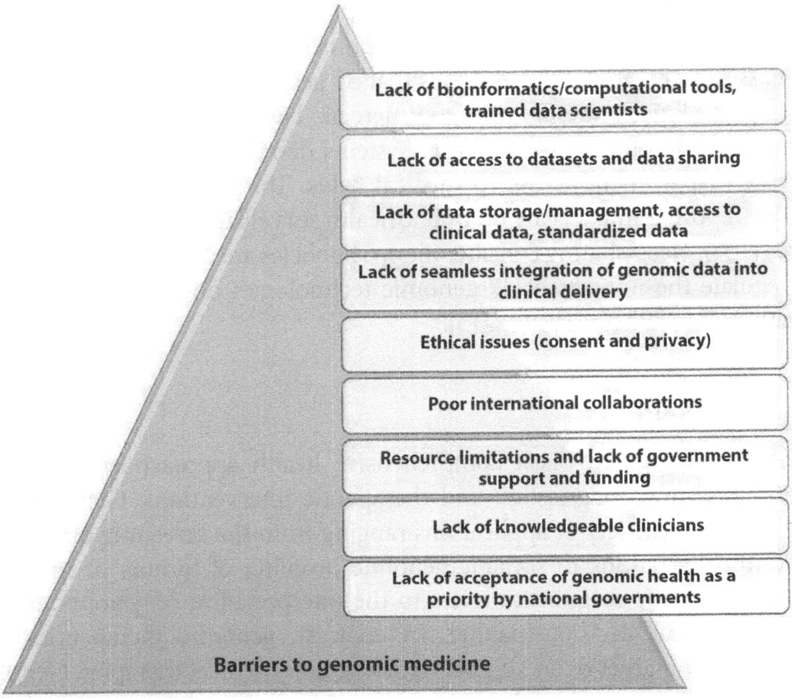

Figure 15.1 Barriers to genomic medicine.

As genomic data is becoming integral to care pathways outside research systems, biomedical scientists and researchers must understand the regulations and compliance associated with handling genomic data in healthcare [5]. Furthermore, with genomics rapidly becoming integral to care pathways, healthcare professionals must keep pace with genomic discoveries and developments through continuous education and training. Genomic healthcare is expected to radically change traditional models of care and patients' engagement in care. There are expected further expansions in genomic healthcare that research and academic institutions, health systems, and governments must proactively consider in their strategic developments, actions, and investments.

15.1.2 Importance and Impact

Recent advances in genomics and biotechnology are converging into a new discipline: genomic healthcare. Genomic healthcare refers to collecting, analyzing, and applying genomic information for healthcare. It encompasses personalized medicine, targeted therapies, hereditary disease prevention, and preventive health management based on genomic information. Genomic healthcare is at the forefront of modern healthcare development and is envisioned to reshape the future of healthcare delivery. Understanding genomic healthcare and its global perspectives is vital for developing effective national policies and strategies.

The genomic revolution has dramatically impacted industries, societies, and daily life. Genomic information, the collection of genes that constitute organisms, determines the nature and characteristics of each living organism. The dramatic decline in DNA sequencing costs has made genomics affordable and applicable to various fields beyond basic research in biology and medicine [6]. The genomic healthcare industry is an emerging market driven by biotechnologies and biomedical applications. Currently, leading genomic companies focus on genomic sequencing services, epigenomic regulation analysis, and the development of artificial intelligence-based genomic data analytics.

15.2 Historical Development of Genomic Medicine

A historical overview of the development of genomic medicine is presented, starting from early genetic studies and leading to the current state of genomic healthcare. The double helix structure of DNA is discussed as a key moment that paved the way for molecular biology. The discovery of how DNA encodes genes and how genes produce proteins that affect phenotypes is explained. The importance of genomics and a genome-centered view of biology is elaborated on, highlighting critical milestones that fueled genomic research and its integration into medicine. The evolutionary steps from conventional to genomic medicine are illustrated, from direct observation-based clinical diagnosis to incorporating genomic data in clinical decision-making. The transition from traditional genetic counseling in a family context to genomic counseling involving complex multilayered data in a clinical context is examined, exploring a prediction-based paradigm with inherent uncertainties. Key historical events are summarized, explaining how the understanding of genetics evolved over decades and shaped current practices in genomic healthcare.

Technological innovations that facilitated the exponential advancement of genetic diagnostics and therapeutics are reviewed, and discoveries that boosted the pace of progress are discussed. The quest to decode the genetic instruction in DNA is presented as the original impetus for genetics, recounting landmark discoveries that shaped basic understanding and the scientific disciplining of heredity, development, and evolution in terms of genes. Current genomic interventions in medicine effectively triumph over prior discoveries in genetics and the resultant technologies. Genomic medicine and healthcare practices are historical outcomes of genetic research and its spiral progression from basic to applied science. An initial impression of the state of genomic healthcare practices is provided and potential historical lessons are highlighted to prevent past mistakes from reoccurring in novel genomic applications. How the past informs the present is deliberated alongside ethical concerns, emotional implications, and misconceptions reminiscent of prior occasions.

15.2.1 Key Milestones

Historical milestones that have defined and shaped genomic medicine and healthcare pathways are discussed, focusing on key events over the past century. Precision medicine, or healthcare tailored to individual characteristics and needs, is greatly facilitated by genomic medicine and is an important societal goal. As a step within genomic medicine, the broad application of specific genomic tests demands a thorough understanding of the key milestones that have added to this knowledge. Consequently, genomic medicine is progressing in routine clinical practice in terms of understanding the genome, employing this understanding in healthcare, and developing healthcare in society. Specific genomic tests enhance the understanding of genetic disorders, some of which are noteworthy discoveries, such as identifying the gene causing Huntington's disease. The first disease-related gene was identified in 1993, and hundreds have been discovered since then. These discoveries have implications for genetic research and therapy. Ethical guidelines and regulatory frameworks were established in the genomic field, as shown in Table 15.1, with the wide application of genomic tests. Innovations are responsible for considering public concerns about, for example, privacy, discrimination, and eugenics. Key milestones are contextualized within broader societal and scientific changes, and public perception of genetics is considered. The past century witnessed important advances in understanding the basis of genetic disorders and the means to combat them. Some milestones of this knowledge accumulation path are discussed, which is a precondition for healthcare genomic services. Furthermore, knowledge accumulation has paved the way for the public and clinical application of genomic health services, and some milestones along this service development path are outlined. Finally, the significance of these key milestones to the novice practice of genomic medicine in healthcare is elucidated.

Table 15.1 Key challenges in genomic healthcare.

Challenge	Impact on global healthcare	Potential solutions	References
Data Privacy & Security	Protection of genetic information from misuse - Risk of data breaches and identity theft	Stronger encryption standards - Comprehensive genetic data protection laws	[7]
Ethical Considerations	- Issues regarding consent, genetic modification, and discrimination - Balancing innovation and ethics	- Development of ethical guidelines - Transparent consent protocols	[8]
Access & Equity	- Disparities in access to genomic healthcare between regions and income groups	- Policy initiatives to reduce healthcare gaps - Mobile genomics clinics	[9]
Regulation & Standardization	- Lack of uniform global standards for genomic testing and treatments - Differences in regulatory policies	- International regulatory bodies - Global agreements on clinical standards	[10]
Workforce Training	- Shortage of skilled professionals to interpret and apply genomic data effectively	- Expansion of genomic education programs - International training initiatives	[11]
Cost & Affordability	- High costs of genomic sequencing and personalized medicine limiting accessibility	- Cost-reduction through technological innovations - Public-private partnerships	[12]

15.3 Current Applications of Genomic Healthcare

In recent years, healthcare has been undergoing a paradigm shift toward precision medicine, primarily spurred by advances in genomics. Put precision medicine refers to using genomic information to guide treatment decisions for individual patients. On average, patients are better off with treatments optimized for their unique genetic profiles than with a randomized approach. Collectively, genomic applications in healthcare can be characterized as "genomic healthcare." Presently, genomic information is routinely used in healthcare delivery in several ways. First, genetic testing has become a standard practice in diagnosing inherited diseases. In many cases, whole-exome or genome sequencing can detect an underlying genetic cause for a patient's symptoms, allowing for earlier detection and intervention. In addition, clinical laboratories routinely assay specific genes associated with heritable cardiac arrhythmias and cardiomyopathies, enabling risk assessment and guiding preventative treatments [13].

Second, a tumor's genomics is routinely profiled in oncology to determine its most effective therapy. Generally, each clinical intervention is based on the specific mutations present in a patient's tumor rather than its tissue of origin. Similarly, there are several genomic applications in pharmacogenomics, where genetic factors influence a drug's efficacy and safety. Genomic variants in target genes, drug transporters, and metabolizing enzymes can alter a drug's intended effect or lead to adverse events and, in some instances, genomic information is already used to select one drug over another. Additionally, genomic risk estimates for complex diseases can guide preventative measures [14]. For example, carriers of specific APOE alleles are at higher risk for Alzheimer's disease, and experimental approaches to mitigate this risk are being considered. However, due to current limitations, routine use of polygenic risk scores for complex diseases is still not implemented. Illustrate the current applications of genomic healthcare; several real-world examples are presented.

15.3.1 Precision Medicine

Precision medicine is at the forefront of genomic healthcare today. Generally, precision medicine entails tailoring preventative or therapeutic interventions to the individual characteristics of each patient, achieved by more thoroughly understanding the underlying causes of diseases and their progression within each individual. In a precision medicine framework, genetic, environmental, and lifestyle factors are jointly considered to create personalized interventions for preventing, managing, or treating a given disease [15]. This growing paradigm shift in modern medicine has

Table 15.2 Impact of genomic healthcare on specific diseases.

Disease	Potential of genomic healthcare	Current developments	References
Cancer	Personalized cancer treatments based on genetic mutations Precision oncology therapies	Liquid biopsy for early detection Targeted therapies (e.g., CAR T-cell therapy)	[17]
Rare Genetic Disorders	Identification of genetic causes for rare diseases Gene therapy as a potential cure	CRISPR-based treatments Gene therapy clinical trials	[18]
Cardiovascular Diseases	Genomic screening to identify genetic risk factors Personalized prevention and treatment plans	Genetic testing for inherited heart conditions Pharmacogenomics for heart disease management	[19]
Neurological Disorders	Insights into genetic causes of neurodegenerative diseases Potential for personalized interventions	Genome-wide association studies (GWAS) for Alzheimer's, Parkinson's	[20]
Infectious Diseases	Genomic surveillance of pathogens Personalized treatment based on genetic resistance and susceptibility	Pathogen genome sequencing for tracking outbreaks Antiviral therapies tailored to genetic makeup	[21]

the potential to significantly increase the effectiveness of treatments while minimizing their adverse effects. Accordingly, patients receiving precision medicine-guided treatments may expect better health outcomes than under conventional "one-size-fits-all" approaches.

There are several real-world implementations of precision medicine. Genomic or other omics data from individual patients have effectively guided the choice of therapies, particularly for oncology and chronic diseases. For instance, enabling clinic genomic screening has made it possible to choose cancer drugs based on identifying specific mutations in tumor cells. In another case of chronic myeloid leukemia (CML), therapy choice was based on the patient's sex and the precise transcriptomic definition of the mutually exclusive presence of BCR-ABL oncogenic gene fusions. Successful genome-guided therapy choices can also lead to serendipitous discoveries of novel treatments, such as one CML patient's off-label imatinib prescription that revealed a more profound understanding of the role of leukemic stem cells in disease initiation and relapse. In addition to addressing patients with particular diseases, there is an increasing number of studies on preventative health plans for high-risk individuals, as shown in Table 15.2. This strategy is directed toward a population in which each individual's genomic data would be used to assess the risk of developing certain diseases, such as diabetes and familial hypercholesterolemia, based on which preventative steps would be taken for risk reduction. Although precision medicine offers great promise for patient well-being, many ethical questions exist, such as whether all patients have equal access to precision medicine facilities and experimentation with genomic data [16]. In principle, an individual's genetic data should always remain his or her own, but in practice, many studies aggregate data from many individuals to achieve more robust results.

15.4 Ethical, Legal, and Social Implications of Genomic Healthcare

Rapid advancements in genomics have yielded numerous benefits for healthcare systems worldwide, yet they also raise several ethical, legal, and social implications. Responsible use of genomic technologies requires carefully weighing their advantages against the risks posed to individual privacy, autonomy, and human rights. Nevertheless, while certain risks might be negligible in some social contexts, they could threaten the very existence of individuals or communities in others. Thus, the precautionary principle suggests that even low-risk actions should be avoided in environments where the stakes are high. Although most genomic advancements are still experimental, epidemiological concerns about technologies applied to large populations could be detrimental. For instance, genetic discrimination has emerged as a relevant concern, particularly in employment and insurance, as access to genomic data could impede individuals' capacity to make informed decisions about risk. In this respect, protective legislation is required, as illustrated by the U.S. Genetic Information Nondiscrimination Act (GINA).

Additionally, ethical dilemmas about informed consent processes for genetic testing and the subsequent use of genomic data for research arise. Transparency is critical, as individuals ought to understand what they authorize [22]. If individuals have been tested due to certain medical conditions, access to their data should be restricted solely to researchers studying those conditions, thereby upholding patients' rights. Moreover, possible scenarios of unequal access to genomic technologies among various populations or using underprivileged groups as a source of genomic information while denying them the benefits of enhanced technologies require consideration. Indeed, the current commodification of genetic information might reinforce preexisting inequities in access to healthcare systems and research benefits. Finally, the societal impact of genomic healthcare technologies, especially regarding public trust in science and scientists' attitudes, is paramount. For instance, while epidemiological studies are generally accepted, the implementation of genome-wide association studies might be met with public resistance and demands for deliberation. Given the coherence of societal impacts complications arising from the public's perspective of science might be detrimental regardless of the epistemic field these concerns should be examined and discussed in parallel with scientific debates. Genomic advancements in biomedicine and public health offer numerous opportunities for design and implementation. However, they also entail several ethical, legal, and social implications that must be resolved before establishing robust policy frameworks.

15.4.1 Privacy and Data Security

As genomic healthcare develops, the key issues of privacy and data security are brought to the forefront. The reason is simple: genetic information is an individual's most private and sensitive information [7]. While the potential of genomic data to deliver better healthcare and new therapeutic solutions is enormous, concerns around privacy and data security loom large. Well-publicized data breaches occurring at private companies and even hospitals have generated concern amongst the public, who are increasingly aware of their digital footprint and the dangers of theft or mishandling of personal data. Genetic data is a particularly contentious issue. The danger of harm is not abstract. Federal regulators accused a genetic-testing firm in 2018 of mishandling data in one case where someone was not adequately warned that data could be sent to law enforcement.

Data breaches or unauthorized access to genetic data can lead to the potential misuse of such data, whether intentionally or unintentionally. Hence, genomic data security is paramount to avoid discrimination against an individual based on their genetic data. Laws and regulations require healthcare institutions to implement robust data security measures, and regulatory frameworks relevant to genomic data security exist. Several organizations have also proposed best practices for genomic data privacy [23].

Nevertheless, genetic data security is an ongoing challenge for the healthcare sector, as past experiences with public genomic databases have reflected that it is challenging to mitigate security risks entirely. However, such databases are expected to be secure. As genomic data privacy advocates stress, ethical considerations must also be considered. Emerging technologies, such as the use of blockchain for secure genomic data sharing, promise solutions to some of these security problems, but new technologies also pose new challenges. For instance, biotechnologies that can transform data into a DNA sequence might complicate the security efforts, as data encoded in DNA would need to protect the substrate itself. Examples from the past have shed light on the real risks involved when genomic data has been mishandled. Since privacy is paramount in genomic healthcare, it is pertinent to discuss trust issues, especially between patients and healthcare service providers. As noted, a long-established ethical principle in medicine is that a patient's information is private and confidential. Genetic data privacy violations could cause a patient to be stigmatized, most likely rendering trust in the healthcare system moot.

15.5 Global Initiatives and Policies in Genomic Healthcare

International cooperation is essential for fostering innovation and the development of new products and services, for responding to health threats, and for the prevention and control of diseases that transcend national borders. Accordingly, the World Health Assembly in 2012 called upon Member States to take actions to enhance the role of research and innovation on genomic and post-genomic technologies in the public health system. In 2021, the World Health Assembly further encouraged Member States to develop national strategies to strengthen the integration of genomics into their public health systems. Several countries are establishing national strategies to incorporate genomics in health care systems, with a focus on opportunities for economic growth and new services. However, due consideration is not always given to the ethical and cultural implications of such developments. Different countries opt for different strategies, reflecting their own cultural and ethical considerations [8]. There is a wide variety of governmental and non-governmental initiatives worldwide that address mutual needs and concerns, including the public sharing of global genomic data. More recently, the COVID-19 pandemic has demonstrated the value of global genomic data sharing in monitoring the epidemiology and evolution of viral pathogens, as well as in tracking the effectiveness of countermeasures and vaccine development. Much can be done to enhance the health and wealth of nations through the cooperative global sharing of genomic data and other resources that underpin scientific research and technological development. However, global initiatives for the cooperative sharing of genomic data would benefit from an analysis of past and present initiatives that illustrate both obstacles and pathways to success. Examples include the Human Genome Project, the 1000 Genomes Project, and initiatives by the World Health Organization (WHO) and the Global Alliance for Genomics and Health (GA4GH) [9]. There exists a plethora of global, regional, and bilateral initiatives that promote the sharing of genomic data and consider the implications of associated new goods and services. Some focus on genomic research per se, while others deal with the wider public health implications of genomics. The WHO plays a prominent role in several initiatives directed at public health issues and the governance of shared biotechnological resources. Other initiatives are coordinated by academic institutions and professional societies. There are many successful examples of cooperation that can be built upon. At the same time, some caution is warranted; there are also examples of failed or ineffectual initiatives. The responsible planning of new cooperative efforts must include a consideration of the prospective regulatory environment and the likely geopolitical interactions [24].

The WHO has an important role in several global initiatives that promote the sharing of genomic data for public health purposes and consider the implications of new goods and technologies. For example, the WHO's Pandemic Influenza Preparedness Framework provides guidance on the equitable sharing of benefits from the use of influenza viruses with pandemic potential. In 2021, the WHO, in collaboration with the GA4GH, also launched a global initiative to enhance the public sharing of genomic sequence data from pathogens that affect human health and to consider the implications of new technologies and goods developed from shared data.

15.5.1 International Collaborations

The implementation of genomic-based research and healthcare approaches at a national or regional level presents significant challenges for many countries. Overcoming these challenges necessitates the reallocation of limited resources or, ideally, the pooling of available resources through international collaborations. Some collaborative partnerships between developing countries and leading research nations emphasize the potential benefits of genomic approaches for countries with diverse ethnic populations, as highlighted in the African post-Human Genome Project perspective. Successful cooperative initiatives that have been effective in genomic population studies and healthcare, particularly in developing or transitional countries, have addressed local needs while involving countries playing a leading role in other research areas [8]. Multinational consortia that combine the resources of several countries can facilitate progress in genomic discoveries and the translation of findings into clinical application in these research areas. Some initiatives pool resources from a number of countries to tackle globally important health issues using genomic approaches. The latter includes "one health" studies in Africa, which focus on the coexistence of human and animal reservoirs of infection and the consideration of environmental aspects in the spread of diseases, particularly using genomic investigative approaches. Collaborative projects that have been implemented will be described below to illustrate how readily global cooperative actions can enhance the efficacy of genomic healthcare actions [25]. Efforts to ensure that the benefits of genomic research reach the entire world population should include agreements on data sharing, ownership, and access, as well as the establishment of common ethical standards to foster trust between stakeholders in different countries [10]. However, the unequal distribution of power and resources in the world makes it difficult to coordinate activities that involve countries with very different regulatory environments and cultural contexts. The challenges and problems faced in genomic population studies that have been implemented or are in progress will also be considered.

15.6 Technological Advancements in Genomic Healthcare

Technological advancements are significantly transforming genomic healthcare, bringing new opportunities but also challenges. In both research and clinical practice, technology-driven innovations are expected to radically change everyday genomic practices. Genomic data analysis and interpretation are rapidly progressing due to the impact of next-generation sequencing technologies on the availability and accessibility of genomic data.

The role of bioinformatics is rapidly evolving as genomic data are routinely generated in clinically relevant settings. Bioinformatics is essential to managing complex genomic datasets and integrating them into healthcare systems. Fantasy bioinformatics tools and platforms are envisioned that impact patient testing, data management, bioinformatics analysis, and interpretation reports.

Genomic datasets will be transferred from laboratories to clinics, with computational analysis carried out by bioinformatics pipelines embedded in laboratory devices. Allied to this, revolutionary genomic technologies such as CRISPR-Cas9 are opening new horizons for gene editing. Such technologies will transform genetic research, development of animal models, applications in healthcare, and the access of research groups to gene editing. *Ex-Vivo* and *In-Vivo* gene therapy represent future applications that may enhance pharmaceutical treatment options for numerous diseases [26].

Genomic technologies will impact the development of clinical diagnostics enhancing monitoring of disease status and risk assessment. Discovery of non-invasive biomarkers will progress late-stage diagnostics development, and genomic methylation will be integrated into the routine cancer testing development shown in Table 15.3. Genomic technologies will dramatically enhance treatment options and the understanding of treatment

Table 15.3 Technological innovations driving genomic healthcare.

Technology	Impact on genomic healthcare	Applications	References
Next-Generation Sequencing (NGS)	Allows for faster, cheaper, and more accurate sequencing Enables personalized medicine	Whole Genome Sequencing (WGS) Liquid biopsy technology	[27]
CRISPR/Cas9 Gene Editing	Potential for correcting genetic disorders Raises ethical questions on gene modification	Clinical trials for gene therapy Research into genetic diseases	[28]
Artificial Intelligence (AI) in Genomics	AI for data analysis, pattern recognition, and predictive analytics Improves efficiency of genomics research	AI algorithms for predicting genetic disease risks AI for drug discovery	[29]
Wearable Genomic Devices	Integration of personal genomics into daily healthcare monitoring Real-time health data collection	Personal genomics testing kits Wearable health-monitoring devices	[30]

selectivity in a precision medicine context, impacting drug development. In a global perspective, genomic technologies and healthcare innovations will reshape academic research, translational research, and pharmaceutical industry interactions. However, fast development generates challenges that need to be addressed, including technological feasibility, clinical validity, and regulatory framework issues. Ethical and societal issues concerning the access and use of genomic data will also remain critical considerations. A vision is presented on the impact of technological advancements on genomic healthcare and the opportunities and challenges. Genomic and healthcare technologies will converge to reshape the future genomic practice and develop innovative solutions to the new challenges.

15.6.1 CRISPR-Cas9 Technology

One of the pivotal advancements in genomic healthcare that can revolutionize the way genomes are studied and edited is the CRISPR-Cas9 technology. CRISPR-Cas9 technology is a groundbreaking and transformative gene-editing technology that allows precise modifications to be made within the genomes. It is based on the natural processes of bacteria, employing a similar adaptive immune strategy to recognize, cut, and destroy invading viruses and other foreign nucleic acids [11]. This technology consists of two key components: a single-guide RNA (sgRNA) that directs the Cas9 nuclease to the target site and the Cas9 protein itself, which creates a double-strand break at the specific location in the genome. Following cleavage, the genome is repaired by either non-homologous end joining or homology-directed repair, resulting in knockout or precise modifications of the target genes [31].

Since its first successful application in eukaryotes, CRISPR-Cas9 technology has been widely and rapidly adopted by researchers. Its simplicity, low cost, and versatility make it an exceptional tool for almost all organisms represented in Figure 15.2. Diverse applications of CRISPR-Cas9 technology have been developed in basic research, agriculture, and biomedicine, including but not limited to loss-of-function gene screening, precise gene knockout, gene knock-in, transgenesis, promoter modification, DNA base editing, epigenetic editing, and transcription regulation. Particularly, the emergence of CRISPR-Cas9 technology has illuminated a new door for the treatment of genetic disorders, cancer, and infectious diseases. Notably, the first ever CRISPR-edited drug treatment, *ex vivo* hematopoietic stem cells edited with the CRISPR-Cas9, received FDA approval in September 2021 and progressed to clinical trials for treating patients with sickle cell disease and β-thalassemia. Currently, there are over 150 clinical trials of CRISPR-Cas9 running worldwide, covering a wide spectrum of disorders.

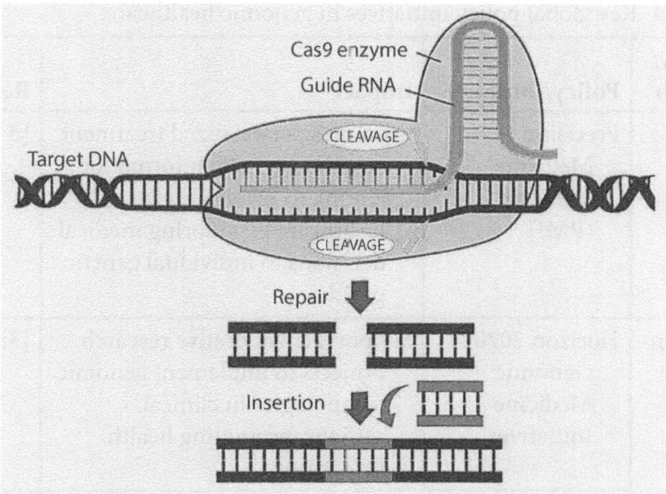

Figure 15.2 CRISPR-Cas9 technology.

15.7 Challenges and Limitations in Genomic Healthcare

A variety of challenges and limitations have emerged that could affect the success and accessibility of genomic healthcare. Foremost among these are challenges related to the equitable access to genomic testing and genomic testing-related therapies. Although rare disease and oncology health systems that include genomic testing and genomic testing-related interventions implemented by drug companies are expanding rapidly, there are likely to be growing disparities in access based on individual, societal, and geographical locational socioeconomic status. Furthermore, testing and intervention access disparities are also possible between countries bordering the global north-south divide. Even where there is access, the actual interpretability of the genome test results can be uncertain due to the varying clinical implications of even benign single nucleotide variations in genetic markers across different populations [32].

Unfortunately, limitations in currently available technological approaches mean that whole genome sequencing data would need to match exome data to provide a full genomic analysis. As a result, exome-based data approaches would likely result in a 20% loss of comprehensive genomic patient care. Similarly, relying only on NGS variants in a transcriptome atlas could miss pathogenic gene variants found outside or regarded as unknown in a reference transcriptome data set shown in Table 15.4.

Table 15.4 Key global policy initiatives in genomic healthcare.

Country/ region	Policy/initiative	Impact	References
United States	Precision Medicine Initiative (PMI)	Promotes personalized treatment based on genetic information, aiming to revolutionize healthcare by tailoring medical decisions to individual genetic profiles.	[34]
European Union	Horizon 2020 – Genomic Medicine Initiatives	Supports collaborative research projects to implement genomic technologies in clinical settings, promoting health innovation.	[35]
India	National Biotechnology Development Strategy (NBDS)	Aims to promote biotechnology research, including genomic medicine, to address public health challenges and improve healthcare access.	[36]
Australia	Genomics Health Futures Mission	Focuses on integrating genomics into healthcare by funding research into genomic diagnostics and treatments to enhance public health outcomes.	[37]

Direct-to-consumer genetic test companies also pose ethical challenges with a range of consent processes and patient education designed to minimize liability for genomic risks uncovered even while avoiding false safety assurances. In general, there is still a need to develop appropriate regulatory frameworks for any currently experimental genomic interventions to guarantee safety and efficacy in their application. Until this is done, it will be difficult to persuade patients to take part in genomic healthcare [33]. Therefore, it will be especially critical to develop mechanisms and frameworks designed to reduce these challenges and limitations. If achieved, the potential benefits of genomic healthcare could be maximized. More generally, efforts will be needed to ensure that any developed genomic research and healthcare resources are accessible to all who might benefit.

15.7.1 Access and Equity

Genomic healthcare is rapidly becoming a reality across different regions of the globe. As the understanding of the human genome continues to grow, the opportunity to integrate genomic data into routine healthcare systems is growing, as is the promise for patients to benefit from the delivery of precise and personalized healthcare interventions. Along with that promise, multiple challenges arise in realizing the goal of genomic healthcare across distinct social, cultural, economic, and political landscapes. Many countries are implementing genomic approaches to healthcare, while others are still in pre-implementation phases, or do not have plans to implement such approaches. Variations also exist within countries, creating disparities in available genomic services. Moreover, many differences exist in how countries are approaching the integration of genomic healthcare into routine practices [38]. Designing action-centered global perspectives on genomic healthcare requires consideration of locally relevant opportunities and challenges. In addition, there is a need to proactively engage with issues related to access and equity, policy and regulation, health literacy and advocacy, social inclusion, and cultural competence as genomic healthcare progresses. These topics highlight the need for global conversations on genomic healthcare, while acknowledging and respecting local diversity and perturbations.

Unequally distributed opportunities exist for patients to access genomic testing and treatment options. Socially vulnerable populations, such as recent immigrants and refugee communities, are often left behind. Those with lower socioeconomic status face barriers to genomic testing and treatment, despite having greater needs. Transgender and non-binary individuals, as well as individuals with rare diseases, often experience inequitable access to genomic healthcare services. Moreover, currently available genomic services are designed to fit majority populations. For example, genomic tests for hereditary breast and ovarian cancer primarily build on knowledge from European ancestry populations and are less effective for individuals originating from other ancestries or populations. Overall, inequitable access to genomic healthcare services compromises health outcomes. On the other hand, there is a danger that the implementation of genomic medicine will perpetuate or exacerbate existing inequalities. Without planning for equitable implementation, disparities may arise in access to and benefit from genomic treatments. Similarly, concerns have been raised about access to pharmacogenomics in clinical care. In general, pharmacogenomics is more likely to be relevant for non-Hispanic white populations, while underrepresented populations may be left without access to relevant genomic information. Efforts are needed to ensure that all individuals have equitable access to the advantages of genomic healthcare. Global approaches can foster knowledge and action sharing, aligning with local settings and needs. For example, the 2023 WHO Genomic and Public Health Meeting emphasized equitable access to the public health benefits of genomics as the priority topic. In recent years, many countries have developed broad national genomic strategies. However, underdeveloped countries often have limited or no genomic strategies, and the ongoing implementation efforts in some countries may jeopardize equity in access to genomic services. For example, engaging countries with low genomic capacities in the implementation of genomic public health services may create a situation where well-resourced countries rapidly upscale genomic public health services while others lag behind.

15.8 Future Directions and Emerging Trends in Genomic Healthcare

As genomic healthcare rapidly expands worldwide, various local adaptations are emerging to meet specific needs. Some health systems embrace cost-efficient, population-wide genomic screening for rare disease carrier status, while others experiment with targeted cancer genome sequencing plus clinical trials for actionable variants. There are places with no genomic healthcare and well-justified reasons to avoid it. For most nations, genomic health brings progress, even if initially a small gain. Timeout should be considered in genomic health investment because expensive technologies might become obsolete soon. What is desired is genomic information decisions are careful today and flexible tomorrow.

Speculation about the future is often pointless, but postulations about potential developments and trends in genomic health progress can be fascinating. Affordable genomic technologies are soaring, accounting for dizzy growth rates in data volume and applications. On the other hand, diligence is strongly needed in handling genomic information because ethical implications arise. Genomic data are so personal that eagerness for information might annihilate privacy. Informed patient consent must practically follow the golden rule: yours should not be shared. Thus, everything data should remain exclusively in the researcher's domain unless reciprocity is ensured. Reciprocities could be impossible for third-world countries because these often lack genomic literacy. Alternatively, public data redistribution rights must be carefully defined [39].

15.8.1 Personalized Therapies

Emerging Therapies Targeted at Specific Genetic Profiles Personalized therapies, an evolution of a more general term in healthcare, are designed to treat based upon a genomic and biomarker profile. Such therapies have a nascent but rapidly growing footprint in healthcare globally. The promise of personalized treatments is apparent, with the potential to radically change treatment approaches from those accustomed to a one-size-fits-all approach to those bespoke for the individual.

Despite the dramatic change in approach, early and emerging evidence points towards personalized treatments offering superior clinical outcomes. Most recently cited has been the dramatic and instantaneous patient recovery of a young girl treated with a bespoke therapy for her leukaemia after previously failing to respond to more conventional approaches [40]. Parallel to this, the emergence of targeted therapies that focus either wholly or partly on a genomic profile hold great promise, particularly in oncology with several high-profile successes already described. Central to the development of targeted therapies is the identification of biomarkers—a biological characteristic that can be measured and evaluated as an indicator of therapeutic response, as well as disease onset, progression, and risk. A number of treatments have already been developed that are either wholly or partly reliant on a biomarker adjustment, especially in oncology, with success stories reported.

For instance, genomic testing can now detect whether a patient with breast cancer over-expresses the HER2 protein. If so, the treatment Herceptin, which specifically targets HER2 over-expressing tumors, can be administered represented in Figure 15.3. Such biomarker-based therapies have been shown to improve clinical outcomes relative to their non-biomedical cohort and often reduce the side effects associated with treatment. Other high-profile examples include the use of K-Ras status in colorectal cancers to determine the suitability of the anti-EGFR monoclonal antibodies cetuximab and panitumumab, as well as the use of BRAF V600E mutation testing to detect eligibility for the use of vemurafenib in treating malignant melanoma. There are currently over 100 FDA-approved targeted therapies, with that number rising rapidly, and alongside those therapies, over 400 genomic tests currently in use, again, with that number predicted to rise significantly. Statistical models predict that by 2020 there could be upwards of 10,000 genomic tests addressing specific molecular genetic traits and potentially thousands of targeted treatments, predominantly in oncology, but also in cardiovascular disease and rare genetic disorders.

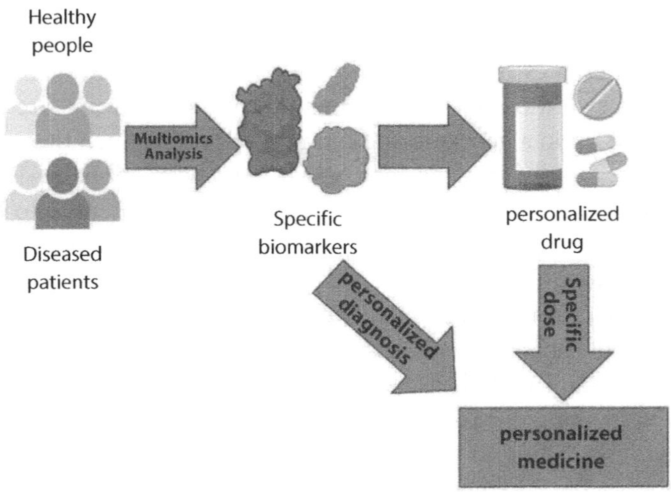

Figure 15.3 Innovations in personalized medicine and gene therapy.

However, despite the apparent clinical benefits of personalized treatments, significant obstacles remain in their widespread adoption. Such therapies require a paradigm shift in how regulatory authorities and public healthcare systems assess and fund treatments. Many currently available targeted treatments were developed at private companies' risk, using their funds, and only later presented to public authorities for scrutiny. Such companies often argue that, because of the investment needed to develop a targeted therapy, diagnostic tests that accompany the treatment should be classed as medical devices, thus enabling the companies to maintain regulatory control over the tests after initial approval. Public authorities often fiercely contested this, claiming that such tests are an intrinsic part of the treatment and, thus, should be tightly regulated to ensure equity of access. Beyond the regulatory hurdle of establishing a positive Treatment Assessment for the therapy and the associated test, *post hoc* concerns from public authorities often center on cost. The rapid rise of high-cost treatments addressing small populations fuels concerns about the affordability of healthcare systems.

15.9 Conclusion

As personalized therapies emerge, there is a risk that the focus on equitable access to these treatments is lost. Most countries, especially developing nations, have minimal capacity and resources to embrace personalized therapies and the attendant diagnostic stratification treatments. Familiar concerns around equity of access are exacerbated when dealing with high-cost precision treatments, often seen as the bastion of wealthier first-world countries. Even in wealthier countries, there is often limited access to many existing genomic tests and a lack of equity in their provision across regional healthcare systems. Thus, parallel to ensuring personalized therapies are robust before deployment, equal emphasis must be placed on ensuring both the therapy and the diagnostic test accompanying it can be equitably accessed, particularly in low- and middle-income countries. Finally, some thoughts on the continued importance of biomarker discovery and how that shapes the future of personalized treatment approaches are also discussed. Personalized therapies are the cutting edge of genomic healthcare, demonstrating how precision medicine redefines treatment paradigms.

Acknowledgment

The authors are thankful to the School of Pharmacy, Guru Nanak Institutions Technical Campus, Hyderabad, Telangana, India, for providing the necessary facilities.

AI disclosure statement: The author acknowledges the use of AI-based tools for minor language editing or grammatical correction. However, all intellectual contributions, analyses, and conclusions presented in this chapter are entirely the author's own.

References

1. Ristevski, B. and Chen, M., Big Data Analytics in Medicine and Healthcare. *J. Integr. Bioinform.*, 15, 20170030, 2018.
2. Thiede, S.N., Snitkin, E.S., Trick, W., Payne, D., Aroutcheva, A., Weinstein, R.A., Popovich, K.J., Genomic Epidemiology Suggests Community Origins of Healthcare-Associated USA300 Methicillin-Resistant Staphylococcus aureus. *J. Infect. Dis.*, 226, 157, 2022.
3. Shieh, J.T.C., Genomic Technologies to Improve Variation Identification in Undiagnosed Diseases. *Pediatr. Neonatol.*, 64, S18, 2023.
4. Reddi, H.V. and Pavlov, Y.I., Editorial: Emerging Talents in Genomic Assay Technology. *Front. Genet.*, 14, 1259011, 2023.
5. Park, S.T. and Kim, J., Trends in Next-Generation Sequencing and a New Era for Whole Genome Sequencing. *Int. Neurourol. J.*, 20, S76, 2016.

6. Uesaka, K., Oka, H., Kato, R., Kanie, K., Kojima, T., Tsugawa, H., Toda, Y., Horinouchi, T., Bioinformatics in Bioscience and Bioengineering: Recent Advances, Applications, and Perspectives. *J. Biosci. Bioeng.*, 134, 363, 2022.
7. Mohammed Yakubu, A. and Chen, Y.P., Ensuring Privacy and Security of Genomic Data and Functionalities. *Brief. Bioinform.*, 21, 511, 2020.
8. Ayesta, A., Bonanad, C., Díez-Villanueva, P., García-Blas, S., Ariza-Solé, A., Martínez-Sellés, M., Ethical Considerations in Elderly Patients with Acute Coronary Syndrome. *Rev. Cardiovasc. Med.*, 23, 55, 2022.
9. Olden, K. and White, S.L., Health-Related Disparities: Influence of Environmental Factors. *Med. Clin. North Am.*, 89, 721, 2005.
10. Marcum, J.A., Nutrigenetics/Nutrigenomics, Personalized Nutrition, and Precision Healthcare. *Curr. Nutr. Rep.*, 9, 338, 2020.
11. Feigin, V.L., Owolabi, M.O., World Stroke Organization–Lancet Neurology Commission Stroke Collaboration Group, Pragmatic Solutions to Reduce the Global Burden of Stroke: A World Stroke Organization-Lancet Neurology Commission. *Lancet Neurol.*, 22, 1160, 2023.
12. Pirmohamed, M., Pharmacogenomics: Current Status and Future Perspectives. *Nat. Rev. Genet.*, 24, 350, 2023.
13. Wang, Y., Liu, B., Zhao, G., Lee, Y., Buzdin, A., Mu, X., Zhao, J., Chen, H., Li, X., Spatial Transcriptomics: Technologies, Applications, and Experimental Considerations. *Genomics*, 115, 110671, 2023.
14. Malone, E.R., Oliva, M., Sabatini, P.J.B., Stockley, T.L., Siu, L.L., Molecular Profiling for Precision Cancer Therapies. *Genome Med.*, 12, 8, 2020.
15. Koulisis, N. and Nagiel, A., Precision Therapy for Inherited Retinal Disease: At the Forefront of Genomic Medicine. *Clin. Lab. Med.*, 40, 189, 2020.
16. Barsoum, I., Tawedrous, E., Faragalla, H., Yousef, G.M., Histo-Genomics: Digital Pathology at the Forefront of Precision Medicine. *Diagn. (Berl.)*, 6, 203, 2019.
17. Roy, P.S. and Saikia, B.J., Cancer and Cure: A Critical Analysis. *Indian J. Cancer*, 53, 441, 2016.
18. Sun, W., Zheng, W., Simeonov, A., Drug Discovery and Development for Rare Genetic Disorders. *Am. J. Med. Genet. A*, 173, 2307, 2017.
19. Halasz, G., Bandera, F., Piepoli, M., Editorial Comments: Focus on Cardiovascular Diseases. *Eur. J. Prev. Cardiol.*, 30, 203, 2023.
20. Pena, S.A., Iyengar, R., Eshraghi, R.S., Bencie, N., Mittal, J., Aljohani, A., Mittal, R., Eshraghi, A.A., Gene Therapy for Neurological Disorders: Challenges and Recent Advancements. *J. Drug Targeting*, 28, 111, 2020.
21. McArthur, D.B., Emerging Infectious Diseases. *Nurs. Clin. North Am.*, 54, 297, 2019.
22. Clayton, E.W., Ethical, Legal, and Social Implications of Genomic Medicine. *N. Engl. J. Med.*, 349, 562, 2003.
23. Thapa, C. and Camtepe, S., Precision Health Data: Requirements, Challenges, and Existing Techniques for Data Security and Privacy. *Comput. Biol. Med.*, 129, 104130, 2021.

24. Bagheri, A. and Delmonico, F.L., Global Initiatives to Tackle Organ Trafficking and Transplant Tourism. *Med. Health Care Philos.*, 16, 887, 2013.
25. Arakawa, N. and Anderson, C., Challenges and Opportunities in Conducting Health Services Research Through International Collaborations: A Review of Personal Experiences. *Res. Soc. Adm. Pharm.*, 16, 1609, 2020.
26. Pushpakom, S., Iorio, F., Eyers, P.A., Escott, K.J., Hopper, S., Wells, A., Doig, A., Guilliams, T., Drug Repurposing: Progress, Challenges, and Recommendations. *Nat. Rev. Drug Discovery*, 18, 41, 2019.
27. McCombie, W.R., McPherson, J.D., Mardis, E.R., Next-Generation Sequencing Technologies. *Cold Spring Harb. Perspect. Med.*, 9, a036798, 2019.
28. Gupta, D., Bhattacharjee, O., Mandal, D., Sen, M.K., CRISPR-Cas9 System: A New-Fangled Dawn in Gene Editing. *Life Sci.*, 232, 116636, 2019.
29. Lipkova, J., Chen, R.J., Chen, B., Lu, M.Y., Artificial Intelligence for Multimodal Data Integration in Oncology. *Cancer Cell*, 40, 1095, 2022.
30. Lotter, W., Hassett, M.J., Schultz, N., AI in Oncology: Current Landscape, Challenges, and Future Directions. *Cancer Discov.*, 14, 711, 2024.
31. Torres-Ruiz, R. and Rodriguez-Perales, S., CRISPR-Cas9 Technology: Applications and Human Disease Modelling. *Brief. Funct. Genomics*, 16, 4, 2017.
32. Tran, K.A., Kondrashova, O., Bradley, A., Deep Learning in Cancer Diagnosis, Prognosis, and Treatment Selection. *Genome Med.*, 13, 152, 2021.
33. Karim, M.R., Islam, T., Shajalal, M., Explainable AI for Bioinformatics: Methods, Tools and Applications. *Brief. Bioinform.*, 24, bbad236, 2023.
34. Sankar, P.L. and Parker, L.S., The Precision Medicine Initiative's All of Us Research Program. *Genet. Med.*, 19, 743, 2017.
35. Kany, S., Jurgens, S.J., Rämö, J.T., Genetic Testing in Early-Onset Atrial Fibrillation. *Eur. Heart J.*, 45, 3111, 2024.
36. Pandey, A., Chawla, S., Guchhait, P., Type-2 Diabetes: Current Understanding and Future Perspectives. *IUBMB Life*, 67, 506, 2015.
37. Lacaze, P., Marquina, C., Tiller, J., Brotchie, A., Kang, Y.J., Merritt, M.A., Green, R.C., Watts, G.F., Nowak, K.J., Manchanda, R., Canfell, K., James, P., Winship, I., McNeil, J.J., Ademi, Z., Combined Population Genomic Screening for Three High-Risk Conditions in Australia: A Modelling Study. *EClinicalMedicine*, 66, 102297, 2023.
38. Chung, C.C.Y., Hong Kong Genome Project, Chu, A.T.W., Chung, B.H.Y., Rare Disease Emerging as a Global Public Health Priority. *Front. Public Health*, 10, 1028545, 2022.
39. Kurnat-Thoma, E., Fu, M.R., Henderson, W.A., Voss, J.G., Hammer, M.J., Williams, J.K., Calzone, K., Conley, Y.P., Starkweather, A., Weaver, M.T., Shiao, S.P.K., Coleman, B., Current Status and Future Directions of U.S. Genomic Nursing Health Care Policy. *Nurs. Outlook*, 69, 471, 2021.
40. Johnson, K.B., Wei, W.Q., Weeraratne, D., Frisse, M.E., Misulis, K., Rhee, K., Zhao, J., Snowdon, J.L., Precision Medicine, AI, and the Future of Personalized Health Care. *Clin. Transl. Sci.*, 14, 86, 2021.

Index

Also of Interest

Check out these other related titles from Scrivener Publishing

Generative Artificial Intelligence in Ophthalmology, Edited by Binod Kumar Mishra, Abhishek Kumar, K. Mariyappan, Vibha Tiwari, Pramod Singh Rathore, and Gaur Hari Das, ISBN: 9781394358175. Empower your clinical practice to lead the future of vision science with this guide to mastering generative AI strategies that transform complex imaging data into high-precision, ethical diagnostic solutions.

Physiotherapy Using Artificial Intelligence: Enhancing Biomechanics for Optimal Rehabilitation, Edited by Abhishek Kumar, T. Ananth Kumar, Sachin Ahuja, J.P. Ananth, and S. Oswalt Manoj, ISBN: 9781394391530. Empower your practice with this definitive resource that bridges the gap between artificial intelligence and biomechanics, providing the essential tools and knowledge to optimize assessments, personalize treatment plans, and predict recovery outcomes in the rapidly evolving landscape of modern physiotherapy.

Environmental Toxicology and Human Health: Current Challenges and Future Perspectives, Edited by Ashish Kumar, Veer Singh, and Maulin P. Shah, ISBN: 9781394399826. Equip yourself to navigate environmental health with this essential resource that integrates fundamentals with the cutting-edge to provide a definitive guide on how chronic exposure to modern pollutants shapes human biological outcomes and regulatory policy.

Advances in Integrative Medicine, Edited by Mohd. Shahnawaz, Sajjid H. Batt, Damodaran Sruthi, Mubashir Javed Mintoo, M.K. Sangale, and Khalid Rehman Hakeem, ISBN: 9781394336777. *Advances in Integrative Medicine* introduces integrative medicine as a holistic approach to health that can effectively address the limitations and side effects of traditional allopathic treatments, providing valuable insights for practitioners and patients alike.

Artificial Intelligence in Healthcare for the Elderly, Edited by Sandeep Sharma, Preeti Nagrath, Bhawna Singla, and Joel J.P.C. Rodrigues, ISBN: 9781394275366. *Artificial Intelligence in Healthcare for the Elderly* provides valuable insights into how artificial intelligence can transform healthcare through personalized monitoring, ethical considerations, and real-world applications.

Design and Forecasting Models for Disease Management, Edited by Pijush Dutta, Sudip Mandal, Korhan Cengiz, Arindam Sadhu, and Gour Gopal Jana, ISBN: 9781394234042. The book provides an essential overview of AI techniques in disease management and how these computational methods can lead to further innovations in healthcare.

Smart Hospitals: 5G, 6G, and Moving Beyond Connectivity, Edited by Arun Kumar, Manoj Gupta, Sanjeev Sharma, Er. Himanshu Sharma, and Khursheed Aurangzeb, ISBN: 9781394275441. Discover the potential of 5G, 6G, and smart hospitals beyond connectivity in Smart Hospitals: 5G, 6G, and Moving Beyond Connectivity and learn how these advancements are revolutionizing healthcare and the digital world.

Sustainable Nanomaterials for Treatment and Diagnosis of Infectious Diseases, Edited by Rajasekhar Chokkareddy, Suvardhan Kanchi and Gan G. Redhi, ISBN: 9781394200016. The book focuses on the design and novel synthetic routes of sustainable nanomaterials in diagnosing and treating infectious diseases offering potential benefits in terms of efficiency, biocompatibility, and environmental impact.

Federated Intelligent System for Healthcare: A Practical Guide, Edited by S. Rakesh Kumar, N. Gayathri and Seifedine Kadry, ISBN: 9781394271351. This practical guide gives valuable insights for integrating advanced technologies in healthcare, empowering researchers to effectively navigate and implement federated systems to enhance patient care.

Internet of Medicine for Smart Healthcare, Edited by Abhishek Kumar, Narayan Vyas, Pramod Singh Rathore, Abhineet Anand, and Pooja Dixit, ISBN: 9781394272235. This book provides in-depth explanations and discussions of the latest applications of AI, machine learning, and the Internet of Medicine, offering readers the cutting edge on this rapidly growing technology that has the potential to transform healthcare and improve patient outcomes.

Viral Vectors for Vaccine Delivery, Edited by Vivek Chavda and Vasso Apostolopoulos, ISBN: 9781394271535. The book is essential for anyone interested in vaccine development, as it highlights the unique advantages of viral vector vaccines in triggering robust, long-lasting immunity and provides an in-depth exploration of the technology and advancements shaping the future of healthcare.

Artificial Intelligence and Cybersecurity in Healthcare, Edited by Rashmi Agrawal, Pramod Singh Rathore, Ganesh Gopal Devarajan, and Rajiva Ranjan Divivedi, ISBN: 9781394229796. *Artificial Intelligence and Cybersecurity in Healthcare* offers a vital exploration of the intersection of artificial intelligence and cybersecurity within healthcare Cyber Physical Systems, equipping readers with knowledge on how to navigate the transformative yet complex technological landscape shaping modern patient care and data protection.

Explainable Artificial Intelligence in the Healthcare Industry, Edited by Abhishek Kumar, T. Ananth Kumar, Prasenjit Das, Chetan Sharma, and Ashutosh Kumar Dubey, ISBN: 9781394249268. Discover the essential insights and practical applications of explainable AI in healthcare that will empower professionals and enhance patient trust with Explainable AI in Healthcare, a must-have resource for anyone looking to navigate the future of medical practices with confidence.

Role of Flavonoids in Chronic Metabolic Diseases: From Bench to Clinic, Edited by Neeraj Mishra, Sumel Ashique, B.H. Jaswanth Gowda, Arshad Farid, Ashish Garg, ISBN: 9781394238040. This comprehensive volume covers the entire field of flavonoids by explaining their complex functions in reducing chronic metabolic illnesses, from the early stages of laboratory research to the development of therapeutic uses.

Viral Vectors for Vaccine Delivery, Edited by Vivek Chavda and Vasso Apostolopoulos, ISBN: 9781394271535. The book is essential for anyone interested in vaccine development, as it highlights the unique advantages of viral vector vaccines in triggering robust, long-lasting immunity and provides an in-depth exploration of the technology and advancements shaping the future of healthcare.

Artificial Intelligence and Cybersecurity in Healthcare, Edited by Rashmi Agrawal, Pramod Singh Rathore, Ganesh Gopal Devarajan, and Rajiva Ranjan Divivedi, ISBN: 9781394229796. *Artificial Intelligence and Cybersecurity in Healthcare* offers a vital exploration of the intersection of artificial intelligence and cybersecurity within healthcare Cyber Physical Systems, equipping readers with knowledge on how to navigate the transformative yet complex technological landscape shaping modern patient care and data protection.

Explainable Artificial Intelligence in the Healthcare Industry, Edited by Abhishek Kumar, T. Ananth Kumar, Prasenjit Das, Chetan Sharma, and Ashutosh Kumar Dubey, ISBN: 9781394249268. Discover the essential insights and practical applications of explainable AI in healthcare that will empower professionals and enhance patient trust with Explainable AI in Healthcare, a must-have resource for anyone looking to navigate the future of medical practices with confidence.

Role of Flavonoids in Chronic Metabolic Diseases: From Bench to Clinic, Edited by Neeraj Mishra, Sumel Ashique, B.H. Jaswanth Gowda, Arshad Farid, Ashish Garg, ISBN: 9781394238040. This comprehensive volume covers the entire field of flavonoids by explaining their complex functions in reducing chronic metabolic illnesses, from the early stages of laboratory research to the development of therapeutic uses.

Printed and bound by CPI Group (UK) Ltd, Croydon, CR0 4YY

17/09/2026

14964266-0007